MATHEMATICS
A Human Endeavor

"This is the part I always hate."

MATHEMATICS
A Human Endeavor

A Book for Those Who Think They Don't Like the Subject

HAROLD R. JACOBS

THIRD EDITION

W. H. FREEMAN AND COMPANY
NEW YORK

Library of Congress Cataloging-in-Publication Data

Jacobs, Harold R.
  Mathematics, a human endeavor: a book for those who think they
  don't like the subject/Harold R. Jacobs.—3rd ed.
  p.   cm.
  Includes index.
  ISBN 0-7167-2426-X
  1. Mathematics—Popular works. I. Title.
QA93.J33   1994
510—dc20   93-37458
                                                        CIP

Printed in the United States of America.

Tenth printing

Contents

3

Functions and Their Graphs   121

4

Large Numbers and Logarithms   183

5

Symmetry and Regular Figures   245

6

Mathematical Curves   327

7

Methods of Counting   401

8

The Mathematics of Chance   447

9

An Introduction to Statistics   525

10

Topics in Topology   601

Appendix: Basic Ideas and Operations   655

Martin Gardner

Photograph by Steve Claris

Foreword

It is not often that one can think of an apt reply to a comment, so I am rather proud of one time that I could. Somebody came up to me after a talk I had given, and said, "You make mathematics seem like fun." I was inspired to reply, "If it isn't fun, why do it?"

—*Ralph P. Boas,*
Professor Emeritus of Mathematics,
Northwestern University.

When Harold Jacobs' *Mathematics: A Human Endeavor* was first published in 1970 by W. H. Freeman and Company, the book astonished the author, publisher, and the entire mathematical community by how quickly it became the nation's most widely adopted high school and college introductory textbook. Why such amazing success?

There are four reasons: The author's choice of exciting topics, with emphasis on their recreational aspects; the author's clear, friendly style; his inclusion of amusing cartoons and comic strips along with other art;

above all, his enthusiasm for mathematics. Other textbooks have since followed Jacobs by imitating the book's format, and by stressing play features. None has topped Jacobs' now classic work in its relevance, its popularity with teachers and professors, and in the delight with which students absorb its captivating content.

I am honored to have been asked for the third time to write a foreword. I want to point out, as I did in the second edition, that the book's great virtue is that Jacobs, amidst all the fun, never loses sight of his main objective: to teach students what mathematics is all about. No games, puzzles, or paradoxes are here merely to entertain or challenge; they are here to draw the reader almost unwittingly into the fundamental ideas of mathematics. You cannot play Sidney Sackson's *Patterns* game without learning what inductive reasoning really is. You cannot understand mind-reading tricks with numbers without learning some elementary number theory. You cannot fathom the mystery of vanishing-area paradoxes without learning some elementary geometry.

There is much talk these days about the "mathematical innumeracy" of most Americans, and how much this contributes to the equally deplorable level of public understanding of science. Compared to Japan and several European countries, it is all too true. Unfortunately, better school buildings and higher educational standards and better wages for mathematics teachers can take us only so far in combating mathematical illiteracy. I am convinced that textbooks written in the spirit of this one are essential if we are to win the battle.

If textbooks such as this had been available thirty years ago, there would be fewer otherwise educated adults to tell you, often with pride, how much they "hated math" in high school. These are the same adults who think accountants are mathematicians, and that mathematics professors are absent-minded old codgers who scratch pancakes and pour molasses on their head. They fail to realize that everything we take for granted in modern technology, a thousand things that would have dumbfounded Isaac Newton, could not have been developed without the creative efforts of mathematicians. Somehow these mathematics haters even manage to get through college without the slightest awareness of mathematical beauty or any inkling of what mathematics is really about.

An old anecdote tells of a meeting of the entire faculty of a major university. The president said his listeners would be amused to know the names of two new students who had flunked freshman classes. A lad named Cicero had failed in Latin. Everybody laughed. Another student named Gauss had failed in mathematics. Only the mathematicians and physicists laughed. This would not have been the case if the members of the liberal arts faculty had explored the world of mathematics with Harold Jacobs.

Photograph by Roy Bishop

A Letter to the Student

In 1623, the great Italian scientist Galileo wrote: "That vast book which stands forever open before our eyes, the universe, cannot be read until we have learned the language and become familiar with the characters in which it is written. It is written in mathematical language, without which means it is humanly impossible to comprehend a single word."

Mathematics has made possible the great advances in science and technology that have occurred since Galileo wrote these words. Indeed, it has become so important to so many fields, including those as diverse as psychology, economics, medicine, linguistics, and even history, that mathematics is now an integral part of most courses of study at universities.

Unfortunately, even though mathematics is a very broad subject, many people leave school with a rather narrow view of it. Someone who has studied only arithmetic may identify mathematics with computation. A student who has taken algebra or geometry may think of it as being limited to solving equations and proving theorems. Even people who have taken advanced courses in mathematics may not be aware of its many applications.

The goal of this book is to give you a much broader view of mathematics by introducing you to areas that you may never have thought about before. Fortunately, you can understand and enjoy mathematics without having to find square roots, or memorize the quadratic formula, or prove geometric facts.

On a recent flight here from Denver, as the pilot announced to passengers that they were about to fly over the Grand Canyon, writer Courtney Anderson, sitting next to the window, offered to change seats with the woman next to him. A woman of about 45, head of an employment bureau in Chicago, she'd told him this was her first trip west.

She looked silently at the great gorge and after a thoughtful moment turned to him and said, "What good is it?" He silently tightened his seat belt and wished he were back home in West L.A.

Courtesy of Matt Weinstock, *Los Angeles Times*

Some of the topics in this book may seem to you to be of little practical use, but the significance of mathematics does not rest only on its practical value. It is hard to believe that someone flying over the Grand Canyon for the first time could remark, "What good is it?" Some people say the very same thing about mathematics. A great mathematician of our century, G. H. Hardy, said, "A mathematician, like a painter or a poet, is a maker of patterns." Some of these patterns have immediate and obvious applications; others may never be of any use at all. But, like the Grand Canyon, mathematics has its own kind of beauty and appeal to those who are willing to look.

Harold Jacobs

Acknowledgments

Since beginning work on the first edition of *Mathematics: A Human Endeavor* some twenty-five years ago, I have benefitted from the comments and suggestions of many people. So many teachers and students have shared their ideas over the intervening years, both through correspondence and at mathematics conferences, that it is impossible for me to individually acknowledge every contribution.

During the time that I have been working on the third edition of this book, colleagues in many parts of the country have served as valued advisors and critics. For their helpful advice, I am grateful to: Peder Bolstad, James Brown, Lou Destito, Nancy A. Freeman, John Glaze, Héctor Hirigoyen, Bernard Hovey, Cheryl Jones, John Leonard, David P. MacAdam, David K. Masunaga, Gary C. Miller, Bernice Nelson, Catherine Pirri, Diana Race, Peter Renz, Carol A. Sipes, Chamille Steiner, Barbara Utter, Michael White, Robert Ross Wilson, and Susan Knueven Wong.

Putting a book such as this together is a formidable task. Without the patience and support of the dedicated staff at W. H. Freeman and Company, it would not have been possible. For their valued efforts, many times above and beyond the call of duty, I would like to thank: Jeremiah Lyons, publisher; Kay Ueno, development editor; Diane Maass, project editor; Travis Amos and Larry Marcus, photo researchers; Scott Zeman, editorial assistant; Paul Rohloff, production coordinator; Nancy Singer, designer; Christine McAuliffe, illustration coordinator; and Lisa Douglas and John Hatzakis, layout artists.

Finally, I would like to thank Martin Gardner for his seemingly endless ideas and inspiring enthusiasm, and acknowledge my indebtedness to Howard Eves, Morris Kline, George Polya, and W. W. Sawyer for demonstrating what good mathematics teaching is all about.

MATHEMATICS
A Human Endeavor

The radio telescope at Arecibo, Puerto Rico

Introduction: Mathematics — A Universal Language

The world's largest radio telescope is located at Arecibo, Puerto Rico. It can communicate with an antenna of equal size anywhere in our galaxy and is being used in NASA's Search for Extra Terrestrial Intelligence Project in the hope of receiving signals from another civilization.

If life does exist on a planet of another star and the beings of that far-off world try to communicate with us, what kind of message would they send? If we send a message to them, what should we say and how should we say it?

We could hardly expect that something like Morse code would work or that any earthly language would make sense. How, then, would it be possible to begin a conversation with another world? Scientists agree that the kind of message most likely to be understood would be a mathematical one.

Here are diagrams of radio signals once suggested by a British physicist as a way of starting a conversation. Each line of pulses represents a mathematical statement. Can you figure out what the statements are?

1.

Hint: This message seems to have three parts, separated by two zigzag patterns. What does each part mean?

2.

3.

4.

5.

6.

7.

8.

A problem somewhat like that of understanding a message from outer space is that of making sense out of the messages left behind by an early civilization here on earth. The earliest known records of mathematics were made by the Babylonians about 4,000 years ago. They left

University Museum, University of Pennsylvania

behind thousands of clay tablets, some of which reveal their number system and their discoveries in algebra and geometry. The photograph at the top of the next page is of a Babylonian tablet of about 1800 B.C. The wedge-shaped writing, called cuneiform, was made by a stylus on wet clay.

A copy of the front and back of another tablet, dug up in the late-nineteenth century, is shown below. Can you translate the groups of wedge-shaped symbols into familiar symbols and explain what the tablet is about? *Hint:* Figure out what all the symbols in the left-hand columns mean before working on the right-hand columns.

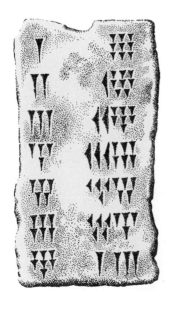
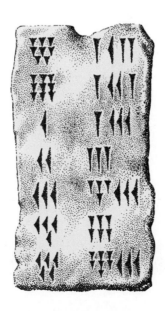

INTERESTING READING

Is Anyone Out There?, by Frank Drake and Dava Sobel, Delacorte Press, 1992.

Cosmos, by Carl Sagan, Random House, 1980: Chapter 12: "Encyclopedia Galactica."

The Search for Extraterrestrial Intelligence, prepared by the National Aeronautics and Space Administration and edited by Philip Morrison, John Billingham, and John Wolfe, Dover, 1979.

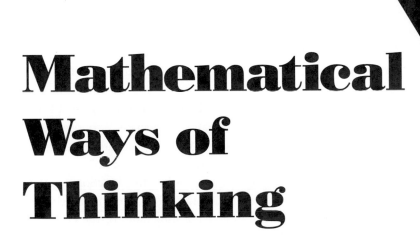

Mathematical Ways of Thinking

LESSON

1

The Path of a Billiard Ball

An expert billiard player's ability to control the path of a ball seems almost miraculous. This is due in part to the fact that the game of billiards can be easily modeled by means of mathematics. The table can be represented by a rectangle, the ball by a point, and its path by a set of line segments.

An ordinary billiard table is twice as long as it is wide (10 feet by 5 feet) and, unlike a pool table, it does not have any pockets. Suppose that a ball is hit from one corner so that it travels at 45° angles with the sides of the table.* If it is the only ball on the table, where will it go?

The first figure at the top of the next page shows the direction that the ball takes as it is hit from the lower-left corner. The second figure shows that the ball hits the midpoint of one of the long sides of the table. In striking the cushion, the ball rebounds from it in a new direction but at

---

*For a discussion of angles and their measurement, see pages 656–657.

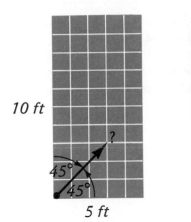
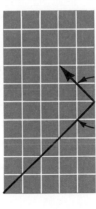
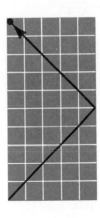

the same angle. (The angles of hitting and rebounding have been marked with curved lines to show that they are equal.) The third diagram shows that the ball goes to the corner at the upper left, and we will assume that the ball stops when it comes to a corner.

What would the ball's path be if the table had a different shape? Suppose that the table was 10 feet by 6 feet and that the ball was again hit from the lower-left corner at 45° angles with the sides as shown in the first figure below. This time, after the first rebound, it would miss the

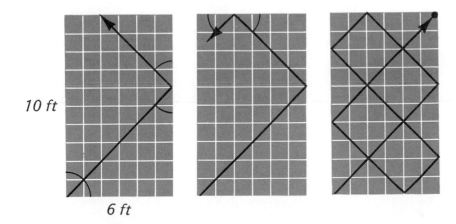

upper-left corner and hit the top side as shown in the second figure. It would rebound from that side in a new direction, but the angles of hitting and rebounding would again be equal. The third figure shows that the ball would rebound several more times before finally ending up in the corner at the upper right.

These two tables suggest several questions about tables of other shapes. Would the ball always end up in a corner? Could it come back to the original corner? If it did end up in a corner, is it possible to predict which one without drawing a figure? Perhaps you can think of other questions as well. We are presented with quite a puzzle. Edward Kasner, in his book *Mathematics and the Imagination* (1941), wrote:

> Puzzles are made of the things that the mathematician, no less than the child, plays with, and dreams and wonders about, for they are made of the things and circumstances of the world he [or she] lives in.

EXERCISES

SET I

On graph paper,* make a diagram of each of the numbered tables below. Be careful to use the same dimensions and write them along the sides of each table as shown.

Draw the path of a ball starting from the lower-left corner of each table, making 45° angles with the sides. (Because of this, the ball travels

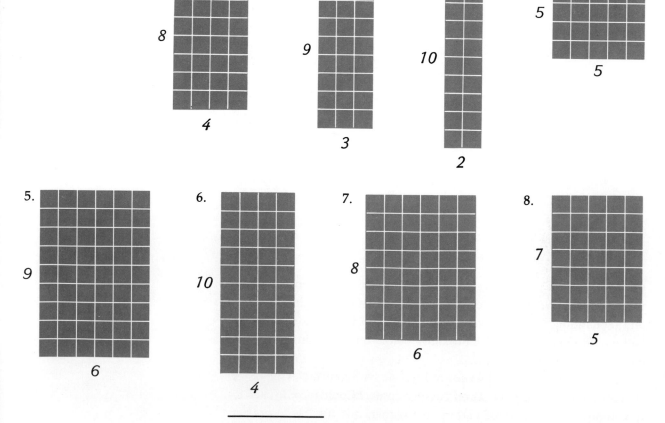

---

*Four or five squares per inch or two squares per centimeter is convenient.

from corner to opposite corner of each square of the grid through which it passes.) Continue each path as far as it can go. If the ball ends up in a corner, mark the corner with a large dot.

In referring to billiard tables in the following exercises, we will call the vertical dimension the *length* and the horizontal dimension the *width.* In stating the dimensions, we will always give the length first. For example, we will refer to the table in exercise 1 as having dimensions "8 by 4."

Length

Width

9. On which table does the ball have the simplest path?

10. What is special about the shape of this table?

11. On which tables does the ball never pass over the same point twice? (In other words, the path does not cross itself.)

12. Judging from the eight tables that you have drawn, do you think that a ball hit from the lower-left corner of a table at 45° angles with the sides will always end up in a corner?

13. Judging from the tables that you have drawn, do you think that a ball hit in this way can end up in *any* of the four corners? Explain.

14. On which table does the ball have the most complicated path?

15. On the table you named in exercise 14, the ball passes over every one of the small squares of the table. Try to draw another table (and the path of the ball) in which the ball passes over every square.

Set II

It is hard, for the most part, to see any pattern in the paths on the tables you have drawn so far because the dimensions of the tables were chosen unsystematically. The following exercises will give you a chance to discover some patterns for tables whose dimensions are related in special ways.

A billiard table with a length of 12 units and a width of 1 unit is shown at the right. Notice that the ball follows a path that consists of 12 segments.

12

1

1. Draw three more billiard tables with lengths of 12 units and widths of 2, 3, and 4 units. Write the dimensions along the sides of each table. Draw the path of a ball starting from the lower-left corner of each table and mark the corner where it ends up with a large dot.

2. Refer to your drawings for exercise 1 to copy and complete this table.

Dimensions of table	No. of segments in path
12 by 1	12
12 by 2	▯▯▯▯▯▯
12 by 3	▯▯▯▯▯▯
12 by 4	▯▯▯▯▯▯

3. How can the number of segments in the paths on these tables be found from the dimensions of the tables?

4. Draw a table with a length of 12 units on which the path of the ball has just two segments.

5. For which tables in Set I does your answer to exercise 3 give the correct number of segments?

Two billiard tables on which the ball travels paths with the same shape are shown at the left.

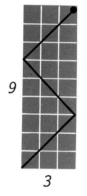

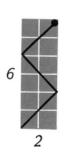

6. Although the tables do not have the same dimensions, there *is* something about their dimensions that is the same. What is it?

7. Draw a billiard table with a length of 6 units and a width of 4 units and a billiard table with a length of 9 units and a width of 6 units. Draw the path of a ball starting from the lower-left corner of each table.

8. What do you notice about the paths on the two tables?

9. Draw a smaller billiard table on which the path of the ball is the same as on the tables you drew for exercise 7. What are its dimensions?

10. Draw a larger billiard table on which the path of the ball is the same. What are its dimensions?

On the set of tables below, the ball travels over every square.

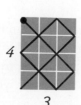

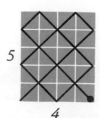

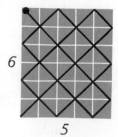

11. Refer to these tables to copy and complete this table.

Dimensions of table	Corner ball ends up in
2 by 1	Upper left
3 by 2	Lower right
4 by 3	▥
5 by 4	▥
6 by 5	▥

12. What do you think are the dimensions of the next table in this set for which the ball ends up in the upper-left corner?

13. Draw the table and show the path of the ball.

14. Use these results to copy and complete this table for two giant billiard tables. (Don't try to actually draw the tables!)

Dimensions of table	Corner ball ends up in
100 by 99	▥
101 by 100	▥

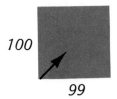

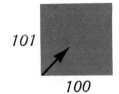

SET III

This picture shows a mirror placed horizontally across the center of the first billiard table you drew in Set I. The reflection in the mirror of the lower half of the ball's path is the same as the actual upper half of the path.

Try putting a mirror horizontally across the centers of the other billiard tables you drew for both Sets I and II of this lesson.

1. How is the corner in which the ball ends up related to whether or not the reflection in the mirror is the same as the actual path?

Try putting a mirror vertically across the centers of the billiard tables you drew for this lesson.

2. What conclusion can you draw from this?

Put the mirror away and try looking at each billiard table upside-down. On some of the tables, the path looks the same rightside-up and upside-down. An example is shown here.

3. What conclusion can you draw about these tables?

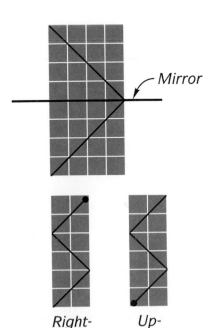

Mirror

Right-
side
up

Up-
side
down

2 More Billiard-Ball Mathematics

W. W. Sawyer, in his book *Mathematician's Delight*, wrote:

> Everyone knows that it is easy to do a puzzle if someone has told you the answer. That is simply a test of memory. You can claim to be a mathematician only if you can solve puzzles that you have never studied before. That is the test of reasoning.

So far, in trying to solve the puzzle of the path of a ball on a billiard table, we have made several discoveries. From the examples that we have considered, it seems that, if the dimensions of the table are whole numbers and the ball is hit from the lower-left corner at 45° angles with the sides, it will end up in one of the other three corners. The path of the ball depends on the shape of the table, and the corner in which it ends up seems to be related to the dimensions of the table in some way.

The simplest possible path is clearly on a table in the shape of a square. On such tables, the ball travels diagonally from one corner to the opposite corner without rebounding from any of the sides. The fact that the path is so simple is related to the fact that the *ratio* of the length to the width of a square table is 1. A ratio is a comparison of two number by division.

The **ratio** of the number x to the number y is the number $\dfrac{x}{y}$.

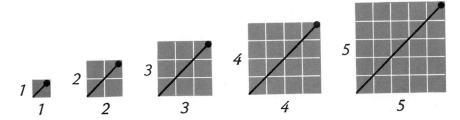

For the square tables shown in the figure above,

$$\frac{1}{1} = \frac{2}{2} = \frac{3}{3} = \frac{4}{4} = \frac{5}{5} = 1.$$

If the length of a billiard table is twice its width, a ball hit from the lower-left corner hits the midpoint of one of the longer sides and then

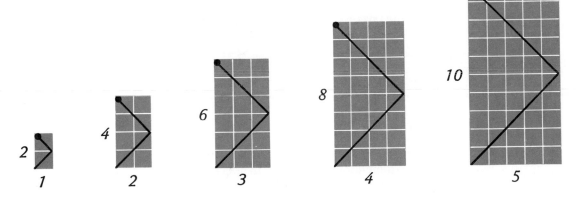

goes to the upper-left corner. Again, the paths on such tables have the same shape because the ratio of the length to the width of each table is the same:

$$\frac{2}{1} = \frac{4}{2} = \frac{6}{3} = \frac{8}{4} = \frac{10}{5} = 2.$$

We have been representing the tables with rectangles.

Rectangles for which the ratios of the lengths to the widths are the same have the same shape and are called **similar**.

Because the paths of the billiard balls on tables that are similar have the same shape, the path on a table with large dimensions can be discovered by reducing the ratio of those dimensions to lowest terms.*

Look, for example, at the two tables below. Because $\frac{12}{8} = \frac{3}{2}$,[†]

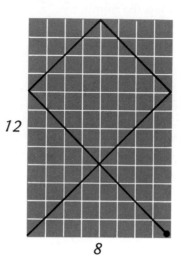

the two tables are similar and hence have paths with the same shape, so that the ball ends up in the same corner. For this reason, we will proceed with the puzzle of predicting the corner in which the ball will end up by reducing the dimensions of some of the tables to lower terms.

EXERCISES

SET I

In the following exercises, we will draw the path of the ball on tables whose dimensions are related in special ways.

1. Draw a set of six billiard tables with widths of 1 unit and lengths of 1, 2, 3, 4, 5, and 6 units. Show the path of the ball on each table, marking the corner in which the ball ends up with a large dot. The first two tables and paths are shown at the left.

---

*The ratio $\frac{x}{y}$ is in lowest terms if there is no whole number larger than 1 that will divide evenly into both x and y.

†Check this with a calculator. Are your ratios 1.5?

Notice that the ball simply zigzags back and forth on these tables, ending up in one of the upper corners.

2. On which tables did it end up in the upper *left* corner?

3. On which tables did it end up in the upper *right* corner?

You know that the numbers

$$1, 3, 5, 7, 9, 11, 13, 15, \ldots \text{ are } odd$$

and that the numbers

$$2, 4, 6, 8, 10, 12, 14, 16, \ldots \text{ are } even.$$

4. On the basis of your answers to exercises 1–3, does the following rule seem to be true?

    On a table whose length is odd and whose width is 1, the ball ends up in the upper-right corner.

5. Make up a similar rule for a table whose length is even and whose width is 1. State your rule as a complete sentence.

6. Draw four different tables whose lengths are even and whose widths are odd numbers other than 1. Show the path of the ball on each table, marking the corner in which the ball ends up with a large dot.

7. On the basis of the tables you drew for exercise 6, make up a corner-predicting rule for a table whose length is even and whose width is odd. State your rule as a complete sentence.

8. Draw four different tables whose lengths are odd and whose widths are even. Show the path of the ball on each table, marking the corner in which the ball ends up with a large dot.

9. On the basis of the tables you drew for exercise 8, make up a corner-predicting rule for a table whose length is odd and whose width is even. State your rule as a complete sentence.

10. Draw four different tables whose lengths are odd and whose widths are also odd numbers other than 1. Show the path of the ball on each table, marking the corner in which the ball ends up with a large dot.

11. On the basis of the tables you drew for exercise 10, make up a corner-predicting rule for a table whose length and width are both odd. State your rule as a complete sentence.

Set II

Here are three tables whose lengths and widths are even.

Table B

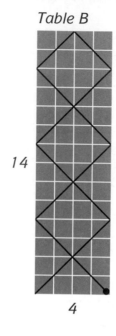

14

Table C

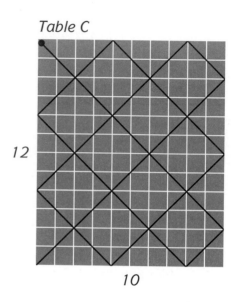

Table A

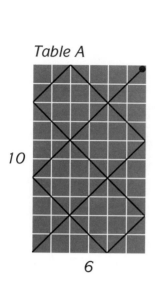

10

6

4

12

10

1. On the basis of these three tables, does there seem to be a corner-predicting rule for a table whose length and width are both even?

The ratio of the length to the width of table A is $\frac{10}{6}$. This ratio can be reduced to $\frac{5}{3}$.

2. Draw a table whose length is 5 and whose width is 3. Show the path of the ball, marking the corner in which it ends up with a large dot.

3. You stated a corner-predicting rule in exercise 11 of Set I for the table you just drew. Does that rule correctly predict the corner in which the ball ended up?

4. The ratio of the length to the width of table B is $\frac{14}{4}$. Reduce this ratio to lowest terms.

5. Draw a table whose length and width are given by the numbers in the reduced ratio. Show the path of the ball, marking the corner in which it ends up with a large dot.

6. Do any of the corner-predicting rules that you wrote in Set I correctly predict the corner the ball ended up in? If so, state the rule.

7. The ratio of the length to the width of table C is $\frac{12}{10}$. Reduce this ratio to lowest terms.

8. Draw a table whose length and width are given by the numbers in the reduced ratio. Show the path of the ball, marking the corner in which it ends up with a large dot.

9. Do any of the corner-predicting rules that you wrote in Set I correctly predict the corner the ball ended up in? If so, state the rule.

10. If the length and width of a billiard table are both even, what should you do before trying to predict the corner in which the ball will end up?

The figures below represent giant billiard tables and are not drawn to scale. In which corner of each table do you think the ball would end up? Tell why you chose each corner.

11.
95

85

12.
105

100

13.
120

110

Set III

As a billiard ball travels around the table, it hits the cushions a number of times. How many times depends on the dimensions of the table. Counting the original and final corner positions as "hits," a square table has two hits. The second table shown has five hits.

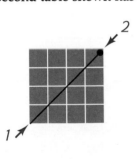

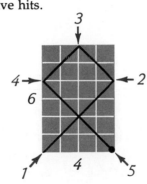

1. Draw two tables, one with length 12 and width 9 and the other with length 4 and width 3. Show the path of the ball on each table and mark the hits.

2. Why do both tables have the same number of hits?

3. From these two tables, try to guess a rule for predicting the number of hits on a table on the basis of its dimensions. Test your rule on some of the other tables of this lesson.

How many hits do you think there would be on each of these giant tables? (They are not drawn to scale.) Explain your thinking.

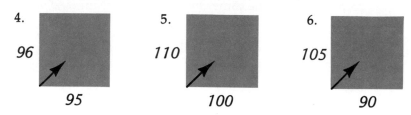

4.   96

      95

5.   110

      100

6.   105

      90

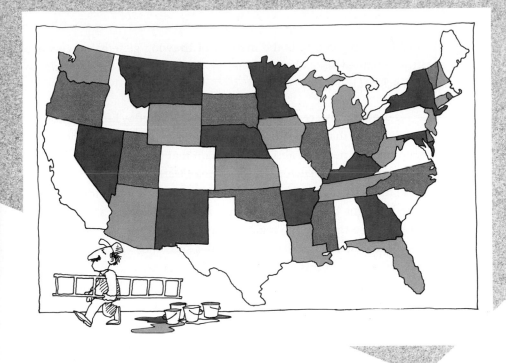

Inductive Reasoning: Finding and Extending Patterns

In his book *Islands Of Truth,* Ivars Peterson wrote:

> Mathematics is really the science of patterns. Mathematical discovery begins with a search for patterns in data—perhaps numbers or scientific measurements. . .

In our study of the path of a ball on a billiard table, we have looked for patterns. First we collected evidence by drawing tables. Next we noticed patterns and drew conclusions from these patterns.

This method of reasoning is used by the scientist who makes observations, discovers regularities, and formulates general laws of nature. In science, this is called the experimental, or scientific, method. In mathematics, it is referred to as reasoning *inductively.*

Inductive reasoning is the method of drawing general conclusions from a limited set of observations. It is reasoning from the *particular* to the *general*.

We use inductive reasoning continually in everyday life. Many of the generalizations that come from this kind of thinking seem highly probable, but we can never be absolutely certain that they are correct. For this reason, scientists call such generalizations *theories*.

An interesting example of arriving at a conclusion by inductive reasoning concerns the coloring of maps. How many colors are needed to color a map so that no two regions sharing a common border have the same color? The map of the United States shown on the previous page has been colored in this way using just four colors.

In 1852, Francis Guthrie, who had recently graduated from University College in London, noticed that it seems to be possible to color *all* maps with four colors or fewer. Since then, so many people have tried without success to draw maps for which more than four colors are needed that many mathematicians became convinced that Guthrie was correct. However, they could not be certain because, although inductive reasoning is of tremendous importance in *developing* mathematical ideas, it cannot *prove* that the ideas are correct. The possibility always exists that additional evidence will reveal that the conclusions are wrong.

In 1976, more than a century after the map-coloring problem had been proposed, two American mathematicians finally figured out how to prove, using a computer, that maps requiring more than four colors do not exist.*

EXERCISES

SET I

$$1 = 1 \times 1$$
$$1 + 3 = 2 \times 2$$
$$1 + 3 + 5 = 3 \times 3$$
$$1 + 3 + 5 + 7 = 4 \times 4$$

Number theory is the branch of mathematics dealing with properties of the whole numbers. An example of a pattern in number theory is shown at the left.

1. Write the next equation in this pattern.

2. Is the equation true?

3. Write what you think the seventh equation of the pattern should be.

---

*An interesting commentary on this appears in the section titled "Maps of a Different Color" in *The Mathematical Tourist* by Ivars Peterson (W. H. Freeman, 1988), pp. 2–4.

4. Is it true?

5. Write what you think the tenth equation of the pattern should be.

6. Is it true?

7. Do you think that the pattern goes on indefinitely?

8. What kind of reasoning are you using?

A different way to multiply that *seems* to give correct answers is shown at the right.

The 36 in the second box comes from 4×9.

9. Where does the 14 in the third box come from?

10. Where does the 63 come from?

11. Where does the 08 come from?

$$\begin{array}{r} 92 \\ \times\ 47 \\ \hline \boxed{63} \\ \boxed{36}\boxed{14} \\ +\ \boxed{08} \\ \hline 4324 \end{array}$$

Use the same method to do each of these problems. Check your answers with a calculator.

12. $\begin{array}{r} 53 \\ \times\ 86 \\ \hline \end{array}$ 13. $\begin{array}{r} 75 \\ \times\ 75 \\ \hline \end{array}$ 14. $\begin{array}{r} 21 \\ \times\ 34 \\ \hline \end{array}$ (Remember: $4 \times 1 = 04$.)

15. If you conclude from these examples that this method always gives the correct answer, what type of reasoning are you using?

Set II

In the late sixteenth century, the great Italian scientist Galileo used inductive reasoning to make several discoveries about the behavior of swinging weights—discoveries that led to the invention of the pendulum clock. One of these discoveries was of a relation between the *length* of the pendulum and the *time* of the swing.

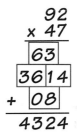

Galileo Galilei

This table lists the swing times of a series of pendulums having different lengths.

Length of pendulum	Time of swing
1 unit	1 second
4 units	2 seconds
9 units	3 seconds
16 units	4 seconds

1. From the pattern in the table, how does the length of the pendulum seem to be related to the time of the swing?

2. What do you think the length of a pendulum with a swing time of 5 seconds would be?

3. What do you think the length of a pendulum with a swing time of 10 seconds would be?

In 1661, the English chemist Robert Boyle did a series of experiments with the pressure of the air. One of his discoveries, through inductive reasoning, was a relation between the volume of a gas and the pressure it exerts. The table below lists some volumes and pressures for a gas at a given temperature.

• Sun
4 • Mercury
7 • Venus
10 • Earth
15 • Mars

Volume	Pressure
1 unit	120 units
2 units	60 units
4 units	30 units
8 units	15 units

4. What happens to the pressure as the volume of the gas gets larger?

5. What happens to the pressure when the volume is multiplied by 2?

6. What happens to the pressure when the volume is multiplied by 4?

52 • Jupiter

7. On the basis of your answers to exercises 5 and 6, what do you think would happen to the pressure if the volume is multiplied by 3?

8. What do you think the pressure would be when the volume is 3 units?

9. What do you think it would be when the volume is 16 units?

In 1772, the German astronomer Johann Elert Bode used inductive reasoning to find a pattern in the distances of the planets from the sun. At that time, only six planets were known. The actual relative distances of the planets from the sun and his pattern are shown at the top of the next page.

96 • Saturn

Planet	Actual distance*	Bode's pattern
Mercury	4	0 + 4 = 4
Venus	7	3 + 4 = 7
Earth	10	6 + 4 = 10
Mars	15	12 + 4 = 16
		▓▓▓ + ▓▓▓ = ▓▓▓
Jupiter	52	48 + 4 = 52
Saturn	96	96 + 4 = 100
		▓▓▓ + ▓▓▓ = ▓▓▓

Notice that there is a pretty good match between the two sets of numbers.

10. What equation do you think belongs between Bode's equations for Mars and Jupiter?

11. What equation do you think belongs after Bode's equation for Saturn?

In 1781, William Herschel discovered Uranus, the next planet beyond Saturn. Because its distance of 192 units comes remarkably close to the number predicted by this equation, astronomers came to the conclusion that the equation between Bode's equations for Mars and Jupiter also must mean something.

12. What do you suppose they thought it meant?

In 1801, the asteroid Ceres was discovered at a distance of 28 units from the sun.

SET III

Patterns, a game invented by Sidney Sackson, is based on inductive reasoning. To play the game, one player creates a pattern that the other players try to discover.

---

*The distances are based on taking the distance from the earth to the sun to be 10 units.

The game is played with four symbols, +, O, ★, and ∴, on a grid of 36 squares. Here are four examples of patterns that might be used in the game. Guess which symbol belongs in the blacked-out square in each pattern.

1.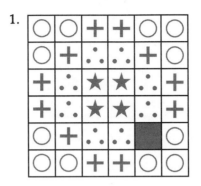

2.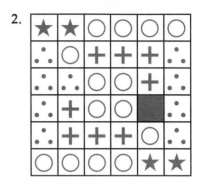

3.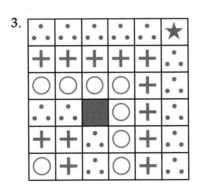

4.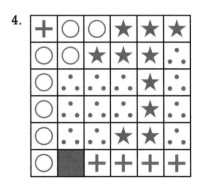

5. Which guess are you least sure of? Why?

"Water boils down to nothing . . . snow boils down to nothing . . . ice boils down to nothing . . . everything boils down to nothing."

The Limitations of Inductive Reasoning

After discovering that water, snow, and ice boil down to nothing, the cave man in this cartoon has concluded that *everything* boils down to nothing. This illustrates the basic weakness of inductive reasoning. Because the conclusions arrived at by this method are drawn from a limited amount of evidence, the possibility always exists that more evidence may be discovered that will prove the conclusions to be incorrect.

An interesting example of this possibility from science concerns the "noble gases." In 1894, a new element was discovered. The element, a gas named argon, had never been found in any compounds because its atoms would not combine with those of any other element. Within the next six years, five more elements of the same type were discovered. Named helium, krypton, neon, xenon, and radon, these elements became known as the "noble gases" because they could not be made to form compounds with other elements. The conclusion that "noble gases cannot form compounds" became generally accepted by chemists and appeared in many books as if it were a statement of fact. In 1962, however, it was proved to be wrong. In that year, a compound of xenon was made for the first time, and since then other compounds formed by these elements that "could not form compounds" have been made.

Conclusions arrived at by inductive reasoning in mathematics are no more certain. Consider, for example, our study of the behavior of a ball on a billiard table. If we had started by drawing a set of tables with lengths of 8 as shown below, we might have noticed that in every diagram the ball ends up in the same place, even after taking widely varying paths. A reasonable conclusion seems to be: "If the length of the table is

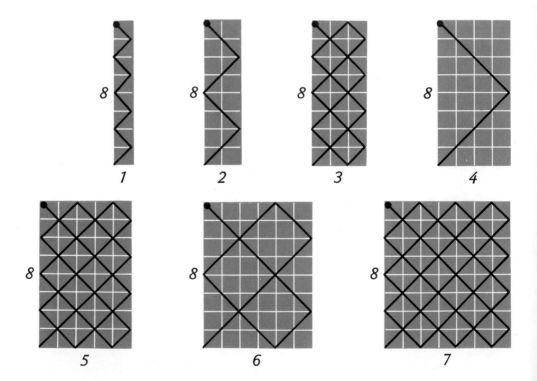

8, the ball will always end up in the upper-left corner." Although this is true for these seven cases, it is revealed to be false by the next case. If the width of the table is also 8, the table is square and the ball ends up in the upper-*right* corner.

Inductive reasoning, then, is of great importance in *suggesting* conclusions in mathematics. To be certain that the conclusions are reliable, however, mathematicians use another method called *deductive reasoning*. Deductive reasoning will be considered in the lessons following this one.

EXERCISES

SET I

The first modern Olympics Games were held in Athens in 1896. The following games were held in 1900, 1904, 1908, and 1912.

    1. What pattern do you see in these dates?

    2. What method of reasoning did you use in arriving at this pattern?

    3. According to the pattern, what year would seem to be the date of the next Olympics after 1912?

    4. Do you think this date is correct?

Do the indicated calculations to find the numbers that will make the following equations true.

    5. $1 \times 9 + 2 = $ ▯▯▯

    6. $12 \times 9 + 3 = $ ▯▯▯

    7. $123 \times 9 + 4 = $ ▯▯▯

    8. $1{,}234 \times 9 + 5 = $ ▯▯▯

    9. Guess the next equation in this pattern.

    10. Check to see whether it is true.

    11. Write what you think the *ninth* equation in this pattern should be.

    12. Check to see whether it is true.

    13. Do you think that the pattern goes on indefinitely?

The numbers in this list are all *prime*. That is, each number is evenly divisible by only itself and 1.

$$31$$
$$331$$
$$3{,}331$$
$$33{,}331$$
$$333{,}331$$
$$3{,}333{,}331$$
$$33{,}333{,}331$$

    14. What is the next number in this list?

    15. If you conclude that it is also prime, what kind of reasoning are you using?

16. Divide the number that you chose in exercise 14 by 17.

17. What does this tell you about the number you chose?

18. What does your result show about inductive reasoning?

Set II

Here is a set of three circles. Two points have been chosen on the first circle and a straight line segment drawn between them. The circle is separated into two regions as a result. Three points were chosen on the second circle and connected with three line segments to form four regions. Four points were chosen on the third circle and, after being connected in all possible ways, eight regions resulted.

Here is a table that includes the results so far.

Number of points connected	2	3	4	5	6
Number of regions formed	2	4	8	|||||||	|||||||

Two more cases have been added.

1. Guess from the pattern in the second line of numbers what the missing numbers are.

2. Draw a pair of large circles and choose five points on one circle and six points on the other. Join the points of each *in every possible way*.

3. How many regions are formed in each?

4. Do both results agree with your guesses?

Cut a sheet of graph paper into two equal pieces. Make neat drawings of the two figures below on one piece of the paper.

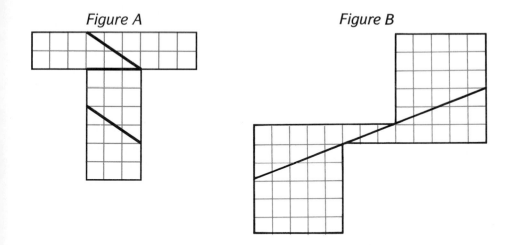

Figure A   *Figure B*

5. Explain how it is possible to tell, without having to count all the small squares, that the area of figure A is 36 square units.

6. What is the area of figure B?

Cut figure A out and into its four separate pieces.

7. Rearrange the pieces to form a square. Make a drawing of the arrangement on the other piece of graph paper.

8. What is the area of the square?

9. Rearrange the pieces again to form a rectangle with a width of 4 units. Make a drawing of the arrangement. (The outline of the rectangle should look like this.)

10. What is the area of this rectangle?

11. Does the area of a figure change when its pieces are arranged in different ways?

Cut figure B out and into its four separate pieces.

12. Rearrange the four pieces to form a square. Make a drawing of the arrangement.

13. What does the area of the square seem to be?

14. Rearrange the pieces again to form a long rectangle with a width of 5 units. Make a drawing of the arrangement.

15. What does the area of this rectangle seem to be?

16. What seems to happen to the area of figure B as its pieces are arranged in different ways?

17. Do you have any ideas about why this happens? Please explain.

Set III

Richard Feynman, winner of the Nobel Prize in physics in 1965 and professor at Caltech for many years, wrote an interesting book about his career, *Surely You're Joking, Mr. Feynman!* (Norton, 1985).

In it, he tells of working on a secret project for the U. S. government at Los Alamos during World War II. In a letter to a friend at that time, he wrote that he had noticed something very peculiar. If you divide 243 into 1, you get

$$0.004115226337 \ldots$$

The letter was returned by the project's censor because the censor thought it might be a code!*

1. What is interesting about the digits in this number?

2. Can you guess any more digits?

3. Divide 243 into 1 as far as you can get to see what happens.

4. Do you think the pattern goes on indefinitely?

---

*Also reported in Chapter 21 of *Penrose Tiles to Trapdoor Ciphers,* by Martin Gardner (W. H. Freeman and Company, 1989).

Deductive Reasoning: Mathematical Proof

The Soma cube is a popular puzzle made from 27 small cubes. Created by Piet Hein, a noted Danish inventor, it consists of seven pieces made by gluing the small cubes together in different arrangements. The object of the puzzle is to put the pieces together to form either a large cube or another specified shape.

Suppose that we have a block of wood in the shape of a cube with which to make a Soma cube. First, we have to cut it up into the 27 small cubes. One way to do this is to make a series of six cuts through the cube while keeping it together in one block. The cuts are indicated by the arrows in the diagrams below.

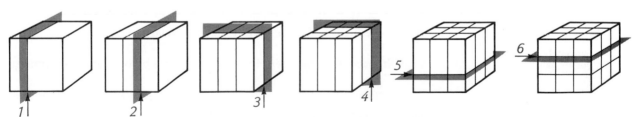

Now suppose that, instead of keeping the cube together in one block, the pieces are rearranged between each cut. Could the cube be cut into the 27 smaller cubes with fewer than six cuts? The two diagrams below show one way in which the first two pieces might be arranged. Notice that the second cut will now cut through more wood than it would have if the piece at the left had not been moved to the bottom.

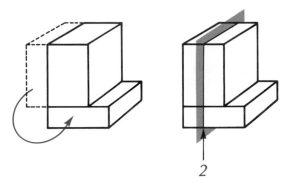

The number of pieces increases with each cut, and there are many ways of rearranging them; so the number of different ways of making the cuts is very large. Suppose that we had a thousand blocks of wood and tried a thousand different ways to cut them up with fewer than six cuts without success. We still would not know for certain that it could not be done because this conclusion is based only on inductive reasoning.

Instead of reasoning inductively by testing many different cases, we can reason *deductively*.

> **Deductive reasoning** is a method of using logic to draw conclusions from statements that we accept as true.

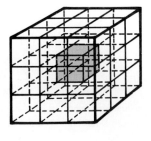

By reasoning deductively, we can prove that the cube cannot be cut into the 27 smaller cubes with fewer than six cuts. The key to the proof is the cube in the center, shown in brown in the figure at the left. Every one of its faces must be created by a separate cut. Because it has *six* faces, *six* cuts are required to create it regardless of the ways in which the other pieces may be rearranged.

In contrast to inductive reasoning, which helps us find what *may* be true, deductive reasoning tells us what *must* be true. Because of this, mathematicians usually use deductive reasoning when they want to prove that something is true. Mathematical statements that are proved true by deductive reasoning are called *theorems*.

> A **theorem** is a statement that is proved true by deductive reasoning.

In 300 B.C., Euclid, in his book the *Elements* (the most widely known textbook ever written), used deductive reasoning exclusively in proving statements about geometric figures. His success in doing so helped to establish deductive reasoning as a fundamental tool of mathematics.

EXERCISES

SET I

Here is a game with a winning strategy that can be discovered by using inductive reasoning. Deductive reasoning proves that this strategy will always win. The game is played on a small checkerboard containing sixteen squares as shown below.

Player A begins by placing pennies on any two of the squares. Player B then places paper clips on the board so that each clip lies on two squares that share a common side. The clips may not overlap each other. To win, player B has to place seven paper clips in this way so that they lie on the fourteen squares not occupied by the pennies. If player B cannot do this, player A wins. The figures below show the outcomes of eight games.

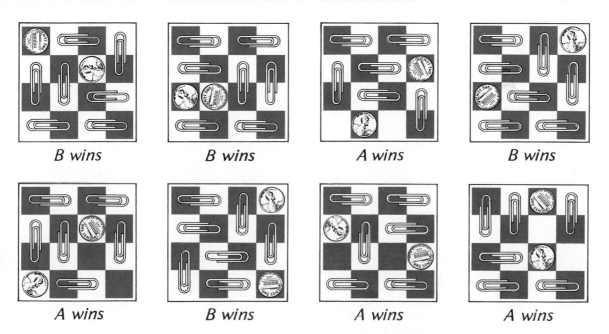

B wins	B wins	A wins	B wins
A wins	B wins	A wins	A wins

1. Look at the squares on which the pennies were placed on each board. How do their colors seem to be related to who wins?

2. If you conclude from these eight examples that your answer to the first question is correct for every possible game, what kind of reasoning are you using?

3. Do the eight examples prove it?

4. According to the rules, each paper clip has to be placed on two squares that share a common side. What can be concluded about the colors of any two such squares?

5. If player B is able to place seven paper clips so that they cover fourteen squares, what must be true about the colors of those fourteen squares?

Of the fourteen remaining squares, how many squares are there of each color if player A begins by placing the pennies on

6. two white squares?

7. two brown squares?

8. a white square and a brown square?

9. What kind of reasoning are you using when you conclude from your answers to exercises 5 through 8 that your answer to the first exercise is correct for every possible game?

Many puzzles, including the following one about marbles in matchboxes, can be solved by reasoning deductively.

The figure at the left shows three matchboxes. One contains two red marbles, one contains two white marbles, and one contains a red marble and a white marble. The labels telling the contents of the boxes have been switched, however, so that *the label on each box is wrong.*

You are permitted to choose one box and open it far enough to see just one marble. The challenge is to explain how you can figure out from this what is in each box.

Suppose that you choose the box labeled "2 red."

10. If you opened it and saw a red marble, what color would the other marble have to be and how do you know?

11. If you opened it and saw a white marble, could the other marble be white? Could it be red?

Suppose instead that you choose the box labeled "2 white."

12. If you opened it and saw a white marble, what could you conclude?

13. If you opened it and saw a red marble, would you know the color of the other marble?

Suppose instead that you choose the box labeled "1 red, 1 white."

14. If you opened it and saw a red marble, what could you conclude?

15. If you opened it and saw a white marble, what could you conclude?

On the basis of your answers so far, one box would be the best choice to look inside.

16. Which one?

Suppose that you see a red marble when you open the box labeled "1 red, 1 white." What can you conclude about

17. the other marble in the box?

18. the marbles in the box labeled "2 white"?

19. the marbles in the box labeled "2 red"?

Set II

Deductive reasoning is a method of drawing conclusions from statements that we accept as true by using *logic*. The following puzzle, from a book titled *100 Games of Logic*, is by Pierre Berloquin, the author of several very successful puzzle books in France.

Andre is a butcher and president of the street storekeepers' committee, which also includes a grocer, a baker, and a tobacconist. All of them sit around a table.

Andre sits on Charmeil's left.
Berton sits at the grocer's right.
Duclos, who faces Charmeil, is not the baker.

1. The puzzle consists of five statements that we accept as true. Use the statements to make a drawing to show who sits where.

2. Use deductive reasoning to figure out what kind of store each person has.

The figures below illustrate the names of different parts of a cube.

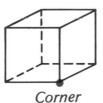   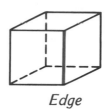   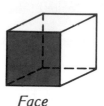

Corner   *Edge*   *Face*

3. How many corners does a cube have?

4. How many edges?

5. How many faces?

Suppose that a wooden cube is painted brown and then cut up into smaller cubes as shown in the first figure at the left. Use deductive reasoning to answer the following questions about the results.

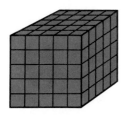

6. How many small cubes are produced? Explain.

The second figure shows the cubes at the corners of the large cube.

7. How many faces of one of these cubes are painted brown?

8. How many of these cubes are there altogether? (Show your method.)

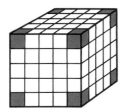

The third figure shows the cubes along the edges between the corner cubes.

9. How many faces of one of these cubes are painted brown?

10. How many of these cubes are there altogether? (Show your method.)

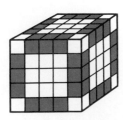

The fourth figure shows the cubes on the faces between the edge cubes.

11. How many faces of one of these cubes are painted brown?

12. How many of these cubes are there altogether? (Show your method.)

Some of the small cubes have no faces painted brown.

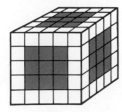

13. Where are these cubes located?

14. How many of these cubes are there altogether?

15. Show that your answers to exercises 8, 10, 12, and 14 account for all of the small cubes.

Suppose that a large wooden cube is painted brown and then cut up into smaller cubes as shown in this figure. Show your methods in finding answers to the following questions.

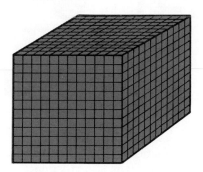

16. How many small cubes are produced?

How many of the cubes have

17. three painted faces?

18. two painted faces?

19. one painted face?

20. no painted faces?

21. Show that your answers to exercises 17–20 account for all of the small cubes.

Every year a calendar different from the year before is needed. This is due to the fact that a year contains 365 days (or 366 days if it is a leap year) and a week contains 7 days.

The year 1999 contains 365 days and begins on Friday.

22. Use deductive reasoning to figure out what day of the week the year 2000 begins on. (Show your method.)

The year 2000 contains 366 days.

23. Use deductive reasoning to figure out what day of the week the year 2001 begins on. (Show your method.)

Every year has a calendar different from the year before.

24. Use deductive reasoning to figure out how many different calendars are possible. (Show your method.)

JANUARY 1999						
S	M	T	W	T	F	S
					1	2
3	4	5	6	7	8	9
10	11	12	13	14	15	16
17	18	19	20	21	22	23
24	25	26	27	28	29	30
31						

Set III

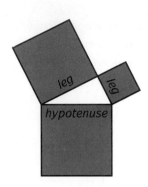

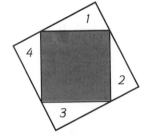

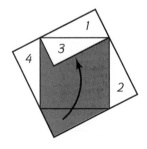

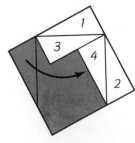

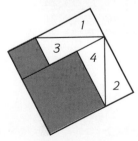

The best-known mathematical theorem is named for Pythagoras, a Greek mathematician of about 500 B.C. The Pythagorean theorem states that, if squares are drawn on the three sides of a triangle that has a right (90°) angle, the largest square is equal in area to the areas of the other two squares put together. The longest side of a right triangle is called the *hypotenuse* and the other two sides are called the *legs.*

An example of this is shown on this postage stamp from Greece. Notice that the square on the hypotenuse of the triangle contains 25 small squares, that the squares on the two legs contain 16 squares and 9 squares, and that $25 = 16 + 9$.

The figure on the postage stamp does not prove that the Pythagorean theorem is true for *all* right triangles because right triangles come in many different shapes. The figures on the left, however, can be used to make a general proof.

The first figure consists of four identical right triangles and a square between them.

1. On which side of the triangles is this square drawn?

The second and third figures show two of the triangles being moved to different positions. The last figure shows the four triangles with two squares.

2. On which sides of the triangles are these squares drawn?

3. What does rearranging the triangles in this way show about the area filled by the large square and the area filled by the two smaller squares put together?

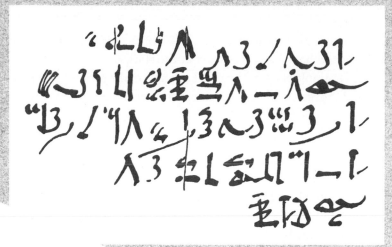

Number Tricks and Deductive Reasoning

In the preface to *The Rhind Mathematical Papyrus,* a translation of the ancient Egyptian document named after A. Henry Rhind, who bought and gave it to the British Museum, Arnold Chace wrote:

> I venture to suggest that if one were to ask for that single attribute of the human intellect which would most clearly indicate the degree of civilization of a race, the answer would be, the power of . . . reasoning, and that this power could best be determined in a general way by the mathematical skill which members of the race displayed. Judged by this standard the Egyptians of the nineteenth century B.C. had a high degree of civilization.

Two of the problems in the papyrus are number tricks. One of them is reproduced at the top of this page; it shows how it was written by a scribe many centuries ago. According to its directions, a number is chosen and then several things are done with it. At the end of the trick, the result is a number that could have been predicted at the start.

Here is an example of a similar trick. Each step of the trick is described at the left, and the results for four numbers chosen at random are shown in the columns at the right.

Choose a number.	4	7	12	35
Add 5.	9	12	17	40
Double the result.	18	24	34	80
Subtract 4.	14	20	30	76
Divide by 2.	7	10	15	38
Subtract the number first thought of.	3	3	3	3
The result is 3.				

The fact that the result is 3 in all four cases shown does not prove that the trick will work for every number. Although we can reason inductively that it may work, we need to use the deductive method to prove that it does.

We will go through the trick again, but this time with one slight change. Rather than choosing a specific number at the start, we will use a box to represent the number originally chosen. Throughout the trick, each box represents this same number. To represent numbers we know, we will use circles.

Choose a number.	□
Add 5.	□ ○○○○○
Double the result.	□□ ○○○○○ ○○○○○
Subtract 4.	□□ ○○○ ○○○
Divide by 2.	□ ○○○
Subtract the number first thought of.	○○○
The result is 3.	

Here we have a proof that the result is *always* 3. Now it is easy to see why the choice of the original number makes no difference in the result.

Boxes and circles are somewhat clumsy, and so mathematicians prefer instead to use a letter of the alphabet to represent the original number and ordinary numerals for the other numbers. Here is the same proof written in the symbols of algebra.

Choose a number.	n
Add 5.	$n + 5$
Double the result.*	$2(n + 5)$ or $2n + 10$
Subtract 4.	$2n + 6$
Divide by 2.	$n + 3$
Subtract the number first thought of.	3
The result is 3.	

EXERCISES

SET I

If the original number in a number trick is represented by $\square$, what would represent

1. the result of doubling the number?

2. the result of adding 2 to the original number?

3. the number one larger than the original number?

If the number resulting from one of the steps of a number trick is represented by $\square \, \square \, \square \, \circ \circ \circ$, what would represent

4. the result of subtracting 3 from this number?

5. the result of dividing this number by 3?

6. the result of adding the original number to this number?

In algebraic symbols, $\square \, \square \, \circ \circ \circ \circ$ would be written as $2n + 4$. Write each of the following in algebraic symbols.

7. $\square \, \square \, \square \, \square \, \circ \circ$

8. $\square \, \circ \circ \circ \circ \circ$

9. $\circ \circ \circ \circ \circ$

10. $\square \, \square \, \square \, \square \, \square$

---

*The principle being used here is called the *distributive rule.* Look on pages 658–659 if you are not familiar with this idea.

The following number trick is illustrated with two different numbers. Proofs that it will always work are shown with boxes and circles, and with algebraic symbols.

	Examples		Proof 1	Proof 2
Choose a number.	5	8	□	n
Multiply by 3.	15	24	□ □ □	$3n$
Add 6.	21	30	□ □ □ ○○○ ○○○	$3n + 6$
Divide by 3.	7	10	□ ○○	$n + 2$
Subtract the number first thought of.	2	2	○○	2
The result is 2.				

Copy each of the following number tricks. *Using the same format as in the example above,* illustrate each trick with two different numbers. Then write two proofs that show that the trick will always work—one proof with boxes and circles, and one with algebraic symbols.

11. Choose a number.
    Add 3.
    Multiply by 2.
    Add 4.
    Divide by 2.
    Subtract the number
      first thought of.
    The result is 5.

12. Choose a number.
    Double it.
    Add 9.
    Add the number
      first thought of.
    Divide by 3.
    Add 4.
    Subtract the number
      first thought of.
    The result is 7.

13. Choose a number.
    Triple it.
    Add the number one larger
      than the number first thought of.
    Add 11.
    Divide by 4.
    Subtract 3.
    The result is the original number.

Set II

The following number tricks are of a different type from those in Set I.

Trick 1

>   Choose any three-digit number.
>   Multiply it by 7.
>   Multiply the result by 11.
>   Multiply the result by 13.

An example of carrying out these steps is shown at the right.

1. Do the trick again with a different three-digit number. Show all four steps.

2. What relation does the final result seem to have to the number that was originally chosen?

3. Suppose that you did this trick with many different three-digit numbers and it always worked. What kind of reasoning would you be using if you concluded that it will work for *every* three-digit number?

4. Find $7 \times 11 \times 13$.

524
$524 \times 7 = 3,668$
$3,668 \times 11 = 40,348$
$40,348 \times 13 = 524,524$

The number 1,001 is the key to this number trick.

5. *Without using a calculator,* multiply the three-digit number that you chose for exercise 1 by 1,001. Show your work.

6. Multiplying a number by 7, 11, and 13 in succession gives the same result as multiplying it by what number?

7. Does this seem to suggest that the trick will work for every three-digit number?

Trick 2

>   Choose any two-digit number.
>   Multiply it by 13.
>   Multiply the result by 21.
>   Multiply the result by 37.

8. Do this trick with a two-digit number. Show all of the steps.

9. Do it again with a different two-digit number.

10. What do you notice about the results?

11. Find $13 \times 21 \times 37$.

12. Why does the trick work?

Set III

Here is an old trick with numbers that has been used in magic shows. A variation of it was included in a book written by the New York magician Al Baker in 1923 that sold for $50 at the time.*

Write down the year of your birth.
Write down a year in which an important event in your life occurred.
Write down your age as of the last day of the current year.
Write down the number of years since the important event occurred.
Add the four numbers together.

1. Try this out. Show your work. What is the result?

2. If possible, try it out on someone else. If possible, choose someone who is not your age. What is the result?

If the steps are carried out correctly, the result can be correctly predicted without knowing any of the four numbers.

3. How do you think this is possible? (In other words, what do you suppose is special about the result?)

4. Can you explain how the trick works?

---

*Reported by Martin Gardner in his *Mathematics, Magic and Mystery.*

Summary and Review

In this chapter we have been introduced to:

Inductive reasoning *(Lessons 1, 2, 3, and 4).* Inductive reasoning is the method of drawing general conclusions from a limited set of observations. It is reasoning from the *particular* to the *general.*

Although inductive reasoning is of great importance in developing *theories,* it cannot prove that they are correct.

Deductive reasoning *(Lessons 5 and 6).* Deductive reasoning is a method of using logic to draw conclusions from statements that we accept as true. It is used for mathematical proofs. Statements that are proved true using deductive reasoning are called *theorems.*

EXERCISES

SET I

A prime number is a number that cannot be divided evenly by any whole number other than itself and 1. The first ten prime numbers are

$$2 \quad 3 \quad 5 \quad 7 \quad 11 \quad 13 \quad 17 \quad 19 \quad 23 \quad 29$$

Each of the first five even numbers larger than 4 can be written as the sum of two prime numbers as shown below.

$$6 = 3 + 3$$
$$8 = 3 + 5$$
$$10 = 3 + 7$$
$$12 = 5 + 7$$
$$14 = 3 + 11$$

1. Try to continue this list by expressing the next ten even numbers as the sum of two prime numbers.

$$16 = \text{▨} + \text{▨}$$
$$18 = \text{▨} + \text{▨}$$
$$20 = \text{▨} + \text{▨}$$

etc.

2. Can you find an even number larger than 4 that *cannot* be written as the sum of two prime numbers?

3. What kind of reasoning suggests that *every* even number larger than 4 can be written as the sum of two prime numbers?

Draw two billiard tables having the dimensions shown in the figures at the left.

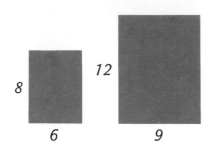

4. Show the path of a ball hit from the *upper-left* corner of each table at 45° angles with the sides.

5. What do you observe about the paths on the two tables?

6. Use the dimensions of the tables to explain why.

Albrecht Dürer, a German artist of the sixteenth century, made a famous engraving titled *Melancholy,* which contains an interesting square of numbers. Two of the numbers in the square have been blacked out in this copy of that engraving.

7. What number do you think appeared in the corner?

8. What number do you think appeared inside?

9. If you conclude from your observations that there is a pattern in the arrangement of the numbers in the square, what kind of reasoning are you using?

10. If you assume the pattern you observed is correct and use it to figure out what the missing numbers are, what kind of reasoning are you using?

Three golfers named Tom, Dick, and Harry are walking to the clubhouse. Tom, the best golfer of the three, always tells the truth. Dick sometimes tells the truth, while Harry, the worst golfer, never does.

11. To figure out who is who, it is best to first determine which one is Tom. Why?

12. How can you deduce which one is Tom from what each golfer says?

13. How can you determine which one is Harry?

14. Is Dick lying or telling the truth?

Set II

Do the following problems on your calculator.

1. 99×99

2. 999×999

3. $9,999 \times 9,999$

The answer to the third problem probably filled the display of the calculator.

4. Guess, on the basis of your three answers, the answer to this problem: $99,999 \times 99,999$

5. On what kind of reasoning is your guess based?

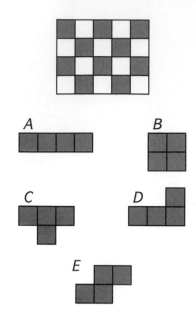

A   *B*

C   *D*

E

Here is a board consisting of 20 small squares and five pieces, each consisting of 4 squares.

Is there a way to arrange the five pieces on the board so that they completely cover it? To answer this, first notice that no matter where piece A is placed on the board, it will cover 2 brown squares and 2 white squares.

6. Copy and complete the following table for this and the other four pieces.

Piece	Squares it would cover
A	2 brown and 2 white
B	|||||||||
C	|||||||||
D	|||||||||
E	|||||||||

7. Can the pieces be arranged so as to completely cover the board? Explain why or why not.

Here is an interesting number trick.

> Choose any three-digit number whose first and last digits differ by more than one.
> Write down the same three digits in reverse order to form another three-digit number.
> Subtract the smaller of your two numbers from the larger and circle your answer.
> Reverse the digits in the number you circled to form another number and circle it.
> Add the two circled numbers together.

An example of carrying out these steps is shown below.

$$175 \qquad \begin{array}{r} 571 \\ -175 \\ \hline 396 \end{array} \qquad 396 \qquad \begin{array}{r} 396 \\ +693 \\ \hline 1{,}089 \end{array}$$

571   693

8. Do the trick with a different three-digit number. Show all of the steps.

9. What do you notice about the results?

10. If you did this trick with many different numbers and it always worked, would that prove that it will work for every such number?

The first step in the trick is to choose a number in which the first and last digits differ by *more than* one. Suppose the first and last digits differed by *exactly* one. An example of what would happen is shown below.

$$\begin{array}{cccc} 483 & 483 & 99 & 99 \\ \times & -384 & \times & +99 \\ 384 & \overline{(99)} & (99) & \overline{198} \end{array}$$

11. Do the trick with a different number whose first and last digits differ by exactly one. Show all of the steps.

12. What do you notice about the results?

13. What would happen if you started with a number in which the first and last digits were the *same*?

This puzzle appeared in an article by Martin Gardner titled "My Ten Favorite Brainteasers."

Miss Green, Miss Black, and Miss Blue are out for a stroll together. One is wearing a green dress, one a black dress, and one a blue dress.

"Isn't it odd," says Miss Blue, "that our dresses match our last names, but not one of us is wearing a dress that matches her own name?"

"So what?" said the lady in black.

14. What color is each lady's dress?

Set III

Here is a puzzle that can be solved by deductive reasoning.

The puzzle is to remove five matches from this figure so that three of the original squares are left with no extra matches.

1. How many matches would be left after the five are removed?

2. Since three squares are supposed to be left, can they share any of these matches as sides?

3. Show a solution to the puzzle.

4. How many solutions are there?

CHAPTER

1

Further Exploration

LESSON 1

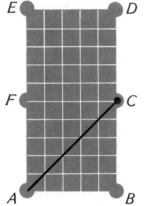

1. A pool table has six pockets: one at each corner and two at the midpoints of the longer sides. The first figure at the left shows a pool table of standard dimensions whose pockets have been labeled for reference. If a ball is hit from the corner labeled A at 45° angles with the sides, it will end up in the pocket labeled C. The second figure shows that, on a table one unit wider, a ball hit from corner A would end up in pocket D instead.

    a. Draw pool tables with other shapes and show the path of a ball from corner A at 45° angles from the sides of each table. Try to find examples in which the ball ends up in each of the other pockets.

    b. If the ball is hit from corner A at 45° angles from the sides, do you think there are any pockets on the table in which it cannot end up? If you do, explain.

2. In 1961, Dr. Andrés Zavrotsky of the University of the Andes in Venezuela patented a device that can be used to find the greatest common divisor of a pair of numbers.* It consists of four adjustable mirrors that can be used to form a rectangle with length and width equal to the two numbers. The figure on the facing page shows an overhead view of the mirrors adjusted to form a rectangle measuring 6 by 14.

    A beam of light is sent through a crack at the lower left corner at 45° angles with the sides of the rectangle. It is reflected from mirror to mirror until it lights up one of the other three corners.

    The greatest common divisor of the numbers under investigation is related to the number of the closest lighted point on mirror A to O, the corner where the light begins. This point is marked with an arrow in the figure.

---

*The greatest common divisor of a pair of numbers is the largest number that can be divided into both of them evenly. For example, the greatest common divisor of 18 and 24 is 6.

    Dr. Zavrotsky's device is described in *Martin Gardner's Sixth Book of Mathematical Games from Scientific American* (W. H. Freeman and Company, 1971).

In the illustration at right, the greatest common divisor of 6 and 14 is 2 and the closest lighted point on mirror A is at 4.

Show the path of the light when the mirrors are adjusted to form rectangles of the following dimensions.

a. 6×12
b. 8×12
c. 9×12
d. 11×12
e. How do you think the greatest common divisor of a pair of numbers can be found from the number of the closest lighted point on mirror A?

Lesson 2

So far in our study of the path of a billiard ball, we have not encountered any situations in which the ball came back to the point from which it was hit. If the ball is hit from a point other than a corner, it is possible for this to happen, in which case the ball travels in an endless loop.

Look, for example, at the four drawings below. If the ball is hit from corner A, it travels diagonally across the table, ending up in the opposite corner. If it is hit from points B, C, or D at a 45° angle from the lower side, however, it could theoretically continue rebounding from the walls forever.

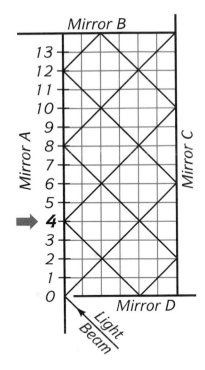

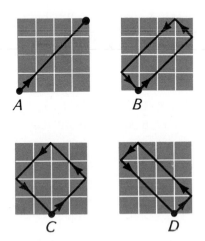

Is such a loop possible on a table that is not square? If it is, can the existence of such a path be predicted from the dimensions of the table?

Make as many drawings as you need to arrive at what you feel are reasonable conclusions. For simplicity, draw only paths that make 45° angles with the sides of the tables.

LESSON 3

1. Here is a remarkable discovery made by Lee Sallows.*

5	22	18
28	15	2
12	8	25

→

Five	Twenty two	Eighteen
Twenty eight	Fifteen	Two
Twelve	Eight	Twenty five

→

4	9	8
11	7	3
6	5	10

a. What is interesting about the square of numbers on the left? The names of the numbers in this square are listed in the figure in the center.

b. How is the square of numbers on the right related to it?

c. What is interesting about the square of numbers on the right?

2. Some fractions have decimal forms that end. For example,

$$\frac{1}{2} = 0.5$$

Other fractions have decimal forms that never end. For example,

$$\frac{1}{3} = 0.3333333 \ldots$$

Ending decimals

$\dfrac{1}{2}$

Non-ending decimals

$\dfrac{1}{3}$

a. Use a calculator to separate the following fractions into two sets: those whose decimal forms end and those whose decimal forms do not seem to end. Start your list with the two examples shown on the left.

$\dfrac{2}{3}$

$\dfrac{1}{6}, \dfrac{5}{6}$

$\dfrac{1}{9}, \dfrac{2}{9}, \dfrac{4}{9}, \dfrac{5}{9}, \dfrac{7}{9}, \dfrac{8}{9}$

$\dfrac{1}{4}, \dfrac{3}{4}$

$\dfrac{1}{7}, \dfrac{2}{7}, \dfrac{3}{7}, \dfrac{4}{7}, \dfrac{5}{7}, \dfrac{6}{7}$

$\dfrac{1}{10}, \dfrac{3}{10}, \dfrac{7}{10}, \dfrac{9}{10}$

$\dfrac{1}{5}, \dfrac{2}{5}, \dfrac{3}{5}, \dfrac{4}{5}$

$\dfrac{1}{8}, \dfrac{3}{8}, \dfrac{5}{8}, \dfrac{7}{8}$

All of the fractions you have tested have been in lowest terms.

b. Does either the numerator or denominator seem to have anything to do with the decimals that end?

---

*Reported in Chapter 5 of *Islands of Truth—A Mathematical Mystery Cruise*, by Ivars Peterson (W. H. Freeman and Company, 1990).

c. Use a calculator to add the following fractions to your list.

$$\frac{1}{11}, \frac{1}{12}, \frac{1}{13}, \frac{1}{14}, \frac{1}{15}, \frac{1}{16}, \frac{1}{17}, \frac{1}{18}, \frac{1}{19}, \frac{1}{20}$$

d. Can you guess a rule for telling whether or not a fraction in lowest terms has a decimal form that ends?

LESSON 4

1. This remarkable poem by Donald Knuth is titled *Disappearances.**

Enchanted words like "hocus pocus" can	I wonder how magicians make their rabbits disappear; not interfere
with laws of science	and facts of mathematics that are clear.
The prestidigitators, making use of devious schemes, (although they never tell you how)	transport things as in dreams:
At times	suspended, banished, null and void—or so it seems.
There must be something secret, yes, a trick that will	involve
—when done with sleight of hand—	a force that's able to *dissolve.*

If the poem is cut into three pieces along the lines indicated and the two right-hand pieces are interchanged, a new poem results.

Enchanted words like "hocus pocus" can	transport things as in dreams:
with laws of science	suspended, banished, null and void—or so it seems.
The prestidigitators, making use of devious schemes,	involve
(although they never tell you how)	a force that's able to *dissolve.*
At times	I wonder how magicians make their rabbits disappear; not interfere
There must be something secret, yes, a trick that will	
—when done with sleight of hand—	and facts of mathematics that are clear.

The original poem has eight lines but the new poem has only seven.

a. Which line of the original poem disappears?

b. What has happened?

**The Mathematical Gardner*, edited by David A. Klarner (Wadsworth, 1981.)

2. John Wilson, an English mathematician who lived in the eighteenth century, discovered something about prime numbers that he was never able to prove.

Number	Pattern		
2	$1 + 1 = 2$	Is 2 a factor of 2?	Yes.
3	$1 \times 2 + 1 = 3$	Is 3 a factor of 3?	Yes.
4	$1 \times 2 \times 3 + 1 = 7$	Is 4 a factor of 7?	No.
5	$1 \times 2 \times 3 \times 4 + 1 = 25$	Is 5 a factor of 25?	Yes.
6	$1 \times 2 \times 3 \times 4 \times 5 + 1 = 121$	Is 6 a factor of 121?	No.
7	$1 \times 2 \times 3 \times 4 \times 5 \times 6 + 1 = 721$	Is 7 a factor of 721?	Yes.
8	$1 \times 2 \times 3 \times 4 \times 5 \times 6 \times 7 + 1 = 5,041$	Is 8 a factor of 5,041?	No.
9	$1 \times 2 \times 3 \times 4 \times 5 \times 6 \times 7 \times 8 + 1 = 40,321$	Is 9 a factor of 40,321?	No.
10	$1 \times 2 \times 3 \times 4 \times 5 \times 6 \times 7 \times 8 \times 9 + 1 = 362,881$	Is 10 a factor of 362,881?	No.

a. Write what you think the next line of this table should be.

b. What connection do you think John Wilson thought existed between the numbers in the left column and the yeses or nos in the right column?

LESSON 5

1. Here is a set of symbols of the sort used by the Japanese to represent poetry as far back as A.D. 1000.*

1	⊦⊦⊦⊦	9	⊓⊓⊓
2	⊓⊦⊦	10	⊓⊓
3	⊓⊦	11	⊓⊓⊦
4	⊓⊦⊦	12	⊓⊓⊦
5	⊦⊓⊦	13	⊓⊓⊓
6	⊦⊓	14	⊦⊓⊓
7	?	15	⊓⊓⊓
8	⊓⊓		

Fractal Music, Hypercards and More, by Martin Gardner (W. H. Freeman and Company, 1992), Chapter 2.

a. One symbol has been omitted. Can you figure out what it is?

Symbol 12 represents the following lines from a poem by Robert Frost.

> Whose woods these are I think I know.
> His house is in the village though;
> He will not see me stopping here
> To watch his woods fill up with snow.
>
> "Stopping by Woods on a Snowy Evening," Robert Frost

b. What is the connection between symbol 12 and the lines by Frost?

Which symbol do you think represents each of the following?

c. Twas the night before Christmas, when all through the house
Not a creature was stirring—not even a mouse;
The stockings were hung by the chimney with care,
In hope that St. Nicholas soon would be there.

> "A Visit from St. Nicholas," Clement Clark Moore

d. It is an ancient Mariner,
And he stoppeth one of three.
"By thy long gray beard and glittering eye,
Now wherefore stopp'st thou me?"

> "The Ancient Mariner," Samuel Taylor Coleridge

e. How do I love thee? Let me count the ways.
I love thee to the depth and breadth and height
My soul can reach, when feeling out of sight
For the ends of Being and ideal Grace.

> *Sonnets from the Portuguese,* Elizabeth Barrett Browning

f. And here's the happy bounding flea—
You cannot tell the he from she.
The sexes look alike, you see;
But she can tell, and so can he.

> "The Flea," Roland Young

g. How many types of "poems" having three lines do you think are possible?

h. Draw their symbols.

2. The desk calendar shown in this photograph contains four cubes.* The two black cubes on the ends show the month and the day of the week, and the two white cubes in the middle show the date.

Each face of each white cube contains a single digit, and the two white cubes can be arranged so that their front faces indicate every date from 01 to 31.

Can you figure out what digits are on each cube given that the one on the right contains the digits 3, 4, and 5? If so, tell what they are.

Lesson 6

1. Here is a very old trick with dice.*

Throw three dice on a table and add the three numbers that turn up. Next, pick up any one of the three dice and add the number on the bottom of it to the previous total. Roll this die again and add the number that turns up to the total.

A magician who is familiar with this trick could now look at the three dice and, without knowing which die you rolled a second time, tell you the final sum.

   a. Try the trick several times. Keep a record of the results in a table like this.

Sum of first three numbers	Your final sum (the sum of all five numbers)	Sum of those numbers showing at end
‖‖‖‖	‖‖‖‖	‖‖‖‖
‖‖‖‖	‖‖‖‖	‖‖‖‖
‖‖‖‖	‖‖‖‖	‖‖‖‖

etc.

*This puzzle is by Martin Gardner and appears in his book titled *Mathematical Circus*.

b. How is the magician able to tell you your final sum?
c. What is it about dice that makes the trick possible?

2. Here is a clever card trick invented by Mark Wilson.

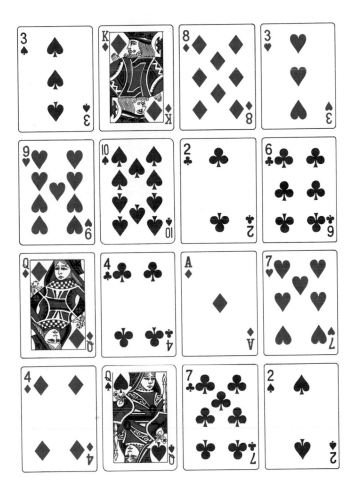

First, pick any brown card shown in this figure and place a coin on it. Now, take the coin and move it left or right to the nearest black card. Next, move the coin vertically up or down to the nearest brown card. Next, move diagonally to the nearest black card. Finally, move down or to the right to the nearest brown card.

a. Try the trick several times. What happens?
b. Can you figure out how the trick works? If so, explain.

Mathematics, Magic and Mystery, by Martin Gardner (Dover, 1956).

Number Sequences

By permission of Johnny Hart and Field Enterprises, Inc.

1

Arithmetic Sequences

A child first becomes aware of numbers through counting. Arranged in order, the counting numbers form a *number sequence:*

$$1 \quad 2 \quad 3 \quad 4 \quad 5 \quad 6 \quad 7 \quad 8 \quad 9 \quad 10 \quad 11 \ldots$$

A **number sequence** is an arrangement of numbers in which each successive number follows the last according to a uniform rule.

For the sequence of counting numbers, the rule is "add 1 to each number to get the next number." The symbol ". . .", called an *ellipsis,* indicates that the sequence continues. A mathematician once said: "A number sequence is like a bus; nobody ever doubts that there is always room for one more."

The numbers in a sequence are called its *terms.* To represent the terms of a number sequence, we will use the notation

$$t_1 \quad t_2 \quad t_3 \quad t_4 \quad t_5 \quad \ldots \quad t_n.$$

The little number on the right of each t is called a *subscript* and is the term number. It keeps track of where we are in the sequence; t_3 (pronounced "tee-sub-three"), for example, represents the third term. The symbol t_n represents the nth term.

Each successive term of the sequence

$$8 \qquad 20 \qquad 32 \qquad 44 \qquad 56 \qquad . . .$$

can be found by *adding the same number* to the preceding term.

$$8 \nearrow 20 \nearrow 32 \nearrow 44 \nearrow 56 \nearrow . . .$$
$$+12 \quad +12 \quad +12 \quad +12 \quad +12$$

Such a sequence is called *arithmetic* (pronounced "arith*me*tic").

> An **arithmetic sequence** is a number sequence in which each successive term may be found by adding the same number.

Another example of an arithmetic sequence is illustrated by this figure. The first term is 2 ($t_1 = 2$) and the number that is added is 3. This number is also called the *common difference* of the sequence because it is the difference between each pair of successive terms:

$$5 - 2 = 3, \qquad 8 - 5 = 3, \qquad 11 - 8 = 3,$$

and so on.

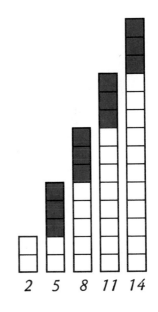

$$2 \quad 5 \quad 8 \quad 11 \quad 14$$

EXERCISES

SET I

In the *Chóu-peï,* an ancient Chinese book on numbers written about 1100 B.C., the following number sequences appear:

$$1 \quad 3 \quad 5 \quad 7 \quad 9 \quad 11 \quad . . .$$

and

$$2 \quad 4 \quad 6 \quad 8 \quad 10 \quad 12 \quad . . .$$

1. Are these sequences arithmetic? Explain why or why not.

2. What kind of numbers make up the first sequence?

3. What kind of numbers make up the second sequence?

The figure at the left shows three stacks of poker chips. The stacks illustrate the first three terms of an arithmetic sequence.

4. What are the terms?

5. What is their common difference?

6. What are the next three terms of the sequence?

When a parachutist jumps from an airplane, the distances in feet traveled during the first few seconds are

$$16 \quad 48 \quad 80 \quad 112 \quad 144 \quad . \, . \, .$$

7. What is the common difference of this sequence?

8. What number follows 144?

9. What do the three dots indicate?

January 1, 2000, falls on a Saturday.

10. List the days of that month that fall on Sunday.

11. What is the common difference of the sequence you wrote?

Copy the following arithmetic sequences, writing in the missing terms.

12. 2   7   12   17   ▓▓▓   ▓▓▓

13. 5   13   21   ▓▓▓   ▓▓▓

14. 11   15   ▓▓▓   23   ▓▓▓

15. ▓▓▓   ▓▓▓   20   29   38

16. 4   ▓▓▓   18   ▓▓▓   32

17. ▓▓▓   33   ▓▓▓   65   ▓▓▓

If a new candle is lit and burns at a steady rate, its height in centimeters at the end of each successive hour might produce the following arithmetic sequence:

$$24 \quad 21 \quad 18 \quad 15 \quad . \, . \, .$$

18. How many centimeters of the candle burn each hour?

19. What is the common difference of this sequence?

20. The first term, 24, represents the height of the candle after it had burned for 1 hour. How tall was the candle before it was lit?

21. Copy and continue the sequence until you get to the term 0.

22. How many hours did the candle last?

Find the missing terms in each of these arithmetic sequences.

23. 10 ▓▓ 70

24. 10 ▓▓ ▓▓ 70

25. 10 ▓▓ ▓▓ ▓▓ 70

26. 10 ▓▓ ▓▓ ▓▓ ▓▓ 70

27. 10 ▓▓ ▓▓ ▓▓ ▓▓ ▓▓ 70

In an auditorium, the numbers of the seats in the rows form an arithmetic sequence. The first row has 16 seats in it and the seventh row has 28 seats.

 16 ▓▓ ▓▓ ▓▓ ▓▓ ▓▓ 28

28. Figure out the common difference of this sequence.

29. Copy the sequence, writing in the missing terms.

30. How many seats would be in the tenth row?

Sandpaper is numbered according to how coarse or fine it is. Five of the grades form an arithmetic sequence whose first term is 240 and whose fifth term is 400.

240 ▓▓ ▓▓ ▓▓ 400

31. Figure out the common difference of this sequence.

32. Copy the sequence, writing in the missing terms.

SET II

The 100th term of the arithmetic sequence

$$1 \quad 2 \quad 3 \quad 4 \quad 5 \quad \cdots$$

is obvious: it is 100. The 100th term of the arithmetic sequence

$$2 \quad 5 \quad 8 \quad 11 \quad 14 \quad \cdots$$

however, is not obvious at all. One way to find out what it is would be to continue writing the sequence until we arrive at it. There is an easier way,

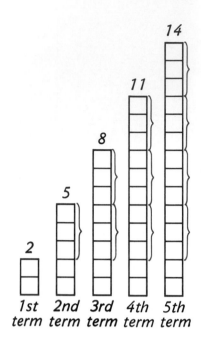

however. Look at the diagram at the left and the accompanying pattern below.

$$\begin{aligned}
\text{The 1st term, } t_1\text{, is} \quad & 2 + 3 \cdot 0 = 2 \\
\text{The 2nd term, } t_2\text{, is} \quad & 2 + 3 \cdot 1 = 5 \\
\text{The 3rd term, } t_3\text{, is} \quad & 2 + 3 \cdot 2 = 8 \\
\text{The 4th term, } t_4\text{, is} \quad & 2 + 3 \cdot 3 = 11 \\
\text{The 5th term, } t_5\text{, is} \quad & 2 + 3 \cdot 4 = 14
\end{aligned}$$

The 100th term must be

$$2 + 3 \cdot 99 = 2 + 297 = 299$$

1. Find the 10th term of the arithmetic sequence

$$3 \quad 10 \quad 17 \quad 24 \quad \ldots$$

by writing down the next six terms.

2. Find the 10th term of the same sequence by using the short-cut suggested by the pattern:

$$\begin{aligned}
\text{The 1st term, } t_1\text{, is} \quad & 3 + 7 \cdot 0 = 3 \\
\text{The 2nd term, } t_2\text{, is} \quad & 3 + 7 \cdot 1 = 10 \\
\text{The 3rd term, } t_3\text{, is} \quad & 3 + 7 \cdot 2 = 17 \\
\text{The 4th term, } t_4\text{, is} \quad & 3 + 7 \cdot 3 = 24
\end{aligned}$$

3. Find the 20th term of the arithmetic sequence

$$11 \quad 15 \quad 19 \quad 23 \quad \ldots$$

by writing down the next 16 terms.

4. Find the 20th term of the same sequence by using the short-cut suggested by the pattern:

$$\begin{aligned}
\text{The 1st term, } t_1\text{, is} \quad & 11 + 4 \cdot 0 = 11 \\
\text{The 2nd term, } t_2\text{, is} \quad & 11 + 4 \cdot 1 = 15 \\
\text{The 3rd term, } t_3\text{, is} \quad & 11 + 4 \cdot 2 = 19 \\
\text{The 4th term, } t_4\text{, is} \quad & 11 + 4 \cdot 3 = 23
\end{aligned}$$

Use shortcuts to find the indicated terms of the following arithmetic sequences.

5. The 11th term of 8   15   22   29   . . .

6. The 25th term of 6   10   14   18   . . .

7. The 16th term of   100   97   94   91   . . .

8. The 17th term of the same sequence.

9. The 40th term of   24   35   46   57   . . .

10. The 71st term of   5   30   55   80   . . .

11. The 72nd term of the same sequence.

The diagram below represents an arithmetic sequence whose first term is t_1 and whose common difference is d. Expressions for the first three terms are t_1, $t_1 + d$, and $t_1 + 2d$, respectively. Write expressions for

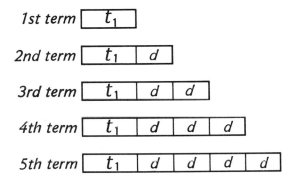

12. the 4th and 5th terms, t_4 and t_5.

13. the 10th term, t_{10}.

14. the nth term, t_n.

Set III

The students at a school in Ontario, Canada, once wanted to build the world's largest human pyramid. To do this, they all lay down on the ground and formed the pyramid shown here.

The numbers of people in the rows of the pyramid formed the arithmetic sequence

1   2   3   4   5   6   7   8   9   10   11   12   13   14   15   16

How many people were there altogether?

Before trying to answer this question, look at the two figures on the next page.

The figure below suggests a shortcut for finding the number of people in a pyramid with 6 rows.

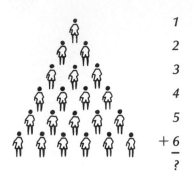

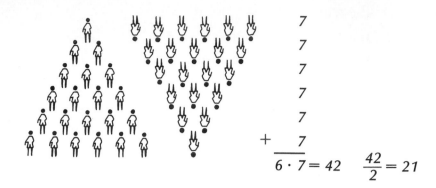

$$6 \cdot 7 = 42 \qquad \frac{42}{2} = 21$$

1. Find the number of people in the pyramid of 16 rows by using the shortcut suggested by this pattern:

1	2	3	4	5	6	7	...	16
+ 16	+ 15	+ 14	+ 13	+ 12	+ 11	+ 10	...	+ 1
17	17	17	17	17	17	17	...	17

2. Check your answer by adding the numbers in order from smallest to largest on a calculator:

$$1 + 2 + 3 + 4 + 5 + 6 + 7 + 8 + 9 + 10 + 11 + 12 + 13 + 14 + 15 + 16.$$

3. Find the sum of the 12 terms shown of this arithmetic sequence by adding them in order from left to right.

$$1 \quad 3 \quad 5 \quad 7 \quad 9 \quad 11 \quad 13 \quad 15 \quad 17 \quad 19 \quad 21 \quad 23$$

4. Find the same sum by using the shortcut suggested by this pattern:

1	3	5	...
+ 23	+ 21	+ 19	...
24	24	24	...

As every bowler knows, the pins are set up in four rows with one pin in the first row, two pins in the second row, three pins in the third, and four in the last. Suppose that a gigantic set of pins was set up in 20 rows, the first row having 1 pin and each successive row having 1 more, so that the last row had 20 pins.

5. How many pins would you have to knock over to make a strike?

6. If pins were set up in the same fashion in 100 rows, how many pins would you have to knock over to make a strike?

Denver Post

Geometric Sequences

In 1935, a chain letter craze started in Denver and swept across the country. It worked like this. You receive a letter with instructions and a list of six names. You send a dime to the person named at the top, cross the name out, and add your own name at the bottom. Then you send a copy of the new list of names to each of five friends with instructions to do the same. When your five friends send 5 letters each, there will be 25 letters in all. If all twenty-five people getting these letters cooperate, 125 letters will be sent, and so on.

A list of these numbers in order, starting with 1 for the letter you receive and ending with the number of lists on which your name appears at the top, is shown here.

$$1 \quad 5 \quad 25 \quad 125 \quad 625 \quad 3{,}125 \quad 15{,}625$$

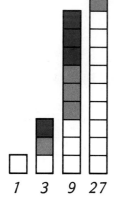

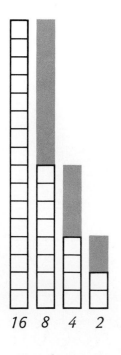

If each of the people receiving one of the 15,625 lists with your name at the top sends you a dime, you will receive $1,562.50. Not bad for an investment of only a dime and a few stamps. It is no wonder that this scheme, even though it was illegal, became popular when the United States was going through the Great Depression.

The terms in the list of numbers of chain letters sent at each stage form a number sequence because each successive number follows the preceding one according to a uniform rule. The rule, however, is to *multiply* each term by 5 to get the next term; so this is not an arithmetic sequence.

$$1 \xrightarrow[\times 5]{} 5 \xrightarrow[\times 5]{} 25 \xrightarrow[\times 5]{} 125 \xrightarrow[\times 5]{} 625 \xrightarrow[\times 5]{} \ldots$$

The sequence is *geometric* and, as the stacks of letters shown on page 67 indicate, it grows faster and faster.

> A **geometric sequence** is a number sequence in which each successive term may be found by *multiplying* by the same number.

If the number by which each term is multiplied is greater than 1, the sequence grows at an increasing rate. The left-hand figure above illustrates a geometric sequence resulting from repeatedly multiplying by 3. If each term is multiplied by a positive number that is less than 1, the sequence shrinks at a decreasing rate. The left-hand figure below illustrates a geometric sequence resulting from repeatedly multiplying by $\frac{1}{2}$.

All ratios of successive terms in a geometric sequence are the same. Called the *common ratio* of the sequence, it is the number by which each term is multiplied to get the next. The common ratio of the left-hand sequence is

$$\frac{3}{1} = \frac{9}{3} = \frac{27}{9} = 3$$

and the common ratio of the right-hand sequence is

$$\frac{8}{16} = \frac{4}{8} = \frac{2}{4} = \frac{1}{2}.$$

EXERCISES

SET I

The following number sequences appear on Babylonian cuneiform tablets of about 1800 B.C.

$$1 \quad 3 \quad 9 \quad 27 \quad 81 \quad . \, . \, .$$

and

$$1 \quad 4 \quad 16 \quad 64 \quad 125 \quad . \, . \, .$$

1. How can successive terms of the first sequence be found?
2. How can successive terms of the second sequence be found?
3. What kind of number sequence are they?

The figure at the right shows three stacks of poker chips. The stacks illustrate the first three terms of a geometric sequence.

4. What are the first three terms?
5. What is their common ratio?
6. What are the next three terms of the sequence?

Copy the following geometric sequences, writing in the missing terms.

7. 4   12   36   ▦   ▦
8. 11   121   ▦   ▦
9. 5   20   ▦   320   ▦
10. ▦   ▦   72   144   288
11. 0   ▦   ▦   ▦
12. 16   40   100   ▦   ▦
13. 81   54   ▦   ▦   ▦

Tell whether each of the following number sequences is arithmetic, geometric, or neither. If the sequence is arithmetic, give the common difference. If it is geometric, give the common ratio.

14. 5   10   15   20   25   . . .
15. 2   8   32   128   512   . . .
16. 1   3   6   10   15   . . .
17. 80   40   20   10   5   . . .

18. 3 20 37 54 71 . . .

19. 2 6 24 120 720 . . .

20. 60 51 42 33 24 . . .

21. 32 48 72 108 162 . . .

When a piano is tuned, the first note to be tuned is the A above middle C. It has a frequency of 440 cycles per second. Then the other seven A's on the keyboard are tuned so that their frequencies form a geometric sequence.

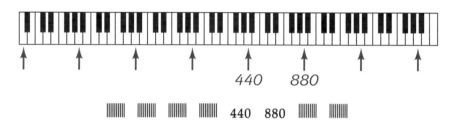

22. What is the common ratio of this sequence?

23. Find the frequencies of the other A's.

24. Write all eight terms of the sequence and then cross out every second term. Do the remaining terms form a geometric sequence?

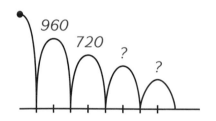

When a rubber ball is dropped from a height of 1280 centimeters, the heights in centimeters of the first few bounces form a geometric sequence:

960   720   ‖‖‖‖‖‖   ‖‖‖‖‖‖

25. What is the common ratio of this sequence?

26. Find the heights of the third and fourth bounces.

Set II

United States currency notes have been printed in the twelve denominations shown on the facing page.

Some of the denominations form the geometric sequence

1   10   100   . . .

1. What is the common ratio of this sequence?

2. How many terms of this sequence are denominations of currency?

3. Find another geometric sequence of currency denominations whose common ratio is 10.

Three of the denominations form the geometric sequence

$$2 \quad 100 \quad 5,000$$

4. What is the common ratio of this sequence?

Find a geometric sequence of currency denominations whose common ratio is

5. 2.

6. 5.

7. 100.

Compare the following patterns of the terms of an arithmetic sequence and a geometric sequence.

An arithmetic sequence: 2   5   8   11   14   . . .

t_1	2	$= 2 + 3 \cdot 0 = 2$
t_2	$2 + 3$	$= 2 + 3 \cdot 1 = 5$
t_3	$2 + 3 + 3$	$= 2 + 3 \cdot 2 = 8$
t_4	$2 + 3 + 3 + 3$	$= \text{▓▓▓▓▓} = 11$
t_5	$2 + 3 + 3 + 3 + 3$	$= \text{▓▓▓▓▓} = 14$

A geometric sequence: 2   6   18   54   162   . . .

t_1	2	$= 2 \cdot 3^0 = 2$
t_2	$2 \cdot 3$	$= 2 \cdot 3^1 = 6$
t_3	$2 \cdot 3 \cdot 3$	$= 2 \cdot 3^2 = 18$
t_4	$2 \cdot 3 \cdot 3 \cdot 3$	$= \text{▓▓▓▓} = 54$
t_5	$2 \cdot 3 \cdot 3 \cdot 3 \cdot 3$	$= \text{▓▓▓▓} = 162$

8. What expressions should be written in the indicated spaces for t_4 and t_5 to complete the pattern in the arithmetic sequence?

9. What expressions should be written in the indicated spaces for t_4 and t_5 to complete the pattern in the geometric sequence?

10. Use the pattern for the arithmetic sequence to write an expression for its 10th term.

11. Find the value of that term.

12. Use the pattern for the geometric sequence to write an expression for its 10th term.

13. Use a calculator to find the value of that term.

The first term of another geometric sequence is 6 and the common ratio of the sequence is 4.

14. Write the first five terms of the sequence.

15. Use the pattern to write an expression for the 100th term of the sequence. (The term has too many digits to find on a calculator. It is 2,410,407,066,388,485,413,312,943,138,511,743, 903,783,304,490,674,189,252,952,064!)

Write expressions for the indicated terms of the following geometric sequences.

16. The 11th term of   7   35   175   875   . . .

17. The 50th term of   2   20   200   2,000   . . .

18. The 123rd term of   3   24   192   1,536   . . .

Suppose the first term of a geometric sequence is t_1 and its common ratio is r. Expressions for its first three terms are t_1, $t_1 \cdot r$, and $t_1 \cdot r^2$, respectively. Write expressions for

19. the 8th term, t_8.

20. the nth term, t_n.

Set III

Although they are illegal, pyramid schemes for making money were a widespread phenomenon in the early 1980's. The operation of a typical pyramid is illustrated in the diagram below.

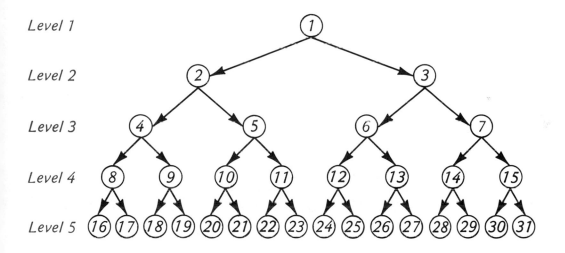

Thirty-one people are represented in the figure. Each person on levels 1 through 4 recruited the two people indicated by the arrows. You have to pay $1,000 to join the pyramid, $500 of it going to the person at the top and the other $500 to the person who recruited you. You immediately get your $1,000 back by recruiting two more people, who pay you $500 each.

At this point in the pyramid, the people on level 5 have just joined, each giving $500 to the person on level 1.

1. How much money did the person at the top get from the people on level 5?

Upon receipt of the money, the person at the top left the pyramid. In order for the two people on level 2 to make money, a sixth level of people had to join the pyramid.

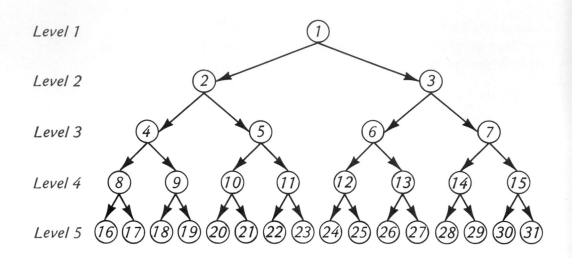

2. How many people were needed for the sixth level?

In order for the people on level 5 to make money, three more levels of people had to join the pyramid.

3. How many people were needed for each of these levels?

4. After these people joined, how many people had joined the pyramid altogether?

5. How many people in the pyramid had made money at this point?

6. How much money had they made altogether?

7. Where did most of this money come from?

8. Through how many levels would the pyramid have to continue for everyone who joined it to make money?

The Binary Sequence

According to legend, the game of chess was invented for a Persian king by one of his servants. The king was so pleased that he asked the servant what he would like as a reward. The man's request seemed very reasonable. He asked that one grain of wheat be placed on the first square of the chessboard, two grains on the second square, four grains on the third, and so on, each square having twice as many grains as the square before. The king was surprised, thinking that the servant had asked for very little. He was even more surprised when he found out how much wheat the man actually wanted.

The numbers of grain of wheat on the squares,

$$1 \quad 2 \quad 4 \quad 8 \quad 16 \quad 32 \quad 64 \quad 128 \quad . . .$$

form a geometric sequence, which grows at an increasing rate. Because there are 64 squares on a chessboard, the inventor's request was for as many grains of wheat as the sum of the first 64 terms of this sequence.

This number,

$$18,446,744,073,709,551,615$$

is equivalent to approximately 175 billion tons of wheat, more than that which has been produced on the earth in recorded history.

The sequence

$$1 \quad 2 \quad 4 \quad 8 \quad 16 \quad . \, . \, .$$

is called the **binary sequence.**

Number	Binary sequence
1	1 2 4 8 . . .
2	1 2 4 8 . . .
3	1 2 4 8 . . .
4	1 2 4 8 . . .
5	1 2 4 8 . . .
6	1 2 4 8 . . .
7	1 2 4 8 . . .
8	1 2 4 8 . . .
9	1 2 4 8 . . .
10	1 2 4 8 . . .

The binary sequence has a remarkable property: every counting number can be expressed as the sum of one or more of its terms. This is illustrated for the numbers from 1 through 10 in the table at the left.

If an electric circuit consisted of a sequence of switches corresponding to the terms of the binary sequence, the counting numbers could be represented by turning the switches on or off as illustrated in the table. For example, to represent the number 3, the switches for 1 and 2 would be turned on. To represent the number 10, the switches for 2 and 8 would be turned on, and so forth.

For this reason, the binary sequence is used in the representation of numbers in the circuits of electronic computers and calculators. The circuits representing the terms of the binary sequence can be seen in the computer-generated plot on paper of a Dynamic Random-Access Memory Chip (DRAM).

To write a number in the form in which it is used in a computer, two digits are used: 1 to show that a switch is on and 0 to show that it is off. Because of this, such numbers are said to be written in *base 2,* or as *binary numerals.*

The binary numerals for the numbers from 1 to 10 are given in the table below. Note that 0's to the left of the first 1 are customarily omitted.

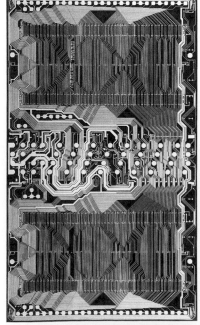

IBM, Burlington, Vermont. Computer generated plot on paper of Dynamic Random-Access Memory Chip (DRAM). 1982. 288,000 transistors. Collection, The Museum of Modern Art, New York. Gift of the manufacturer, courtesy of the IBM Corporation.

Number	Binary numeral					Number	Binary numeral			
	8	4	2	1			8	4	2	1
1				1		6		1	1	0
2			1	0		7		1	1	1
3			1	1		8	1	0	0	0
4		1	0	0		9	1	0	0	1
5		1	0	1		10	1	0	1	0

EXERCISES

SET I

The photograph at the right shows the pattern in the lengths of the marks along 1 inch of a rule. Along the edge, there are 16 small units, which have been numbered.

1. Look at the increasingly longer marks, shown in color. What are the numbers above these marks?

2. What sequence do they form?

3. Write the next five terms of this sequence.

Every counting number can be expressed as the sum of one or more terms of the binary sequence in which no term appears more than once. For example,

$$19 = 16 + 2 + 1$$

and

$$100 = 64 + 32 + 4$$

Write each of the following numbers as the sum of one or more terms of the binary sequence in which no term appears more than once.

4. 5.

5. 20.

6. 42.

7. 71.

8. 95.

Because every counting number can be expressed in terms of the binary sequence, every counting number can be written as a *binary numeral*. The 1's tell us to include the number and the 0's tell us not to include the number. Look at the examples in this table.

Number	Binary numeral						
	64	32	16	8	4	2	1
19			1	0	0	1	1
100	1	1	0	0	1	0	0

Copy the table and then add the following numbers to it, expressing each as a binary numeral.

9. 5.   10. 20.   11. 42.   12. 71.   13. 95.

The largest number that can be expressed as a three-digit binary numeral is 7 because

$$111 \text{ means } 4 + 2 + 1 = 7$$

What is the largest number that can be expressed as

14. a four-digit binary numeral?

15. a five-digit binary numeral?

16. a six-digit binary numeral?

To change a binary numeral such as 11001 into decimal form (our usual base 10 number system), write the place value of each digit above it:

$$
\begin{array}{ccccc}
16 & 8 & 4 & 2 & 1 \\
1 & 1 & 0 & 0 & 1
\end{array}
$$

Adding the place values of the digits that are 1's, we get

$$16 + 8 + 1 = 25$$

Use the same method to change each of the following binary numerals into decimal form:

17. 10100.

18. 101000.

19. 1101001.

20. 11010010.

Adding a zero to the end of a number in decimal form is equivalent to multiplying the number by ten. Compare, for example, 57 and 570.

21. What is adding a zero to the end of a binary numeral equivalent to? (Look at exercises 17 and 18 and your answers, as well as exercises 19 and 20.)

In 1605, the English statesman Francis Bacon devised a code for sending secret diplomatic messages. Each letter of the alphabet was represented by a five-letter group of *a*'s and *b*'s as shown in the table at the top of the next page. (The letters *j* and *v* were not in the alphabet at that time.)

A	B	C	D	E	F
Aaaaa	aaaab	aaaba.	aaabb.	aabaa.	aabab.

G	H	I	K	L	M
aabba	aabbb	abaaa.	abaab.	ababa.	ababb.

N	O	P	Q	R	S
abbaa.	abbab.	abbba.	abbbb.	baaaa.	baaab.

T	U	W	X	Y	Z
baaba.	baabb.	babaa.	babab.	babba.	babbb.

Courtesy of F. G. Heath

If the *a*'s are changed to 0's and the *b*'s to 1's, the five-letter groups in Bacon's code become five-digit binary numerals. For example, the five-letter group for D, *aaabb*, becomes 00011, which is 3 in decimal form.

22. What numbers in decimal form do the five-letter groups for the rest of the alphabet become?

23. How many five-letter groups of *a*'s and *b*'s are possible in all? (Hint: The first letter group is *aaaaa* and the last is *bbbbb*.)

Set II

The number of grains of wheat said to have been requested by the inventor of chess is the sum of the first 64 terms of the binary sequence. The following figures suggest a way to find that sum without doing any addition.

2

?

?

?

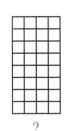

?

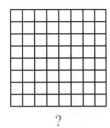

?

?

1

1 + 2 = ?

1 + 2 + 4 = ?

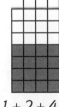

1 + 2 + 4 + 8 = ?

1 + 2 + 4 + 8 + 16 = ?

1 + 2 + 4 + 8 + 16 + 32 = ?

1. Copy and complete the following sequence of the numbers of small squares in the first row of figures:

2

2. Copy and complete the following sequence of the numbers of small squares in the second row of figures:

1

3. How are the numbers in your answer to exercise 2 related to the numbers in your answer to exercise 1?

4. Copy and complete this list to show the first 12 terms of the binary sequence.

$$
\begin{array}{ccccccc}
t_1 & t_2 & t_3 & t_4 & t_5 & t_6 & t_7 \quad \cdots \\
1 & 2 & 4 & 8 & 16 & 32 & 64 \quad \cdots
\end{array}
$$

5. Find the sum of the first seven terms of the binary sequence by adding them:

$$1 + 2 + 4 + 8 + 16 + 32 + 64$$

6. Show a way to find the sum of the first seven terms without doing any addition.

7. What is the sum of the first 10 terms of the binary sequence?

8. Which term of the binary sequence would you need to know in order to quickly find the sum of the first 64 terms?

The value of that term is 18,446,744,073,709,551,616.

9. What is the sum of the first 64 terms of the binary sequence?

The martingale system of betting calls for doubling your bet after any losing bet. If you bet $1 and lose, then according to the system you should bet $2.

Suppose you are playing a game in which, if you win, you get back what you bet plus an equal amount. For example, if you bet $5 and win, you get back the $5 plus $5 more.

Suppose that in playing this game, you start with a bet of $1 and lose five bets in a row:

$$\$1 + \$2 + \$4 + \$8 + \$16$$

10. How much money have you lost altogether?

"What I can't understand—why doesn't *everyone* just keep doubling their bets?"

11. According to the martingale system, how much money should your sixth bet be?

12. If you win the sixth bet, how much money ahead or behind are you for the six bets altogether?

Suppose you start with a bet of $1 and lose seven bets in a row.

13. How much money have you lost altogether?

14. How much money should your eighth bet be?

15. If you win the eighth bet, how much money ahead or behind are you for the eight bets altogether?

Suppose you are so rich that you never have to worry about running out of money and that you start with a bet of $1. Suppose, furthermore, that you are extremely unlucky and lose 64 bets in a row!

16. How much money have you lost altogether? (Think of the chessboard.)

17. How much money should your 65th bet be?

18. If you win the 65th bet, how much money ahead or behind are you for the 65 bets altogether?

19. If you lose the 65th bet, how much money behind are you altogether?

SET III

After thinking that he had invented the binary numerals, the seventeenth-century German mathematician Gottfried Leibniz was astonished to find that an ancient Chinese book, the *I Ching,* contained a set of numbered figures, called hexagrams. Each hexagram consists of six lines, each of which is either solid or broken. The first eight, together with their names, are shown below.

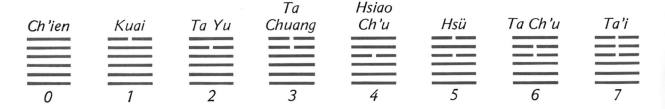

1. The hexagrams are related to the binary numerals in a simple way. What is it?

2. How many such hexagrams are possible? Explain.

3. The hexagrams corresponding to the numbers 10, 20, 30, 40, and 50 have the names K'uei, Chia jên, I, Sung, and Lü, respectively. If you know how they are related to the binary numerals, you should be able to draw them.

University Museum, University of Pennsylvania

The Sequence of Squares

Among the many clay tablets made by the Babylonians about 4,000 years ago are some that contain number sequences. In addition to tablets containing arithmetic and geometric sequences, one tablet shows the first 60 terms of the sequence that begins

$$1 \quad 4 \quad 9 \quad 16 \quad 25 \quad \ldots$$

These numbers come from multiplying each counting number by itself:

$$1 \times 1 = 1, \quad 2 \times 2 = 4, \quad 3 \times 3 = 9, \quad 4 \times 4 = 16, \quad 5 \times 5 = 25, \ldots$$

Multiplying a number by itself is called *squaring* the number and is represented by writing a small 2 at the upper right of the number being squared. The 2 is called an *exponent.* For example, 5×5 can be written as 5^2, which is read as "5 squared."

The **sequence of squares** is

$$1^2 \quad 2^2 \quad 3^2 \quad 4^2 \quad 5^2 \quad \ldots$$

or, equivalently,

$$1 \quad 4 \quad 9 \quad 16 \quad 25 \quad \ldots$$

In about 600 B.C., mathematicians in Greece began representing numbers with dots arranged in geometric shapes. For example, the figures at the left show how they represented the sequence of square numbers. These figures reveal how the square numbers got their name. The square numbers are the numbers that can be represented by dots in square arrays. The number 25, for example, is a square number because it is the number of dots in a square array with 5 dots per side. In addition to calling 25 the *square* of 5, we call 5 a *square root* of 25. In symbols, $25 = 5^2$ and $5 = \sqrt{25}$.

The sequence of squares has an interesting relation to an arithmetic sequence — the sequence of odd numbers:

$$1 \quad 3 \quad 5 \quad 7 \quad 9 \quad 11 \quad \ldots$$

Any sum of consecutive odd numbers starting with one is a square.

$$
\begin{aligned}
1 &= 1 = 1^2 \\
1 + 3 &= 4 = 2^2 \\
1 + 3 + 5 &= 9 = 3^2 \\
1 + 3 + 5 + 7 &= 16 = 4^2 \\
1 + 3 + 5 + 7 + 9 &= 25 = 5^2
\end{aligned}
$$

A visual demonstration of this relationship is shown at the left.

EXERCISES

SET I

The first six terms in the sequence of squares are

$$1 \quad 4 \quad 9 \quad 16 \quad 25 \quad 36$$

1. Copy the six terms and continue the sequence by writing the next six terms.

2. Find the difference between each pair of consecutive terms in your answer to exercise 1 as shown below.

3. What do you notice about the resulting sequence?

4. What is the common difference between each pair of consecutive terms in the new sequence?

This figure represents a ball rolling down a ramp 25 meters long. It takes 5 seconds for the ball to roll from the top to the bottom, and the positions of the ball at the end of each second are shown.

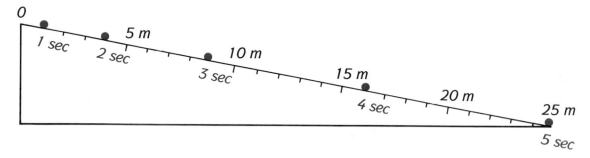

5. Write a number sequence representing the total distances traveled by the ball during the first second, during the first 2 seconds, during the first 3 seconds, during the first 4 seconds, and during the first 5 seconds.

6. What is the name of this sequence?

7. Write a number sequence representing the distances traveled by the ball during the first second, during the second second, during the third second, during the fourth second, and during the fifth second.

8. What is the name of this sequence?

9. The common difference in the sequence that you wrote for exercise 7 is called the *acceleration* of the ball in "meters per second per second." What is the acceleration of the ball?

The *perimeter* of a square is the number of linear units around its border. The figures below illustrate the perimeters of a series of squares having sides of lengths 1, 2, 3, 4, 5, and 6. The perimeter of the first square is

$$1 + 1 + 1 + 1 = 4$$

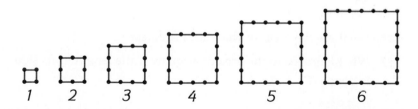

and the perimeter of the second square is

$$2 + 2 + 2 + 2 = 8$$

10. Copy and complete the following number sequence of the perimeters of these squares.

4 8

11. What kind of sequence do the perimeters form?

The *area* of a square is the number of square units inside. The figures below illustrate the areas of the same squares.

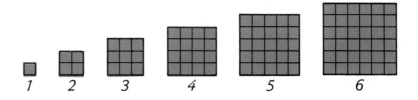

1 2 3 4 5 6

12. Copy and complete the following number sequence of the areas of these squares.

1 4

13. What kind of sequence do the areas form?

These figures illustrate a sequence of squares in which the length of the side is successively doubled.

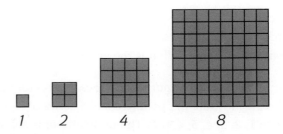

1 2 4 8

14. What are the perimeters of these four squares?

15. What happens to the perimeter of a square if the length of its side is doubled?

16. What are the areas of these four squares?

17. What happens to the area of a square if the length of its side is doubled?

At the beginning of a game of pool, the 15 numbered balls are arranged in the triangular pattern shown here.

For this reason, the number 15 is called a *triangular* number. The first terms in the sequence of triangular numbers are illustrated by the figures below.

18. Write the five numbers illustrated and continue the sequence to show the next five terms.

19. Copy and complete the following pattern.

$$
\begin{aligned}
1 &= 1 \\
1 + 2 &= \text{\tiny{||||||}} \\
1 + 2 + 3 &= \text{\tiny{||||||}} \\
1 + 2 + 3 + 4 &= \text{\tiny{||||||}} \\
1 + 2 + 3 + 4 + 5 &= \text{\tiny{||||||}}
\end{aligned}
$$

20. What kind of numbers are the numbers in the right column?

21. Write the first 10 triangular numbers and then add each pair of consecutive numbers as shown below.

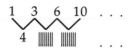

22. What do you notice about the resulting sequence?

23. What connection between the square numbers and the triangular numbers do the figures below illustrate?

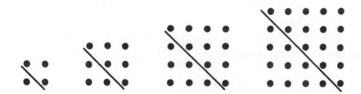

Set II

A table of the squares of the numbers from 1 through 40 is shown below.

No.	Square	No.	Square	No.	Square	No.	Square
1	1	11	121	21	441	31	961
2	4	12	144	22	484	32	1,024
3	9	13	169	23	529	33	1,089
4	16	14	196	24	576	34	1,156
5	25	15	225	25	625	35	1,225
6	36	16	256	26	676	36	1,296
7	49	17	289	27	729	37	1,369
8	64	18	324	28	784	38	1,444
9	81	19	361	29	841	39	1,521
10	100	20	400	30	900	40	1,600

Refer to the table to answer the following questions.

1. Can the square of an even number be odd?
2. Can the square of an odd number be even?

The fourth line of this table.

4 ⟶ 16   14 ⟶ 196   24 ⟶ 576   34 ⟶ 1,156

suggests that if the last digit of a number is 4, the last digit of its square is 6.

3. If the last digit of a number is 7, what is the last digit of its square?
4. For which digits is the last digit of a number the *same* as the last digit of its square?
5. Can a square number end in *any* digit? If not, in which digits can it not end?

One of the following numbers is a square and the others are not:

1,372   2,137   3,721   7,213

6. Which number do you think is the square? Why?

7. Can a square number of two or more digits have all even digits? If so, give an example.

8. Do you think that a square number of two or more digits can have all odd digits? If so, give an example.

9. Make a table of squares of the numbers from 51 to 59.

10. Do you notice a pattern in your table? If so, describe it.

The last digits of the squares of the numbers from 1 through 9 form a pattern.

11. Copy and complete this list of the last digits of the squares of the numbers from 1 through 9.

1   4   9   6   5   ▦   ▦   ▦   ▦

The pattern of numbers in exercise 11 and words such as RADAR, PEEP, and HANNAH are examples of *palindromes.*

12. On the basis of these examples, how would you define *palindrome?*

The *digital root* of a number is found by adding its digits, adding the digits of the resulting number, and so forth, until the result is a single digit. For example, the steps in finding the digital root of 529 are:

$$5 + 2 + 9 = 16; \quad 1 + 6 = 7$$

The digital root of 529 is 7.

13. Copy and complete the following table of the digital roots of the squares of the numbers from 1 through 10.

Number	Square	Digital root of square*	Number	Square	Digital root of square*
1	1	1	6	▦	▦
2	4	4	7	▦	▦
3	9	9	8	▦	▦
4	16	7	9	▦	▦
5	▦	▦	10	▦	▦

*Remember that the digital root of a number is always a *single* digit.

14. Construct a similar table for the digital roots of the squares of the numbers from 11 through 20.

15. If a number is a square, do you think its digital root can be any number? If not, what numbers can be its digital root?

16. What pattern do you notice in the digital roots of the squares of the numbers from 1 through 8?

SET III

Number sequences sometimes appear in science in mysterious ways. An interesting example is the relation of the elements in chemistry.

The elements can be listed in a table according to the arrangements of the electrons in their atoms so that elements with similar properties appear in columns. The first six rows of this table, called the periodic table, are shown below.

Subgroups of elements within the rows are named *s*, *p*, *d*, and *f*. Only the sixth row has elements in all four subgroups.

Notice that each *s* subgroup contains only two elements.

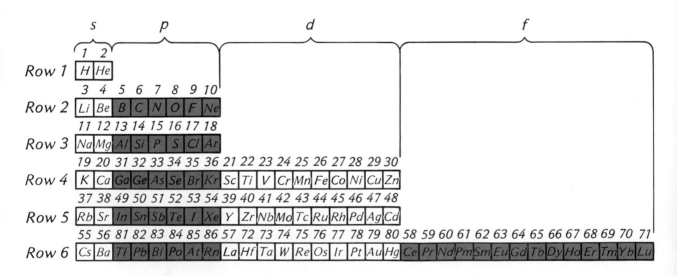

1. Copy and complete the following table of the possible numbers of elements in each subgroup.

Subgroup	*s*	*p*	*d*	*f*
Number of elements	2	▊▊▊	▊▊▊	▊▊▊

2. If there were another subgroup following *f*, how many elements do you think it could contain?

3. Copy and complete the following table of the numbers of elements in each row of the periodic table.

Row	1	2	3	4	5	6
Number of elements	2	8	▓	▓	▓	▓

Notice that all of the numbers in the second row of the table for exercise 3 are *even*.

4. Divide each number in the second row in half to make a new table as shown below.

Row	1	2	3	4	5	6
Half of number of elements	1	4	▓	▓	▓	▓

5. What do the numbers in the second row of your new table have in common?

At present, 105 elements are known, of which 7 have been created in the laboratory since 1950. If the number of new elements to be created were to increase indefinitely, how many elements would you predict for

6. the seventh row of the periodic table?

7. the eighth row of the periodic table?

By permission of Johnny Hart and Field Enterprises, Inc.

5

The Sequence of Cubes

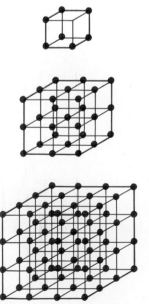

In addition to picturing the square numbers as square arrays of dots, the ancient Greeks pictured other numbers with cubic arrays of dots. The number 8, for example, was pictured as a cube having 2 dots on each edge. Eight can also be represented as $2 \times 2 \times 2$ or 2^3, which is read as "two cubed." Cubes having 3 and 4 dots on each edge contain $3 \times 3 \times 3 = 3^3 = 27$ dots and $4 \times 4 \times 4 = 4^3 = 64$ dots, respectively. For this reason, 27 and 64 are called cube numbers.

The **sequence of cubes** is

$$1^3 \qquad 2^3 \qquad 3^3 \qquad 4^3 \qquad 5^3 \qquad \ldots$$

or, equivalently,

$$1 \qquad 8 \qquad 27 \qquad 64 \qquad 125 \qquad \ldots$$

In the same way that square numbers are related to the *area* of a square (that is, the amount of *surface* that it occupies), cube numbers are related to the *volume* of a cube (the amount of three-dimensional *space* that it occupies). A square measuring 4 centimeters along each side has an area of $4^2 = 16$ square centimeters; a cube measuring 4 centimeters along each edge has a volume of $4^3 = 64$ cubic centimeters.

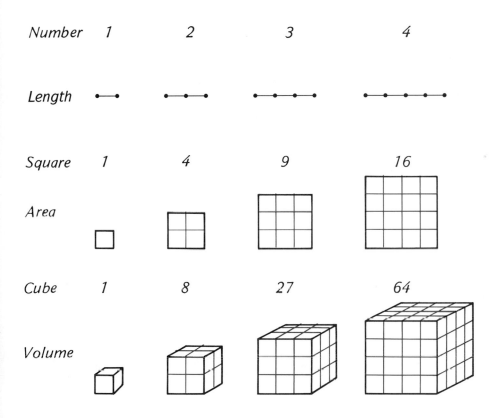

Because the sequence of squares can be written as

$$1^2 \quad 2^2 \quad 3^2 \quad 4^2 \quad 5^2 \quad . \, . \, .$$

and the sequence of cubes as

$$1^3 \quad 2^3 \quad 3^3 \quad 4^3 \quad 5^3 \quad . \, . \, .$$

they are also referred to as the sequences of second and third powers, respectively. We have seen that second powers can be pictured in two dimensions using squares and third powers in three dimensions using cubes. Although we cannot picture sequences of powers higher than the third with arrays of dots (since this requires more than three dimensions), such sequences are both interesting and useful.

The **sequence of fourth powers,**

$$1^4 \quad 2^4 \quad 3^4 \quad 4^4 \quad 5^4 \quad \ldots$$

or

$$1 \quad 16 \quad 81 \quad 256 \quad 625 \quad \ldots$$

can be applied, for example, to relating the luminosity of a star to its temperature.*

EXERCISES

SET I

The first five terms in the sequence of cubes are

$$1 \quad 8 \quad 27 \quad 64 \quad 125$$

1. Copy the five terms and continue the sequence by writing the next five terms.

These figures illustrate pennies arranged in the form of six-sided figures called *hexagons.* Use your imagination to "see" the hexagons. (The single penny will probably require the most imagination.)

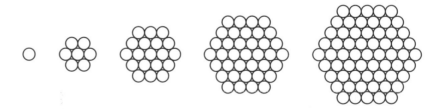

2. Copy and complete the following sequence of the numbers of pennies in each figure.

$$1 \quad 7$$

There is 1 penny in the first figure and there are $1 + 7 = 8$ pennies altogether in the first two figures. How many pennies are there altogether

3. in the first three figures?

4. in the first four figures?

*Stars, by James B. Kaler (Scientific American Library, 1992), p. 72.

5. in the first five figures?

6. What do you notice about these numbers?

The figures below illustrate cubes having edges of lengths 1, 2, 3, 4, and 5.
The *surface area* of a cube is the number of square units on its six faces.

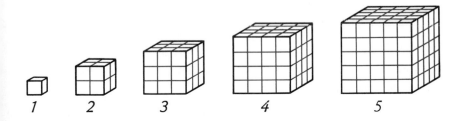

The surface area of the first cube is

$$6 \times 1 = 6$$

and the surface area of the second cube is

$$6 \times 4 = 24$$

7. Copy and complete the following number sequence of the
   surface areas of the cubes shown.

$$6 \quad 24 \quad \text{▓▓▓} \quad \text{▓▓▓} \quad \text{▓▓▓}$$

The *volume* of a cube is the number of cubic units that it contains.

8. Copy and complete the following number sequence of the
   volumes of the cubes shown.

$$1 \quad 8 \quad \text{▓▓▓} \quad \text{▓▓▓} \quad \text{▓▓▓}$$

The following table represents a sequence of cubes in which the length of
the edge is successively doubled.

Length of edge	1	2	4	8	16
Surface area	6	24	▓▓▓	▓▓▓	▓▓▓
Volume	1	8	▓▓▓	▓▓▓	▓▓▓

9. Copy and complete the table.

10. What happens to the surface area of a cube if the length of its
    edge is doubled?

11. What happens to the volume of a cube if the length of its
    edge is doubled?

The weight that a column can support is related to the fourth power of its diameter.

12. If the diameter of the column is multiplied by 3, it can support 3^4 times as much weight. How many times is that?

13. Copy and complete the following sequence of fourth powers.

$$1^4 \quad 2^4 \quad 3^4 \quad 4^4 \quad 5^4 \quad 6^4$$
1 ▨ ▨ ▨ ▨ ▨

Each of the fourth powers you have written can also be written as a square. For example,

$$4^4 = 4 \times 4 \times 4 \times 4 = 16 \times 16 = 16^2$$

14. Copy and complete the following pattern by writing each fourth power as a square.

$$1^4 \quad 2^4 \quad 3^4 \quad 4^4 \quad 5^4 \quad 6^4$$
▨ ▨ ▨ 16^2 ▨ ▨

The amount of power needed by a boat traveling at a high speed is related to the seventh power of its speed.

15. If the speed of a boat is multiplied by 2, the amount of power the boat needs is multiplied by 2^7. Find the value of 2^7.

16. Copy and complete the following sequence of seventh powers.

$$1^7 \quad 2^7 \quad 3^7 \quad 4^7 \quad 5^7$$
1 ▨ ▨ ▨ ▨

Recall that the digital root of a number is found by adding its digits, adding the digits of the resulting number, and so forth, until the result is a single digit.

Number	Cube	Digital root of cube
1	1	1
2	8	8
3	27	9
4	▨	▨
5	▨	▨
6	▨	▨
7	▨	▨
8	▨	▨
9	▨	▨
10	▨	▨
11	▨	▨
12	▨	▨

17. Copy and complete the table at the left of the digital roots of the cubes of the numbers from 1 through 12.

18. What pattern do you notice?

$$1 \qquad\qquad = 1$$
$$3 + 5 \qquad\quad = ▨$$
$$7 + 9 + 11 \quad\; = ▨$$
$$13 + 15 + 17 + 19 = ▨$$

Set II

The following exercises refer to the pattern at the left.

1. Copy it, filling in the missing numbers.

2. What sequence do the numbers on the left sides of the equations form?

3. What sequence do the numbers on the right sides of the equations form?

4. Write the next two lines of the pattern.

5. Are they also true?

The following exercises refer to this pattern.

$$
\begin{aligned}
1 &= 1 \\
1 + 8 &= \text{|||||||} \\
1 + 8 + 27 &= \text{|||||||} \\
1 + 8 + 27 + 64 &= \text{|||||||}
\end{aligned}
$$

6. Copy it, filling in the missing numbers.

7. To what sequence do the numbers on the left sides of the equations belong?

8. To what sequence do the numbers on the right sides of the equations belong?

9. Rewrite the pattern in exercise 6, using exponents. Write the first line as $1^3 = 1^2$.

10. What is a shortcut for finding the last number on each line?

11. Write the next two lines of the pattern.

12. Are they also true?

The following exercises refer to this pattern.

$$
\begin{aligned}
3^2 + 4^2 &= \text{|||||||}^2 \\
3^3 + 4^3 + 5^3 &= \text{|||||||}^3
\end{aligned}
$$

13. Copy it, filling in the missing numbers.

14. Write what you think is the next line of the pattern.

15. Is it also true? Show why or why not.

Set III

Several science-fiction films and television programs have been based on the theme of human beings changing size.* If people could become larger

---

*Among them are *The Incredible Shrinking Woman*, *The Amazing Colossal Man*, "Land of the Giants," and *Honey, I Shrunk the Kids*.

From the motion picture Dr. Cyclops, 1939; courtesy of Universal Pictures.

or smaller, their physical characteristics would not change at the same rate. Their weight, for example, would vary with the *cube* of their height, whereas the strength of their bones would vary with its *square*. This means that someone who became 2 times his or her normal height would have bones $2^2 = 4$ times as strong and be $2^3 = 8$ times as heavy.

1. What would happen to the strength and weight of a man who became 12 times his normal height?

2. A person's thighbones can support as much as 10 times his weight. Why would the legs of a man who became 12 times his normal height break when he stood up?

3. What would happen to the strength and weight of a woman who became one-tenth her normal height?

4. To become stronger in proportion to his or her weight, which should a person do: grow or shrink? Explain.

Leonardo of Pisa, who was called Fibonacci
Courtesy of Columbia University Libraries

The Fibonacci Sequence

One of the greatest mathematicians of the Middle Ages was Leonardo of Pisa, called Fibonacci. The construction of the famous Leaning Tower in that city was begun during his lifetime but was not completed for nearly two centuries. Fibonacci studied in North Africa, where he learned mathematical works available only in Arabic. In 1202, he wrote a book on arithmetic and algebra titled the *Liber Abaci.* This book was influential in introducing into Europe the Hindu – Arabic numerals — 0, 1, 2, 3, 4, 5, 6, 7, 8, and 9 — with which we now write numbers.

One of the many interesting problems in this book was about rabbits. It went like this.

A pair of rabbits one month old are too young to produce more rabbits, but suppose that in their second month and every month there-after they produce a new pair. If each new pair of rabbits does the same, and none of the rabbits die, how many pairs of rabbits will there be at the beginning of each month?

The figure below illustrates what happens in the first six months. The lines in color indicate the births of new pairs of rabbits.

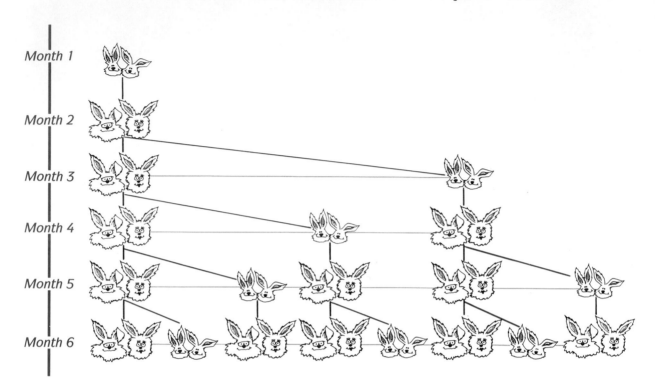

The numbers of *pairs* of rabbits at the beginning of each month form the sequence

$$1 \quad 1 \quad 2 \quad 3 \quad 5 \quad 8 \quad 13 \quad \ldots$$

This sequence is now known as the *Fibonacci sequence.* Its terms follow a simple pattern:

$$
\begin{aligned}
1 \quad & 1 \\
1 + 1 &= 2 \\
1 + 2 &= 3 \\
2 + 3 &= 5 \\
3 + 5 &= 8 \\
5 + 8 &= 13 \quad \ldots
\end{aligned}
$$

The **Fibonacci sequence** is the sequence of numbers

$$1 \quad 1 \quad 2 \quad 3 \quad 5 \quad 8 \quad 13 \quad \ldots$$

in which the first two terms are 1 and each successive term is the sum of the preceding pair of terms.

Since its original appearance in Fibonacci's problem about the rabbits, the Fibonacci sequence has turned up in an amazingly wide variety of creations. One fascinating aspect of mathematics is that ideas from one area of study frequently apply to other, apparently unrelated areas. Among the many subjects in which the Fibonacci sequence has been found are musical scales, pine cones, data sorting, Roman poetry, sunflowers, and the reproduction of bees.

EXERCISES

SET I

The first eight terms of the Fibonacci sequence are listed below.

t_1	t_2	t_3	t_4	t_5	t_6	t_7	t_8
1	1	2	3	5	8	13	21

1. Copy this list and continue it to the 15th term.

2. The terms of the sequence that are even numbers form a pattern. The pattern has to do with their position in the sequence. What is the pattern?

3. The terms that are evenly divisible by 5 form a similar pattern. What is it?

In 1963, someone used deductive reasoning to prove that there are only two numbers in the Fibonacci sequence that are squares.

4. One of the numbers is 1. What is the other?

There are also only two numbers in the sequence that are cubes.

5. What are those numbers?

The figure at the top of the next page shows the family tree of a male bee. A male bee has only one parent, his mother, whereas a female bee has both a father and a mother. In the tree, each male is represented by the symbol ♂ and each female by the symbol ♀.

6. Use the figure to find the numbers of bees in the fourth, fifth, and sixth generations back.

7. What do you notice about the sequence of numbers of bees in successive generations of ancestors?

8. How many bees do you think would be in the seventh, eighth, and ninth generations back?

Generation back		Number of bees in each generation

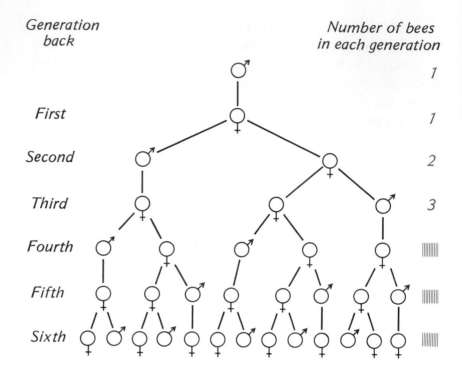

Compare the family tree of a male bee with the rabbits figure on page 100.

9. What does each pair of baby rabbits correspond to on the bee's family tree?

10. What does each pair of adult rabbits correspond to on the bee's family tree?

The ancestors in the sixth row of the bee's family tree can be related to the keys of a piano.

11. Which keys do the female bees correspond to?

The 13 keys shown are one octave of the *chromatic* scale. The white keys in this octave are the notes of a C *major* scale.

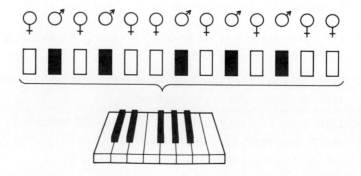

12. How many white keys are in this octave?

13. Is this number a term of the Fibonacci sequence?

14. What is another word beginning with *oct-* whose meaning is related to that of *octave?*

The black keys in an octave are an example of the notes of a *pentatonic* scale.

15. How many black keys are in the given octave?

16. Is this number a term of the Fibonacci sequence?

17. What is another word beginning with *pent-* whose meaning is related to that of *pentatonic?*

Set II

The following exercises refer to this pattern.

Pattern A

$$1 + 1 = 2$$
$$1 + 1 + 2 = 4$$
$$1 + 1 + 2 + 3 = $$
$$1 + 1 + 2 + 3 + 5 = $$
$$1 + 1 + 2 + 3 + 5 + 8 = $$
$$1 + 1 + 2 + 3 + 5 + 8 + 13 = $$
$$1 + 1 + 2 + 3 + 5 + 8 + 13 + 21 = $$

1. Copy it, filling in the missing numbers.

2. Compare the sums with the list that you made for exercise 1 of Set I. What do you notice?

3. Use what you noticed to guess the sum of the first 10 terms of the sequence without adding them.

The following exercises refer to the pattern at the right.

4. Copy it, filling in the missing numbers.

5. What do you notice?

6. Write the next line of the pattern.

Pattern B

$$1^2 + 1^2 = 2$$
$$1^2 + 2^2 = 5$$
$$2^2 + 3^2 = $$
$$3^2 + 5^2 = $$
$$5^2 + 8^2 = $$
$$8^2 + 13^2 = $$

The following exercises refer to this pattern.

Pattern C

$$1^2 + 1^2 \qquad\qquad = 2 = 1 \cdot 2$$
$$1^2 + 1^2 + 2^2 \qquad\quad = 6 = 2 \cdot 3$$
$$1^2 + 1^2 + 2^2 + 3^2 \quad = 15 = 3 \cdot 5$$
$$1^2 + 1^2 + 2^2 + 3^2 + 5^2 = 40 = 5 \cdot 8$$

7. Write the next line of the pattern.

8. Use the pattern to guess the sum of the squares of the first 10 terms of the Fibonacci sequence without adding them.

The following exercises refer to this pattern.

Pattern D

$$1^3 + 2^3 - 1^3 = 8$$
$$2^3 + 3^3 - 1^3 = 34$$
$$3^3 + 5^3 - 2^3 = \text{▊▊▊}$$
$$5^3 + 8^3 - 3^3 = \text{▊▊▊}$$

9. Find the missing numbers.

10. What do you notice?

11. Which one of the preceding patterns—A, B, or C—do these figures illustrate?

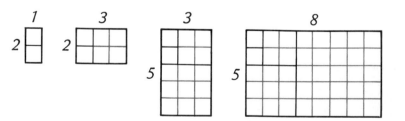

12. Draw the next figure of the pattern.

The first six terms of the Fibonacci sequence and the ratios of consecutive terms are shown below.

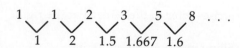

For example, $\dfrac{3}{2} = 1.5$, $\dfrac{5}{3} = 1.667$ (rounded to the nearest thousandth), and $\dfrac{8}{5} = 1.6$.

13. Copy and complete the list of ratios for the following successive terms of the sequence. (Use a calculator and round each ratio to the nearest thousandth.)

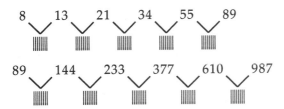

14. What do you notice about the ratios?

15. Does the Fibonacci sequence seem more like an *arithmetic* sequence or a *geometric* sequence? Why?

SET III

This figure shows a sand dollar, a sea creature often washed up on beaches in the United States and Japan.

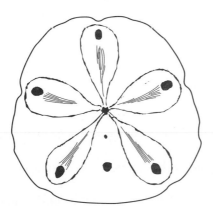

It has five evenly spaced holes that, if connected in every possible way, form the figure shown on the next page. The figure has been drawn so that the distance from A to B is about 34 millimeters. Check this with a ruler.

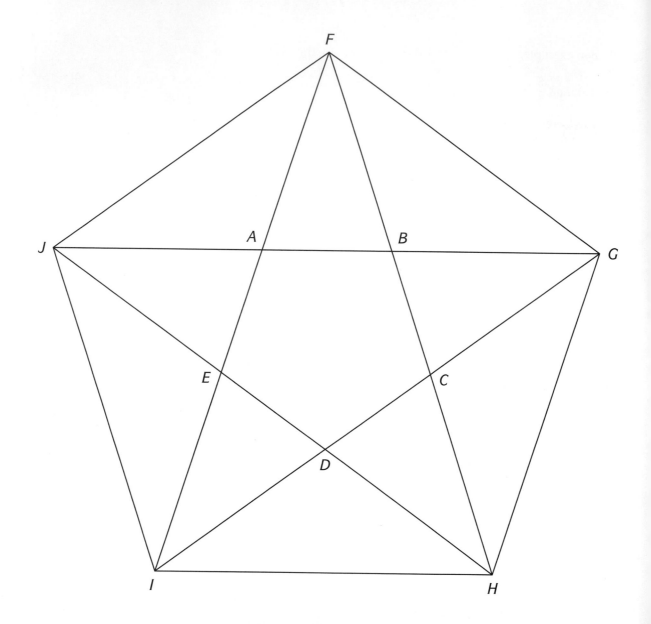

1. How many lengths in the figure are 34 millimeters?

The number 34 is one of the terms of the Fibonacci sequence.

2. Use a ruler to measure other lengths in the figure. What do you notice?

2

Summary and Review

In this chapter we have studied the following number sequences:

Arithmetic sequences *(Lesson 1).* Each successive term is found by adding the same number. This number is called the *common difference* of the sequence.

An example of an arithmetic sequence is

$$1 \quad 5 \quad 9 \quad 13 \quad 17 \quad \ldots$$

Its common difference is 4.

Geometric sequences *(Lessons 2 and 3).* Each successive term is found by multiplying by the same number. This number is called the *common ratio* of the sequence.

An example of a geometric sequence is

$$2 \quad 6 \quad 18 \quad 54 \quad 162 \quad \ldots$$

Its common ratio is 3.

The **binary sequence** is a special geometric sequence:

$$1 \quad 2 \quad 4 \quad 8 \quad 16 \quad \ldots$$

Power sequences *(Lessons 4 and 5).* The terms are found by raising the consecutive counting numbers to the same power. Two examples of power sequences are the *sequence of squares,*

$$1^2 \quad 2^2 \quad 3^2 \quad 4^2 \quad 5^2 \quad \text{or}$$
$$1 \quad 4 \quad 9 \quad 16 \quad 25 \quad \ldots$$

and the *sequence of cubes,*

$$1^3 \quad 2^3 \quad 3^3 \quad 4^3 \quad 5^3 \quad \text{or}$$
$$1 \quad 8 \quad 27 \quad 64 \quad 125 \quad \ldots$$

The **Fibonacci sequence** *(Lesson 6).* The first two terms are 1 and each successive term is the sum of the preceding pair of terms:

$$1 \quad 1 \quad 2 \quad 3 \quad 5 \quad 8 \quad \ldots$$

Courtesy of the Computer Technique Group, Tokyo

These figures, each showing a square transformed into the profile of a woman and then back into a square, were drawn with the help of a computer. The first was programmed according to an arithmetic sequence and the second according to a geometric sequence.

EXERCISES

SET I

PEANUTS reprinted by permission of UFS, Inc.

The number sequence

$$6 \quad 6 \quad 6 \quad 6 \quad 6 \quad 6 \quad \ldots$$

can be thought of as either arithmetic or geometric.

1. As an arithmetic sequence, what is the common difference?
2. As a geometric sequence, what is the common ratio?

Find the missing term in each of the following number sequences.

3. 2   6   10   ▓▓▓   18
4. 9   16   25   ▓▓▓   49
5. 4   12   36   ▓▓▓   324
6. 1   8   27   ▓▓▓   125
7. 8   13   21   ▓▓▓   55
8. 1   3   6   ▓▓▓   15

The three chemical elements lithium, sodium, and potassium are very much alike. They are soft, light metals that will burn your fingers if you touch them. Their atomic weights are 7, 23, and 39.

    9. What kind of number sequence is this?

The Rhind Papyrus, written in Egypt in the seventeenth century B.C., contains the list shown at the right.

Household	7
Cats	49
Mice	343
Barley	2,301
Hekats	16,807

    10. What kind of number sequence does this seem to be?

    11. There is a mistake in the sequence as it appears in the papyrus. What is it?

Among the terms sometimes used by the book industry to indicate the size of a book's pages are *folio, quarto,* and *octavo.* These words refer to the number of pages that can be obtained from large printer's sheets by folding them as shown by the brown lines in the figures at the right. Smaller pages obtained from the large sheets are referred to as *16 mo, 32 mo,* and *64 mo.*

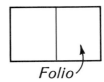

Folio

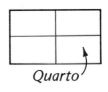

Quarto

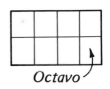

Octavo

    12. What kind of number sequence do these terms suggest?

    13. Why does the folding of the sheets result in this particular sequence?

The pattern below is illustrated by these figures.

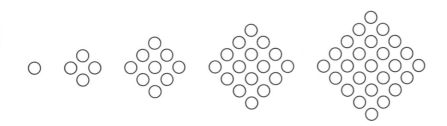

    14. Copy and complete the pattern.

$$
\begin{array}{rcl}
1 & = & 1 \\
1 + 2 + 1 & = & \text{||||||} \\
1 + 2 + 3 + 2 + 1 & = & \text{||||||} \\
1 + 2 + 3 + 4 + 3 + 2 + 1 & = & \text{||||||} \\
1 + 2 + 3 + 4 + 5 + 4 + 3 + 2 + 1 & = & \text{||||||}
\end{array}
$$

    15. What sequence do the numbers in the right column of the pattern form?

    16. How do the figures illustrate this?

Number	Cube
1	1
2	8
3	27
4	64
5	125
6	216
7	343
8	512
9	729
10	1,000
11	1,331
12	1,728

Set II

The brilliant Indian mathematician Srinivasa Ramanujan is pictured on this postage stamp. He specialized in the study of numbers and knew their characteristics in the same way that a baseball fan might know a vast number of statistics about the game. One time a friend, making small talk, said "I came here in a taxi with a most uninteresting number, 1,729." Ramanujan immediately replied: "1,729 is a very interesting number; it is the smallest number expressible as the sum of two cubes in two *different* ways."

1. Use the table of cubes at the left to complete the following equations based on this fact.

$$1{,}729 = \rule{0.6cm}{0.4pt}^3 + \rule{0.6cm}{0.4pt}^3$$
$$1{,}729 = \rule{0.6cm}{0.4pt}^3 + \rule{0.6cm}{0.4pt}^3$$

2. The number 50 is the smallest number expressible as the sum of two *squares* in two different ways. What are they?

The full spectrum of colors is created in a television picture from three primary colors: red, green, and blue. Eight basic colors can be produced from the three shown in this chart, in which 1 means that the color is used and 0 means that the color is not used.

Red	Green	Blue	
0	0	0	Black
0	0	1	Blue
0	1	0	Green
0	1	1	Cyan
1	0	0	Red
1	0	1	Magenta
1	1	0	Yellow
1	1	1	White

3. The "three-digit numerals" in this list are related to a number sequence we have studied. Which one is it?

The number 0 corresponds to black in this list.

4. To what color does the decimal number 4 correspond?

5. To what decimal number does the color white correspond?

The first two terms of an arithmetic sequence are 5 and 17. Both 5 and 17 are prime.*

6. Write some more terms of this sequence.

7. Do you think every term in the sequence is prime?

What kind of number sequence the following sequence is depends on what numbers are chosen for the missing terms.

1  9

8. Copy and complete the sequence so that it is *arithmetic*.

9. Copy and complete it so that it is *geometric*.

10. Copy and complete it so that it is a *power* sequence.

11. Which one of the three sequences grows the fastest?

12. Which one grows the slowest?

The first ten terms of the Fibonacci sequence are

$$1 \quad 1 \quad 2 \quad 3 \quad 5 \quad 8 \quad 13 \quad 21 \quad 34 \quad 55$$

13. Copy and complete the following equations.

$$5^2 = \rule{1cm}{0.4pt} \qquad 3 \cdot 8 = \rule{1cm}{0.4pt}$$
$$8^2 = \rule{1cm}{0.4pt} \qquad 5 \cdot 13 = \rule{1cm}{0.4pt}$$
$$13^2 = \rule{1cm}{0.4pt} \qquad 8 \cdot 21 = \rule{1cm}{0.4pt}$$
$$21^2 = \rule{1cm}{0.4pt} \qquad 13 \cdot 34 = \rule{1cm}{0.4pt}$$

14. How does the square of any term of the Fibonacci sequence seem to compare with the product of the term before it and the term after it?

15. Write the next pair of equations in the pattern.

The following exercises refer to the pattern at the right.

16. Copy it, writing in the missing numbers.

17. Write what you think the next line of the pattern is.

18. Show whether it is true.

19. Write the line of the pattern that starts with 10^2.

20. Show whether it is true.

$$1^2 + 2^2 + 2^2 = \rule{1cm}{0.4pt}^2$$
$$2^2 + 3^2 + 6^2 = \rule{1cm}{0.4pt}^2$$
$$3^2 + 4^2 + 12^2 = \rule{1cm}{0.4pt}^2$$

---

*Remember that a prime number is a number that cannot be divided evenly by any whole number other than itself and 1.

SET III

"I asked you a question, buddy . . . What's the
square root of 5,248?"

The little guy in the bar has been asked an unexpected question and it looks like he may be in trouble if he doesn't come up with an answer.

1. What does the question mean? In other words, what is meant by the *square root* of a number?

Suppose the little guy guesses 74.

2. Show why this guess is too big.

3. Can you find a better answer without using a calculator? If so, what is it?

Try finding the answer *with* a calculator.

4. What answer does the calculator give?

5. Do you think the answer given by the calculator is correct? Tell why or why not.

Further Exploration

Lesson 1

1. The 17-year locust is named for the length of its lifespan. The locust is seen only in the last month of its life, the rest of its existence being spent underground. In Oklahoma, Kansas, and Missouri, it appeared in 1981.

   a. Write a number sequence listing the years of all of its appearances in the nineteenth and twentieth centuries.

   Suppose that the locust had a predator with a life cycle of six years (during most of which it is in larval form) and that the predator appeared as an adult in 1801.

   b. Write a number sequence listing the years of all of the predator's appearances in the nineteenth and twentieth centuries.
   c. In what years did both the locusts and their predators appear?
   d. Do you see a relation between the interval between these years and the lifespans of the locusts and their predators? If so, what is it?

   Suppose that the locust had a lifespan of *16* years rather than 17 and that it appeared in 1981.

   e. Write a number sequence listing the years of all of its appearances in the nineteenth and twentieth centuries.
   f. Which of these years coincide with the years in which the predator appeared?

   Suppose that the locust had a lifespan of *18* years and that it appeared in 1981.

   g. Write a number sequence listing the years of all of its appearances in the nineteenth and twentieth centuries.
   h. Which of these years coincide with the years in which the predator appeared?
   i. Do you see any relations between the years in parts f and h and the lifespans of the locusts and their predators? If so, what are they?

"I can't *stand* this waiting. Couldn't we be *six*-year locusts?"
Copyright © Sydney Harris, 1977

2. Mrs. Marva Drew, a resident of Waterloo, Iowa, typed the numbers 1 to 1,000,000 on a manual typewriter. She worked over a period of six years doing this and used 2,473 sheets of paper.*

   a. How many digits did Mrs. Drew type? Show how you got your answer. (Don't count any *commas*—just the *digits*; 1,000,000, for example, has seven digits.)

   b. What is the *sum* of all the digits Mrs. Drew typed? Show how you got your answer. (*Hint:* Look at the following pattern.)

   0 and 999,999
   1 and 999,998
   2 and 999,997 . . .

LESSON 2

1. The successive terms of the following sequence are found by multiplication, but the sequence is not geometric.

   t_1 t_2 t_3 t_4 t_5 t_6
   1   2   6   24   120   720 . . .

   a. What do you think is the seventh term of the sequence?

   The numbers in the sequence are called *factorials*. Factorial numbers are used in counting numbers of arrangements. For example, the number of ways that five cards in a poker hand can be arranged is the fifth term, t_5, of this sequence, and is called "5 factorial," which equals 120.

   b. The 13 cards in a bridge hand can be arranged in "13 factorial" ways. How many is that?

2. Suppose that each time a pair of blue jeans is washed, they lose some color and that the amounts of color left after successive washings form a geometric sequence.†

   Starting with a new pair of jeans having 100% color, suppose that the first three terms of the sequence are

   95%   90.25%   85.7375%   . . .

   a. Find the common ratio of this sequence.

   b. Use a calculator to find the next seven terms of the sequence.

Guinness Book of World Records.

†Adapted from a problem created by Paul Foerster.

The first three terms of the sequence rounded to the nearest whole number are

95%   90%   86%   . . .

c. Write the seven terms you found for exercise 2, each rounded to the nearest whole number.
d. After how many washings do the jeans have about 90% of their original color?
e. After how many washings do they have about 70% of their original color?

Lesson 3

1. Some of the familiar units for measuring volume, cup, pint, quart, and gallon are part of a larger set of units used in Colonial America. As you can see from the table below, each unit was doubled to find the next.

1 jigger = 2 mouthfuls
1 jack = 2 jiggers
1 jill = 2 jacks
1 cup = 2 jills
1 pint = 2 cups
1 quart = 2 pints
1 pottle = 2 quarts
1 gallon = 2 pottles
1 peck = 2 gallons
1 pail = 2 pecks
1 bushel = 2 pails
1 strike = 2 bushels
1 coomb = 2 strikes
1 cask = 2 coombs
1 barrel = 2 casks
1 hogshead = 2 barrels
1 pipe = 2 hogsheads
1 tun = 2 pipes

a. According to the table, which unit is bigger and by how much: a bushel or a peck?
b. How many cups are in a gallon?
c. How many jiggers are in a pottle?
d. Which unit is the smallest and which is the largest?
e. How many of the smallest unit would be needed to make the same volume as the largest unit?

2. Experiment: *A Card Sorting System*

One application of the binary sequence is in data-processing systems. From the following experiment, you will see how it can be used to sort cards automatically.

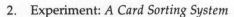

Cut 8 file cards in half as shown in the figure above. Take one of the 16 cards produced and punch a row of four holes below a longer edge as shown in the figure below. The holes should be spaced about 1.5 centimeters apart.

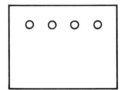

Using this card as a stencil, punch the other 15 cards to match it. Three cards can easily be punched at a time.

Number the cards from 0 through 15. The four holes represent the first four numbers of the binary sequence in reverse order:

$$8 \quad 4 \quad 2 \quad 1$$

Write the number 1 above the appropriate holes of each numbered card to represent the matching binary numeral. The first four cards are illustrated below.

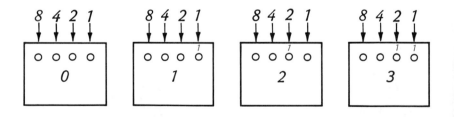

Along the top edge of each card, cut out the space above each hole marked with a 1.

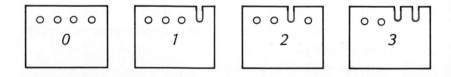

The cards are now ready to use. Shuffle them, being careful not to turn any of them over or upside down.

Make a hook out of a paper clip, something like the one illustrated at the right.

Holding the cards loosely together with one hand, put the hook through the 1-holes and lift up, shaking the hook rapidly so that the cards with notches above the 1-holes remain behind. Slide the cards that came up off the hook and place them on top of the other cards.

Next, put the hook through the 2-holes and carry out the same procedure. Be sure to place the cards that come up on *top* of the cards that remain behind. Repeat with the remaining holes (going from right to left.) When you have done this, the cards should be in correct order from 0 through 15.

After sorting the cards by this method, you might print the following words on them, reshuffle them, and sort them again.

0	This	4	cards	8	sorted	12	of
1	pack	5	has	9	automatically	13	the
2	of	6	just	10	by	14	binary
3	sixteen	7	been	11	means	15	sequence.

a. How many cards could be sorted if each one has five holes rather than four? Explain your reasoning.

b. If each card had 10 holes, more than 1,000 cards could be sorted in just 10 steps. Explain why.

LESSON 4

1. How many squares are there on a checkerboard? If you count only the small squares, there are 64. As the figure at the right shows, however, there are squares of other sizes as well.

Before trying to figure this out, warm up with these smaller boards. A 2 × 2 board has 5 squares: 4 small ones and 1 large one.

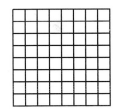

a. A 3 × 3 board has 14 squares. Explain why.

b. How many squares does a 4 × 4 board have? Explain.

c. How many squares does the full-size 8 × 8 checkerboard have?

2. The driver of a car glanced at the odometer and saw that it read 15,951 miles.* The driver thought: "That's interesting. The mileage is a palindrome; it reads the same backward as forward. It will be a long time before that happens again."

---

*This puzzle is from a popular Russian book of mathematical puzzles written by Boris A. Kordemsky. Its English edition is titled *The Moscow Puzzles* (Scribner's, 1972).

After driving for just two hours, however, the mileage shown on the odometer was a new palindrome. How fast was the car going in those two hours?

LESSON 5

1.   According to the *Guinness Book of World Records*, the largest prize ever offered in mathematics was for the first person who could prove Fermat's Last Theorem. The prize was offered in 1908 and resulted in a large number of proofs being submitted, all of them incorrect.

Fermat's Last Theorem is about equations of the form

$$a^n + b^n = c^n$$

in which a, b, c, and n are counting numbers. An example of such an equation is

$$20^2 + 21^2 = 29^2$$

The theorem is named for the seventeenth-century French mathematician Pierre de Fermat, who made a note in the margin of one of his books that equations of this form can be found for squares only. He wrote that he had discovered a proof of this but that the margin of the book was too narrow to contain it. No one to this day has been able to figure out what Fermat's proof was.

The March 7, 1938, issue of *Time* magazine contained an article reporting that a Mr. Krieger had discovered an equation that supposedly disproved the theorem.* The equation was

$$1,324^n + 731^n = 1,961^n$$

in which n was a counting number larger than 2 and which Mr. Krieger refused to disclose. A reporter for the *New York Times* proved that Krieger was mistaken. How did the reporter do it? (*Hint:* What digits can each power in the equation end in?)

2.   Early in the seventeenth century, the astronomer Johannes Kepler made an interesting discovery about the distances of the planets from the sun and their periods, the time that it takes them to travel once around the sun.

The distances in the table on the next page are given in the *astronomical unit*, which is the distance of the earth from the sun. The periods are

---

*Martin Gardner, *Wheels, Life and Other Mathematical Amusements* (W. H. Freeman, 1983), p. 76.

given in terms of years. For example, Neptune's distance from the sun is about 30.1 times the earth's distance and it takes Neptune 165 years to travel once around the sun.

Planet	Distance			Period		
	d	d^2	d^3	p	p^2	p^3
Earth	1	1	1	1	1	1
Mars	1.52	▨	▨	1.88	▨	▨
Jupiter	5.2	27	140	11.9	140	1,700
Saturn	9.54	▨	▨	29.5	▨	▨
Uranus	19.2	▨	▨	84	▨	▨
Neptune	30.1	▨	▨	165	▨	▨
Pluto	39.5	▨	▨	248	▨	▨

a. Copy the table, leaving each of the indicated spaces blank. Then use a calculator to find the squares and cubes of the distances and periods. Round each number to two significant digits.* The calculations for Jupiter are shown at the right as an example:

$5.2 \times 5.2 = 27.04 \approx 27$
$27.04 \times 5.2 = 140.608 \approx 140$
$11.9 \times 11.9 = 141.61 \approx 140$
$141.51 \times 11.9 = 1685.159 \approx 1700$

b. When you have completed the table, see if you notice anything interesting about any of the results. If you do, describe in words what you noticed.

LESSON 6

1. This figure shows part of a honeycomb with a bee in the cell numbered 1.† Suppose that the bee moves to the other cells, always traveling

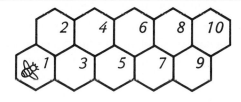

---

*Some examples of numbers rounded to two significant digits are: 3.7, 140, 86,000, and 5,200,000. The significant digits are underlined in each example.

†Martin Gardner, *Mathematical Circus* (Knopf, 1979), p. 162.

Cell 2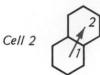

to either neighboring cell to the right. The possible paths of the bee to cells 2, 3, 4, and 5 are shown in the figures below.

Cell 3

Cell 4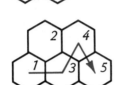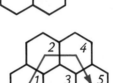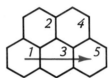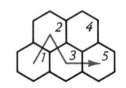

Cell 5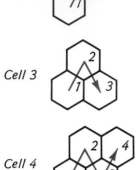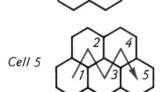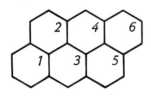

a. How does the number of possible paths to a given cell seem to be related to the number of the cell?

b. How many paths to Cell 6 do you think are possible?

c. Trace the figure at the left as many times as necessary and draw all of the possible paths.

d. How many paths to Cell 10 do you think are possible?

2. The following number trick is based on properties of the Fibonacci sequence. Ask a friend to choose two numbers at random and write one number below the other. Have your friend add them to get a third number. Add the second and third numbers to get a fourth number. Add the third and fourth numbers to get a fifth number. Ask your friend to continue in this fashion as far as he or she likes.

Now have your friend draw a line between any two numbers. You can quickly tell the sum of all of the numbers above the line by simply subtracting the second number in the list from the second number below the line.

a. Try this trick.

b. Represent the two originally chosen numbers by a and b and continue the table at the left. Then prove that the trick will work for any two numbers and a list of as many as twelve numbers.

Example:

12
7
19
26
45
71
⋮

Numbers	Sums
a	a
b	$a + b$
$a + b$	$2a + 2b$
$a + 2b$	$3a + 4b$

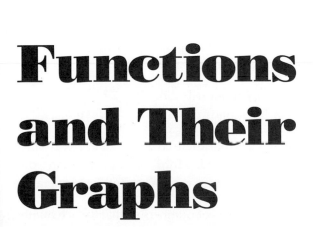

Functions and Their Graphs

Runk/Schoenberger. From Grant Heilman Photography, Inc.

1

The Idea of a Function

A familiar sound in the country on a warm summer evening is the chirping of crickets. The rate at which crickets chirp depends on the temperature: the warmer it is, the more they chirp in any given time. The table below shows how the rate and temperature are related.

Temperature in degrees Fahrenheit	50	60	70	80	. . .
Number of chirps in 15 seconds	10	20	30	40	. . .

To each temperature in this table, there corresponds a rate: 10 corresponds to 50, 20 corresponds to 60, and so forth. A mathematician would say that the rate at which crickets chirp is a *function* of the temperature.

A **function** is a pairing of two sets of numbers so that to each number in the first set there corresponds exactly one number in the second set.

One way to represent a function is with a *table,* as was done on the facing page. Another way is to write a *formula.* From the table, we see that each number on the second line is 40 less than the corresponding number on the first line. If we let F represent the temperature in degrees Fahrenheit and n represent the number of chirps in 15 seconds, we can write

$$n = F - 40$$

The two letters in this formula are *variables:* as F varies in value, so does n. For example, if $F = 55$, then $n = 55 - 40 = 15$, and if $F = 90$, then $n = 90 - 40 = 50$.

If we know the formula for a function, we can create a table by choosing numbers that we want for the first line and then substituting them into the formula to find the corresponding numbers of the second line. For example, a formula for the temperature in degrees Celsius, C, as a function of the number of chirps that a cricket makes in 15 seconds, n, is

$$C = 0.6n + 4$$

To create a table for this function, we first choose some numbers for n:

$$n \quad 0 \quad 10 \quad 20 \quad 30$$

We then substitute these numbers into the formula to find the numbers for the second line:

If $n = 0$, then
$C = 0.6(0) + 4 = 0 + 4 = 4$
So $C = 4$ when $n = 0$

If $n = 10$, then
$C = 0.6(10) + 4 = 6 + 4 = 10$
So $C = 10$ when $n = 10$

If $n = 20$, then
$C = 0.6(20) + 4 = 12 + 4 = 16$
So $C = 16$ when $n = 20$

If $n = 30$, then
$C = 0.6(30) + 4 = 18 + 4 = 22$
So $C = 22$ when $n = 30$

The table is

$$
\begin{array}{ccccc}
n & 0 & 10 & 20 & 30 \\
C & 4 & 10 & 16 & 22
\end{array}
$$

EXERCISES

SET I

One way to represent a function is with a table. Copy and complete the tables shown for the functions having the following formulas:

1. Formula: $y = x + 4$

x	0	1	2	3	4
y	4				

2. Formula: $y = 7x$

x	0	1	2	3	4
y		7			

3. Formula: $y = 8 - x$

x	0	1	2	3	4
y	8				

4. Formula: $y = \dfrac{12}{x}$

x	1	2	3	4	5
y	12				

5. Formula: $y = 11x + 1$

x	1	2	3	4	5
y	12				

6. Formula: $y = 2(x - 5)$

x	5	6	7	8	9
y	0				

7. Formula: $y = 0x + 3$

x	1	2	3	4	5
y	3				

8. Formula: $y = x^2$

x	1	2	3	4	5
y					

9. Formula: $y = x^2 + 2x + 1$

x	1	2	3	4	5
y	4				

10. Formula: $y = (x - 1)^3$

x	2	3	4	5	6
y					

11. Formula: $y = 2^x$

x	2	3	4	5	6
y	4	8			

The second line of numbers in the table for exercise 2 is

y	0	7	14	21	28

These numbers are part of an arithmetic sequence because each number can be found by adding 7 to the preceding number.

12. In which of the other tables in exercises 1 through 11 are the numbers in the second line in arithmetic sequence?

13. The second line of one of those tables is part of a geometric sequence. Which table is it?

14. In which tables are the numbers in the second line squares?

A function is *increasing* if, as one variable increases, the other variable also increases.

15. Which of the functions in exercises 1 through 11 are increasing?

A function is *decreasing* if, as one variable increases, the other variable decreases.

16. Which of the functions in exercises 1 through 11 are decreasing?

Set II

To guess a formula for the function represented by this table,

x	11	12	13	14	15
y	1	2	3	4	5

compare each y-number with the x-number above it.

x	11	12	13	14	15
y	1	2	3	4	5

Each y-number can be found by subtracting 10 from the corresponding x-number. This can be written as the formula

$$y = x - 10$$

Use the same method to guess a formula for the function represented by each of the following tables. Begin each formula with $y =$.

1.
x	3	4	5	6	7
y	6	8	10	12	14

2.
x	2	3	4	5	6
y	10	11	12	13	14

3.
x	7	8	9	10	11
y	4	5	6	7	8

4.
x	3	4	5	6	7
y	9	16	25	36	49

5.
x	2	3	4	5	6
y	22	33	44	55	66

6.
x	5	6	7	8	9
y	51	61	71	81	91

(*Hint:* Multiply and then add.)

7.
x	6	7	8	9	10
y	28	33	38	43	48

(*Hint:* Multiply and then subtract.)

8.
x	1	2	3	4	5
y	1	8	27	64	125

9.
x	1	2	3	4	5
y	2	9	28	65	126

(*Hint:* Compare this table with the preceding one.)

10.
x	1	2	3	4	5
y	2	6	12	20	30

The length of anchor line needed by a boat is a function of the depth of the water.

Depth of water in feet, x	10	20	30	40	50
Length of line in feet, y	70	140	210	280	▨

11. What is the missing number in this table?

12. How can this number be found from the previous y-number, 280?

13. How can the y-numbers in the table be found from the corresponding x-numbers?

14. What is a formula for this function?

According to an experimental study, the lifespan of a hamster is a function of the time the hamster spent hibernating.

Percent of lifetime in hibernation, x	0	10	20	30
Expected lifespan in days, y	▓	▓	▓	▓

A formula for this function is

$$y = 18x + 660$$

15. Use the formula to find the missing numbers in the table.

16. What kind of number sequence do the y-numbers form?

17. If the amount of its lifetime that a hamster spends in hibernation increases by 10%, by how many days does its expected lifespan increase?

Your ideal heart rate during aerobic exercise is a function of your age.

Age in years, x	20	30	40	50	60
Heart rate in beats per minute, y	200	190	180	170	▓

18. What is the missing number in this table?

19. How can this number be found from the previous y-number?

20. How can the y-numbers in the table be found from the corresponding x-numbers?

21. What is a formula for this function?

The fuel efficiency of a car is a function of the number of cylinders in its engine. A representative formula for this function is

$$y = 44 - 2.5x$$

in which x is the number of cylinders in the engine and y is the number of miles per gallon that the car gets in highway driving.

22. Use the formula to copy and complete the following table.

Number of cylinders, x	4	6	8
Numbers of miles per gallon, y	▓	▓	▓

23. What happens to the number of miles per gallon as the number of cylinders increases by 2?

SET III

If you could jump as high as you wanted, the time you would be in the air would be a function of the height to which you jumped.

Broadway dancers.

Height of jump in feet, x	1	4	‖‖‖‖	16	25
Time in air in seconds, y	0.5	1	1.5	2	‖‖‖‖

1. What kind of numbers are the x-numbers in this table?

2. What kind of sequence do the y-numbers form?

3. What are the missing numbers in the table?

4. Can you think of a way to get each y-number from the corresponding x-number? If so, what is it?

A stamp issued by France on the 300th anniversary of Descartes' invention of coordinate geometry

Descartes and the Coordinate Graph

One of the greatest mathematical achievements of all time was the invention of coordinate geometry by René Descartes. It made possible a new method of studying geometric figures and the relations between them, relations first proved by the ancient Greeks. The new method used algebra, a subject developed many centuries after geometry, and marked the beginning of modern mathematics. With this method, lines, circles, and other figures could be related to equations, so that algebra and geometry were combined into a new subject more powerful than either of its separate parts.

René Descartes was born in France in the late sixteenth century. In the age in which Descartes lived, Europe was in political and religious turmoil, yet there was also great intellectual progress. In England, Shakespeare was writing his plays; great scientific discoveries were being made by Galileo in Italy; and the French mathematicians Fermat and Pascal were developing another new branch of mathematics called probability theory.

As a young student Descartes began to question the truth of much of what he was being taught. The subject of mathematics, however, appealed to him because its methods of reasoning seemed universal and without fault. He decided that mathematics, to quote his words, "is a more powerful instrument of knowledge than any other that has been bequeathed to us by human agency." This belief led him to apply deductive reasoning, the mathematical method developed by the Greeks so many centuries earlier, to other areas of study. In 1637, his book titled *A Discourse on the Method of Rightly Conducting the Reason and Seeking Truth in the Sciences* established Descartes as the "father of modern philosophy." The book ended with a section on coordinate geometry, his great contribution to the subject of mathematics.

Descartes' invention was clever, yet, like many important discoveries in mathematics, it was very simple. The idea was that the location of a point in a plane can be described by giving its distances from a pair of perpendicular lines. The lines are called the *x-axis* and the *y-axis*, and the point at which they intersect, labeled O, is called the *origin*. The axes are numbered at equal intervals in each direction from the origin. On the *x*-axis, positive numbers are used to the right of the origin and negative numbers* are used to the left. On the *y*-axis, positive numbers are used above the origin and negative numbers are used below it.

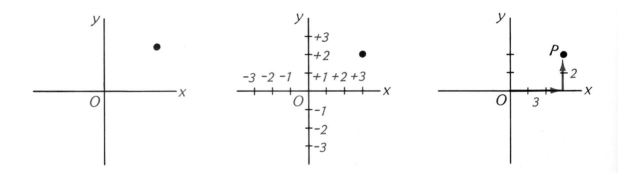

To locate a point, we first move along the *x*-axis until we are directly above or below the point, counting the units as we go. Then we move directly up or down to the point itself, again counting the units along the way. These two numbers are called the *coordinates* of the point and are written in parentheses like this: (3, 2). The first number is the *x-coordinate* and the second number is the *y-coordinate*.

Other examples of how the coordinates of a point are found are shown in the figure on the facing page.

---

*Look on page 659 if you are not familiar with negative numbers.

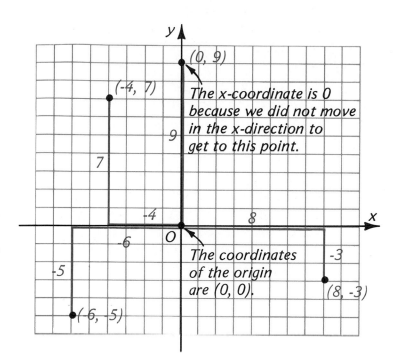

EXERCISES

SET I

The curve in this graph is a circle. Which point on the circle has each of the following coordinates?

1. (4, 3).

2. (3, 4).

3. (−3, −4).

4. (4, −3).

5. (−4, 3).

6. (5, 0).

7. (−5, 0).

8. (0, 5).

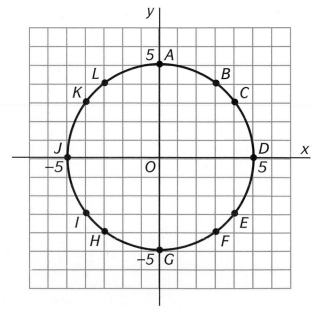

Which point on the circle has

9. the same x-coordinate as C?

10. the same y-coordinate as H?

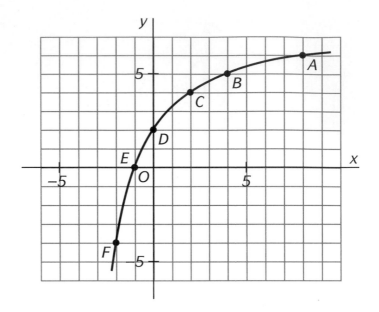

This graph shows part of a curve called a *hyperbola.*

> 11. Write the coordinates of each lettered point on the hyperbola. For example, the coordinates of point A are (8, 6).

The curve in this graph is called a *parabola.*

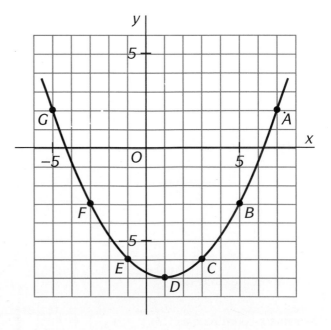

> 12. Write the coordinates of each lettered point on the parabola.

On graph paper, draw a pair of axes extending from −8 to +8 on each axis.

13. Plot and label the following points: A(−5, 5), B(−2, 3), C(1, 2), D(4, 3), E(6, 5), F(6, 7), G(4, 7), H(2, 5), I(1, 2), J(2, −1), and K(4, −4). Join the points in order with a smooth curve. The result looks like a figure called the *folium of Descartes.*

Set II

A function can be pictured with a graph by using the pairs of numbers in its table as coordinates of points.

For example, for the table

x	1	2	3	4
y	3	5	7	9

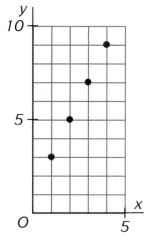

the points (1, 3), (2, 5), (3, 7), and (4, 9) can be plotted as shown in the figure at the right.

On graph paper, draw and label six pairs of axes as shown in the figure below. Use the pairs of numbers in the tables for the following functions to plot points.

1. Function A

x	1	2	3	4	5
y	4	5	6	7	8

2. Function B

x	1	2	3	4	5
y	6	5	4	3	2

3. Function C

x	1	2	3	4	5
y	7	7	7	7	7

4. Function D

x	1	2	3	4	5
y	1	2	4	8	16

5. Function E

x	1	2	3	4	5
y	3	6	9	12	15

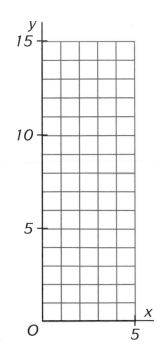

6. Function F

x	1	2	3	4	5
y	12	6	4	3	2.4

In answering the following questions, use the letters naming the functions to identify them.

7. For which functions do the points lie on a straight line?

8. For which functions do the y-numbers form arithmetic sequences?

9. For which function do the y-numbers form a geometric sequence?

10. Which functions are *increasing*?

11. Which functions are *decreasing*?

12. For which function do the points lie on a horizontal line?

13. For which function does each pair of coordinates add up to 7?

14. For which function does each pair of coordinates multiply to give 12?

15. For which function is each y-coordinate 3 more than the corresponding x-coordinate?

16. For which function is each y-coordinate 3 times the corresponding x-coordinate?

SET III

The number of words in a child's vocabulary is a function of the child's age. A typical child at the age of 20 months has a vocabulary of 300 words. At the age of 50 months, the child's vocabulary has increased to 2,100 words.

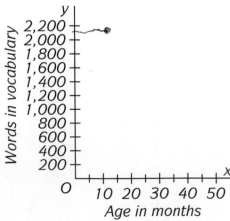

1. On graph paper, draw and label a pair of axes as shown on the previous page. Plot two points: one showing that the vocabulary of a 20-month-old child is 300 words and the other showing that the vocabulary of a 50-month-old child is 2,100 words. Draw a straight line through the two points.

Use your graph to estimate how many words

2. a 30-month-old child knows.

3. a 40-month-old child knows.

4. If the graph is correct, how many new words does a typical child learn each month from the age of 20 months to the age of 50 months?

Draw the line until it meets the x-axis.

5. Could the graph be used to estimate how many words a 10-month-old child knows? Explain.

3

Functions with Line Graphs

One of the tallest persons who ever lived was Robert Wadlow, born in Alton, Illinois, in 1918. The photograph shows him standing behind a man of average height when he was 21.

By the age of 5 he was already 5 feet 4 inches tall. The table below shows his height in inches as a function of his age.

Age	5	8	10	20
Height in inches	64	72	77	103

The table contains *pairs of numbers,* a height number for each age number. These pairs of numbers can be used as coordinates of points to make a graph of the function.

To graph the function, we first plot the points: (5, 64), (8, 72), (10, 77), and (20, 103). The first coordinate of each point is the age and the second coordinate is the height, and so we will name the axes a and h. Because the height coordinates grow much faster than the age coordinates, we choose scales on the two axes that allow room to show all four points.

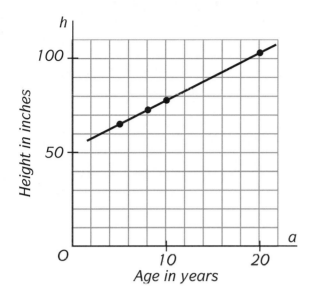

After plotting the points, we see that they seem to lie along a straight line. It makes sense to draw this line because there are *other ages and heights between those listed in the table.* For example, from the graph it appears that when he was 15, Robert was about 90 inches tall. The formula of the function is

$$h = 2.6a + 51$$

and the line is called its graph.

To graph a function for which we are given a formula, it is first necessary to make a table. For example, to make a table for the function

$$y = 2x - 1$$

we first choose some numbers for x: 1, 2, 3, 4, and 5 are convenient. Then we substitute each number into the formula to find the corresponding number for y, as shown below.

Let $x = 1$:
$y = 2(1) - 1 = 2 - 1 = 1$
So $y = 1$ when $x = 1$.

Let $x = 2$:
$y = 2(2) - 1 = 4 - 1 = 3$
So $y = 3$ when $x = 2$.

Let $x = 3$:
$y = 2(3) - 1 = 6 - 1 = 5$
So $y = 5$ when $x = 3$.

Let $x = 4$:
$y = 2(4) - 1 = 8 - 1 = 7$
So $y = 7$ when $x = 4$.

Let $x = 5$:
$y = 2(5) - 1 = 10 - 1 = 9$
So $y = 9$ when $x = 5$.

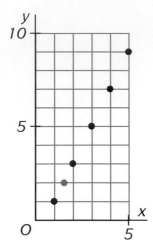

The resulting table,

x	1	2	3	4	5
y	1	3	5	7	9

contains the coordinates of five points. Plotting them on a pair of axes reveals that they also lie along a straight line. Other numbers chosen for x also result in points on this line. For example,

$$\text{if} \quad x = 1.5$$
$$\text{then} \quad y = 2(1.5) - 1 = 3 - 1 = 2$$

The point (1.5, 2) also lies on the line. Because of this, it makes sense to draw a line through the points to get the graph shown in the second figure.

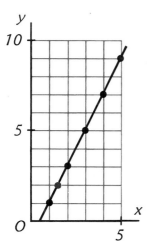

EXERCISES

SET I

A certain function has the formula

$$y = 2x + 3$$

1. Copy and complete the following table for this function.

x	0	1	2	3	4
y	▓	▓	▓	▓	▓

2. How many points are included in the table?

3. Graph the function by drawing a pair of axes, plotting these points, and drawing a line through them.

A certain function is represented by this table.

x	2	3	4	5	6
y	6	7	8	9	10

4. What is a formula for this function? (Begin your formula with $y =$.)

5. Graph the function by drawing a pair of axes, plotting the points in the table, and drawing a line through them.

6. At what value of y does the line meet the y-axis?

A certain function has the graph shown here.

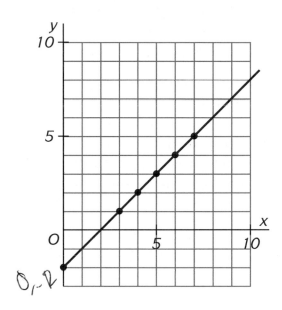

7. Where does the line meet the y-axis?

8. Copy and complete the following table for this function.

x	3	4	5	6	7
y	1				

9. What is a formula for this function?

Three functions have the following formulas:

$$\text{Function A:} \quad y = 2x + 1$$
$$\text{Function B:} \quad y = 3x + 1$$
$$\text{Function C:} \quad y = 4x + 1$$

10. Make three tables, one for each function. Let $x = 0, 1, 2, 3,$ and 4 in each table.

11. Draw a pair of axes as shown in the adjoining figure. Graph all three functions on this pair of axes. Draw a line through each set of points and label each line with its formula.

12. What point do all three lines go through?

13. What number do all three formulas have in common?

14. Which line is steepest?

15. How does the steepness of the line change as the number by which x is multiplied in the formula gets larger?

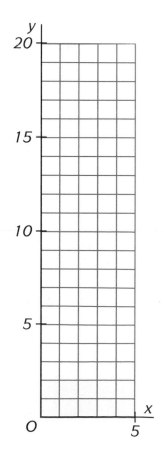

Three functions have the following formulas:

Function D: $y = x + 3$
Function E: $y = x + 5$
Function F: $y = x + 8$

16. Make three tables, one for each function. Let $x = 0, 1, 2, 3,$ and 4 in each table.

17. Draw a pair of axes like the one you drew for exercise 11. Graph all three functions on this pair of axes. Draw a line through each set of points and label each line with its formula.

18. What do you notice about these three lines?

19. At what value of y does each line meet the y-axis?

20. Write a formula for the function whose graph is a line in the same direction and meeting the y-axis at 12.

Two functions have the following formulas:

Function G: $y = 6 + x$
Function H: $y = 6 - x$

21. Make a table for each function, letting $x = 0, 1, 2, 3,$ and 4 in each table.

22. Graph both functions on one pair of axes.

23. What angle do the two lines seem to make with each other?

SET II

The amaryllis grows from a bulb to a plant with flowers in just a few weeks. Part of a graph of the height of the plant as a function of time is shown here.

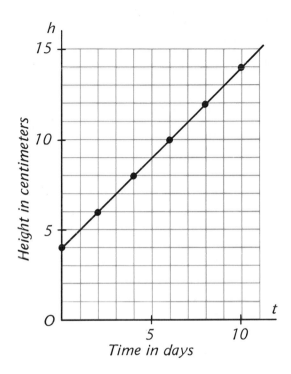

1. Refer to this graph to copy and complete the following table.

t	0	2	4	6	8	10
h	▊▊	▊▊	▊▊	▊▊	▊▊	▊▊

2. Write a formula for this function.

3. If the amaryllis continues to grow at the same rate, how tall would it be after 20 days?

Before the invention of mechanical clocks, candles were sometimes used to measure the passage of time. A formula for the height of such a candle as a function of time is

$$h = 10 - 2t$$

in which h represents the height of the candle in centimeters and t represents the time in hours that the candle has been burning.

4. Use this formula to copy and complete the following table.

t	0	1	2	3	4	5
h	▓	▓	▓	▓	▓	▓

5. Graph the function, letting the horizontal axis represent time and the vertical axis represent height.

6. What was the height of the candle before it was lit?

7. What does the 10 in the formula represent?

8. How much does the height of the candle change each hour?

9. What does the 2 in the formula represent?

10. How many hours does the candle burn?

The amount of time that you can spend in the sun without burning is related to the number of the sunscreen lotion you use. The table below is for a person who can stay in the sun without any sunscreen lotion for only 15 minutes without burning.

Sunscreen number, x	8	10	12	14
Time in minutes, y	120	150	180	210

11. How does increasing the sunscreen number by 2 change the time that can be spent in the sun?

12. Draw a graph for this function, letting 5 units on the x-axis represent 10 and 5 units on the y-axis represent 100.

13. Write a formula for this function.

How far a baseball travels depends, in part, on the speed of the bat. A typical formula for this function is

$$y = 6x - 40$$

in which x represents the speed of the bat in miles per hour and y represents the distance the ball travels in feet.

14. Use this formula to copy and complete the following table.

Speed of bat, x	50	60	70	80	90
Distance ball travels, y	260	▓	▓	▓	▓

15. How does the distance traveled by the ball change as the speed of the bat increases by 10 miles per hour?

16. How far do you think the baseball would travel if the speed of the bat were 100 miles per hour?

17. Draw a graph of this function. Let 1 unit on the *x*-axis represent 10 and 2 units on the *y*-axis represent 100.

Temperatures in the metric system are measured in degrees Celsius. A table expressing Fahrenheit temperatures as a function of Celsius temperatures is shown here.

Degrees Celsius, *x*	0	10	20	40	80	100
Degrees Fahrenheit, *y*	32	50	68	104	176	212

18. How does increasing the Celsius temperature by 20 change the Fahrenheit temperature?

19. Draw a graph for this function, letting 2 units on the *x*-axis represent 10 and 2 units on the *y*-axis represent 20.

20. Use your graph to estimate the Fahrenheit temperature that corresponds to 60°C.

21. Show how you could figure out the Fahrenheit temperature that corresponds to 60°C from the table.

Set III

In 1900, the typical surfboard was 16 feet long. Since then, surfboards have become shorter and shorter.

Bishop Museum.

Tom Blake with six of his surfboards.

A formula that gives a pretty good estimate of the length of a surfboard as a function of the year is

$$y = 263 - 0.13x$$

in which x is the year and y is the length of the surfboard in feet.

1. Show that this formula gives a length of 16 feet for the year 1900.

2. Use the formula to copy and complete the following table.

Year, x	1900	1930	1970	1980
Length of surfboard, y	16	▨	▨	▨

3. Draw and label a pair of axes as shown in the figure below and graph the function.

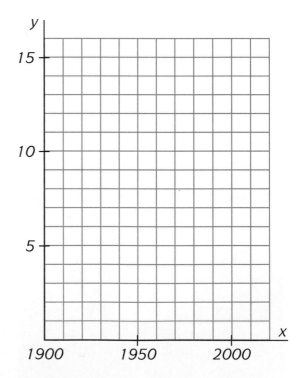

4. Use your graph to estimate the length of surfboards in 1950.

5. Do you think your graph would continue to give reasonable lengths of surfboards for years following 1980?

By permission of Johnny Hart and Field Enterprises, Inc.

Functions with Parabolic Graphs

If it takes 16 seconds for a rock to hit the bottom of a well, can anything be concluded about the depth of the well? Suppose, for example, that a heavy rock and a light rock were thrown in at the same time. Would they hit the bottom at the same time?

The Greeks thought that the heavy rock would hit the bottom first. They reasoned that if one object is heavier than another, it is because it is more strongly attracted to the earth, and the more strongly the earth attracts an object, the faster it will fall. But the Greeks were wrong.

In the seventeenth century, the great Italian scientist Galileo discovered that the speed at which an object falls does not depend on its weight. Drop a small stone and a large rock from the same height and they will hit the ground at the same time.

Galileo knew that the distance an object falls is a function of time. If the time is measured in quarter-seconds and the distance in feet, the two variables are related in an especially simple way. This table shows what it is.

Time in quarter-seconds, t	0	1	2	3	4	5	. . .
Distance in feet, d		0	1	4	9	16	25	. . .

The formula for this function

$$d = t^2$$

no doubt delighted Galileo.

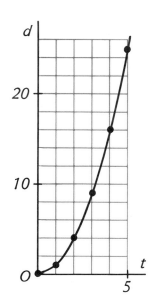

The graph of this function is shown at the left. Unlike those of the graphs of functions in the last lesson, the points in this graph do not lie on a straight line. They can be connected, instead, with a smooth curve to show the times and distances between those listed in the table. The graph shows part of a curve called a *parabola*.

Although the graph does not extend far enough, the formula for the function can be used to find out how many feet deep B.C.'s well is. Noting that 16 seconds is the same as 64 quarter-seconds and substituting, we get

$$
\begin{aligned}
d &= t^2 \\
&= (64)^2 \\
&= 4{,}096
\end{aligned}
$$

The well is apparently more than 4,000 feet deep!

EXERCISES

SET I

A certain function is represented by this table of numbers.

x	0	1	2	3	4
y	0	1	4	9	16

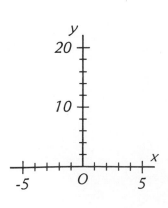

1. What is a formula for this function? Begin your formula with $y =$.

2. Graph the function by copying the pair of axes shown here, plotting the points included in the table, and drawing a smooth curve through them.

Because the graph is a curve, a more complete picture of it can be obtained by including some points with negative coordinates.

3. Copy and complete the following table for the same function.

x	-4	-3	-2	-1
y	16*	‖‖‖‖‖	‖‖‖‖‖	‖‖‖‖‖

Add the points to your graph and extend the curve through them.

The curve you drew in exercises 2 and 3 is a parabola. Two more functions whose graphs are parabolas have the following formulas:

$$\text{Function A:} \quad y = x^2 + 2$$
$$\text{Function B:} \quad y = x^2 + 10$$

4. Make two tables, one for each function. Let $x = -3, -2, -1, 0,$ 1, 2, and 3 in each table.

5. Draw a pair of axes with the same scales as those in exercise 2. Graph both functions on this pair of axes. Draw a curve through each set of points and label each curve with its formula.

6. What do you notice about these two curves?

7. At what values of y do the curves meet the y-axis?

8. Compare the curve shown here with the two you have just drawn. What do you think is the formula for the function shown?

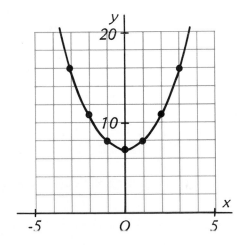

---

*Because the product of two negative numbers is positive, the square of a negative number is always positive. See pages 660–661 if this is not clear.

A function has the formula

$$y = 12 - x^2$$

9. Copy and complete the following table for this function.

x	−3	−2	−1	0	1	2	3
y							

10. Graph it on a pair of axes with the same scales as those in exercise 2.

11. In what way is the curve you drew different from the ones in the previous exercises?

12. At what value of y does it meet the y-axis?

13. Compare the graph shown here with the one you have just drawn. What do you think is the formula of the function shown?

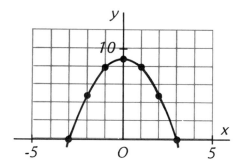

A function has the formula

$$y = (3 - x)^2$$

14. Copy and complete the following table for this function.

x	0	1	2	3	4	5	6
y							

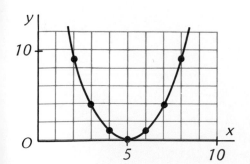

15. Graph it on a pair of axes with the same scales as those in exercise 2.

16. At what value of x does the curve touch the x-axis?

17. At what value of y does it meet the y-axis?

18. Compare the graph shown at the left with the one that you drew for exercise 15. What do you think is the formula of the function shown?

SET II

The collision impact* of an automobile is a function of its speed. For a certain automobile, it is given by the formula

$$I = 2s^2$$

in which I represents the collision impact and s represents the speed in kilometers per minute.

1. Use this formula to copy and complete the following table. (The expression $2s^2$ means to square s first and then multiply by 2. For example, if $s = 3$, $2(3)^2 = 2 \cdot 9 = 18$.)

s	0	1	2	3	4
I					

2. What happens to the collision impact of the automobile if its speed is doubled?

3. What happens to the collision impact of the automobile if its speed is tripled?

4. Graph this function, labeling the axes as shown here.

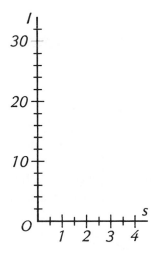

---

*Collision impact is a measure of the damage that a moving automobile can cause when it hits an object.

A cable television company's total income depends on how many sub-scribers it has. The number of subscribers depends on how high the monthly fee is.*

A typical table showing how the number of subscribers is related to the monthly fee is shown below.

Monthly fee in dollars, x	0	10	20	30	40	50	60
Thousands of subscribers, y	12	10	8	6	4	2	0

5. Graph this function. Let 2 units on the x-axis represent 10.

6. What happens to the number of subscribers as the monthly fee increases?

The cable company's total income can be found by multiplying the monthly fee by the number of subscribers. For example, from the table above, if the monthly fee is $20, the company would have 8 thousand subscribers: $(20)(8) = 160$. The company's monthly income would be $160,000.

7. Copy and complete the following table by multiplying the numbers in the table above.

Monthly fee in dollars, x	0	10	20	30	40	50	60
Total income in thousands of dollars, y	▓▓▓	▓▓▓	160	▓▓▓	▓▓▓	▓▓▓	▓▓▓

8. Graph the function. Let 2 units on the x-axis represent 10 and 5 units on the y-axis represent 100.

9. What are the coordinates of the highest point on your graph?

Refer to this point to answer the following questions.

10. What is the highest total monthly income that the cable company could have?

11. What monthly fee would produce this income?

The cliff divers in Acapulco, Mexico, hold the record for the highest head-first dives. They have to leap forward as they dive to avoid hitting rocks below.

A formula expressing their distance above the water, y, in feet is

$$y = 88 - 0.12x^2$$

in which x is how far forward they have moved in feet.

---

*Based on an example in *Economics*, 13th edition, by Paul A. Samuelson and William D. Nordhaus (McGraw-Hill, 1989, pp. 575–577).

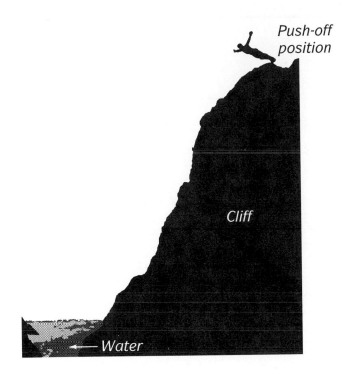

Push-off position

Cliff

Water

Joe Viesti, Viesti Associates, Inc.

Cliff diving, Acapulco, Mexico.

For example, if $x = 5$,

$$y = 88 - 0.12(5)^2 = 88 - 0.12(25) = 88 - 3 = 85$$

So the diver is 85 feet above the water when he has moved 5 feet forward.

12. Copy and complete the following table for this function.

x	0	5	10	15	20	25	27
y	▓	85	▓	▓	▓	▓	▓

13. Graph this function. Let 2 units on each axis represent 10 feet.

Your graph is a picture of the diver's path. The point where the curve meets the y-axis shows the position of the diver at the beginning of the dive.

14. What does the point where the curve meets the x-axis represent?
15. How high is the diver above the water at the beginning of the dive?
16. How far forward does the diver move during the dive?

SET III

"Of course you smell pizza pie—it's right there in front of you!"

The price of a pizza depends on its diameter. A typical formula is

$$y = 0.07x^2 - x + 8.25$$

in which x is the diameter of the pizza in inches and y is its price in dollars.

1. Use this formula to copy and complete the following table for this function, which includes the usual diameters of small, medium, large, and extra-large pizzas.

Diameter in inches, x	10	12	15	18
Price in dollars, y	▨	▨	▨	▨

2. Graph this function. Let 5 units on the x-axis represent 10 inches and 5 units on the y-axis represent $5. Draw a curve through the four points.

3. According to the formula, what price would you expect for a pizza with a diameter of only 4 inches?

4. According to the formula, what price would you expect for a pizza with a diameter of 1 inch?

5. Do the points corresponding to your answers to exercises 3 and 4 seem as if they would lie on the curve you drew in exercise 2?

6. Does the formula seem to give reasonable prices for pizzas of all possible diameters?

Courtesy of Elmer Atkins

"Well, finally! I thought this thing would never end."

More Functions with Curved Graphs

The time that you have to wait at a train-crossing for a train of a given length to pass by depends on the speed of the train. The slower the speed of the train, the longer you have to wait.

A typical formula for this function is

$$y = \frac{120}{x}$$

in which y is the waiting time in minutes and x is the speed of the train in miles per hour. The table below shows the waiting times for several different speeds.

Speed in miles per hour, x	10	20	30	40	50	60
Waiting time in minutes, y	12	6	4	3	2.4	2

Speed in miles per hour, x	10	20	30	40	50	60
Waiting time in minutes, y	12	6	4	3	2.4	2

If the points having these coordinates are plotted on a graph, they lie on a curve, but the curve is not a parabola. It gets closer and closer to the

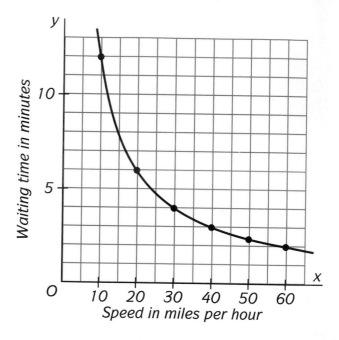

x-axis as we look toward the right, corresponding to the fact that as the speed of the train becomes greater and greater, the waiting time becomes smaller and smaller.

If the train could go 600 miles per hour, then according to the formula, it would pass the crossing in only

$$y = \frac{120}{600} = 0.2 \text{ minute}$$

or 12 seconds! This means that if the graph were extended to the right, at $x = 600$, the curve would be very close to the x-axis.

On the other hand, if the train went only 1 mile per hour, then it would take it

$$y = \frac{120}{1} = 120 \text{ minutes}$$

or 2 hours to pass the crossing! This means that if the graph were extended upward, at $y = 120$, the curve would be very close to the y-axis. A view of the graph extended this far in each direction is shown on the next page.

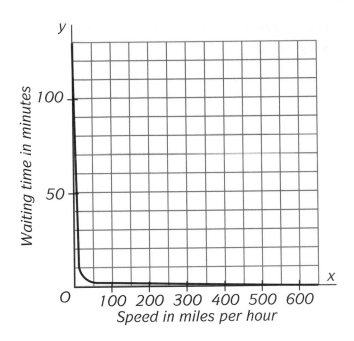

Speed in miles per hour

EXERCISES

SET I

The "waiting time for a train" function is one member of a family of related functions. A similar function is

$$y = \frac{6}{x}$$

1. Copy and complete the following table for this function.

x	1	2	3	4	5	6
y	▓▓▓	▓▓▓	▓▓▓	1.5	▓▓▓	▓▓▓

2. Graph the function by copying the pair of axes shown here, plotting the points included in the table and drawing a smooth curve through them.

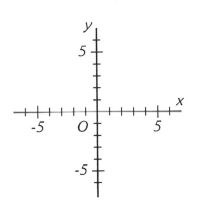

3. What happens to the values of y as the values of x get larger?

4. Use the formula to find the value of y when $x = 100$.

5. Do you think there is a value of x for which $y = 0$?

It is impossible to find a value for y if $x = 0$ because $\frac{6}{0}$ is not equal to any number. This means that there is no point on the curve for which $x = 0$. There are points for which x is a negative number, however.

6. Copy and complete the following table of negative numbers for the function.

x	−6	−5	−4	−3	−2	−1
y	−1	−1.2				

Plot the points included in this table on the pair of axes that you have already drawn and connect them with a smooth curve.

Another function related to the function that you have just graphed is shown in this graph.

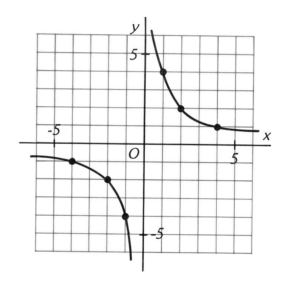

7. Refer to the graph to copy and complete the following table.

x	−4	−2	−1	1	2	4
y	−1					

8. Its formula is similar to the formula of the function that you graphed. What do you think it is?

Another function with a curved graph is

$$y = x^3$$

9. Copy and complete the following table for this function. [Notice that the cube of a negative number is also negative. For example, $(-4)^3 = (-4)(-4)(-4) = (+16)(-4) = -64$.]

x	−4	−3	−2	−1	0	1	2	3	4
y	−64								

10. Graph the function, letting 1 unit on the y-axis represent 10.

Compare the graph of the function

$$y = x^3 + 10$$

shown at the right with the graph of the function

$$y = x^3$$

that you have just drawn.

11. In what way are the two graphs alike?

12. In what way are they different?

Graphs of the functions $y = x^4$, $y = x^5$, $y = x^6$, $y = x^7$, and $y = x^8$ are shown below.

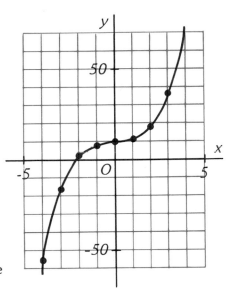

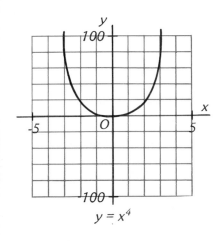

$$y = x^4$$

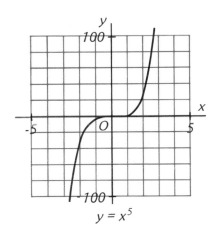

$$y = x^5$$

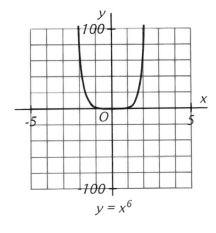

$$y = x^6$$

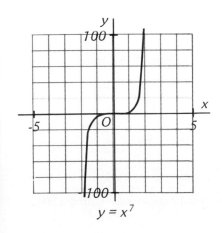

$$y = x^7$$

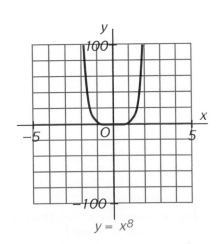

$$y = x^8$$

13. Which of the other functions have graphs similar to the graph of $y = x^4$?

14. What do their formulas have in common?

15. Which functions have graphs similar to the graph of $y = x^3$? (See your graph for exercise 10.)

16. What do their formulas have in common?

Set II

The length of a sound wave is a function of its frequency; it is given by the formula

$$y = \frac{330}{x}$$

in which y is the wavelength in meters and x is the frequency in Hz.*

1. The human ear can hear sound waves ranging from 20 Hz to 20,000 Hz. Use the formula to find the lengths of sound waves having these frequencies.

2. What happens to the wavelength of a sound as its frequency increases?

3. Copy and complete the following table for this function.

Frequency in Hz, x	20	40	60	80	100	150
Wavelength in meters, y						

4. What happens to the wavelength of a sound when its frequency is doubled?

5. Graph the function, letting 5 units on the x-axis represent 50 and 5 units on the y-axis represent 5.

The volume of a bird egg is a function of its length. A formula for this function is

$$v = 2.2\ell^3$$

in which v is the volume in cubic centimeters and ℓ is the length in centimeters.

---

*Hz stands for hertz, a unit of frequency named for the scientist Heinrich Hertz. A hertz is 1 cycle per second.

6. Use this formula to copy and complete the following table. For example, if $\ell = 1$, $v = 2.2(1)^3 = 2.2(1) = 2.2$.

Length in centimeters, ℓ	1	5	8	10	12	15
Volume in cubic centimeters, v	2.2	▦	▦	▦	▦	▦

7. A typical chicken egg is 5 centimeters long and a typical hummingbird egg is 1 centimeter long. What is the volume of each egg in cubic centimeters?

8. How many times the volume of a hummingbird egg is the volume of a chicken egg?

9. A typical ostrich egg is 15 centimeters long. How many times the length of a chicken egg is the length of an ostrich egg?

10. How many times the volume of a chicken egg is the volume of an ostrich egg?

11. Graph the function, letting 5 units on the ℓ-axis represent 5 and 2 units on the v-axis represent 1000. Mark the three points that represent a typical hummingbird egg, a typical chicken egg, and a typical ostrich egg.

If the moon were much closer to the earth than it is, it would look much larger. A formula relating apparent area and distance is

$$y = \frac{57,600}{x^2}$$

in which y is the apparent area of the moon and x is its distance from the earth in thousands of miles.

12. Copy and complete the following table for this function.

Distance in thousands of miles, x	15	30	60	120	240
Apparent area of moon, y	▦	▦	▦	▦	▦

The table shows that when the moon is 240 thousand miles from the earth (its actual distance), its apparent area is 1 unit (its "actual apparent size.")

13. What happens to the apparent area of the moon each time the distance is cut in half?

14. Graph this function, letting 5 units on each axis represent 50 units.

"Maynard, I do think that just this once you should come out and see the moon!"

SET III

As a piece of iron is heated, a blacksmith can judge its approximate temperature from its color. At first it glows a dull red, then bright red, and then a bright yellow-white.

The color of the iron changes as the heat radiated by it increases. A typical table for this function is

Color	Black	Dull red	Bright red	Bright orange	Yellow-white
Temperature in thousands of degrees Kelvin, x	1	2	3	4	5
Heat radiated in thousands of calories per minute, y	1	16	81	256	625

1. What happens to the heat radiated when the temperature is doubled?

2. Guess a formula for this function.

3. Graph it, letting 2 units on the x-axis represent 1 and 2 units on the y-axis represent 100.

4. The temperature at the surface of the sun is about 6000°K. If the iron heated by the blacksmith were that hot, how much heat would it radiate?

Courtesy of the National Oceanic and Atmospheric Administration

Interpolation and Extrapolation: Guessing Between and Beyond

Mark Twain once remarked that in eternity he planned to spend 8 million years on mathematics. In his book *Life on the Mississippi,* Twain used some mathematics to make a strange prediction about the future of the Mississippi River. The river is extremely crooked with many curves, some of which are shown in the photograph above. From time to time, the river changes its course from a wide bend to a more direct path, called a cutoff. As a result, the length of the Mississippi is becoming shorter and shorter. Twain gave some figures:

The Mississippi between Cairo and New Orleans was 1,215 miles long 176 years ago. It was 1,180 after the cutoff of 1722 . . . its length is only 973 miles at present [1875].

Because the length of the Mississippi is a function of time, a table can be made from these numbers. Rounding them slightly, we get:

Time (in years), t	1700	1720	1875
Length in miles, ℓ	1,215	1,180	975

Although we have no formula for this function, the points in the table can be graphed. The result, shown below, suggests that they might lie on a straight line.

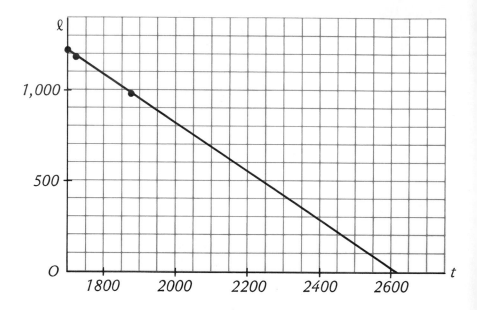

If a line is drawn through the three points, we can get additional information from it. For example, it appears that in 1800 the river was approximately 1,100 miles long. This information, found by reading between the values that we know, was obtained by *interpolation*.

> To **interpolate** is to guess other values of a variable *between* values that are known.

If the line representing the length of the Mississippi is extended until it intersects the time axis, it looks as if the river will disappear sometime about the year 2600. Twain jokingly made this prediction:

> Any calm person, who is not blind or idiotic, can see . . . that 742 years from now the Lower Mississippi will be only a mile and three-quarters long. . . . There is something fascinating about science. One gets such wholesale returns of conjecture out of such a trifling investment of fact.

What's wrong with this reasoning? The trouble is that, although the three points may seem to lie along a straight line, the graph must actually curve later on. The river cannot become any shorter after all the bends are gone. In drawing the line past the third point, we are *extrapolating*.

> To **extrapolate** is to guess other values of a variable *beyond* those that are known.

If a function continues to behave in the same way, then our guess may be very close; otherwise, you can see what may happen.

EXERCISES

SET I

When chocolate fudge cools after it has been cooked, its temperature is a function of time. A table is shown here.

Time in minutes, x	10	20	40	70	80
Temperature in °C, y	95	77	52	28	23

1. Draw a pair of axes, letting 1 unit on each axis represent 10. Plot the five points in the table.

2. Do they seem to lie along a *straight line* or a *curve*? Connect them accordingly.

3. Use your graph to estimate the temperature of the fudge when it has cooled for 50 minutes.

4. Did you *interpolate* or *extrapolate* in making this estimate? Explain.

5. Estimate the temperature of the fudge when it has cooled for 90 minutes.

6. Did you *interpolate* or *extrapolate* in making this estimate? Explain.

7. What happens to the rate at which fudge cools as time passes?

In 1787, the French physicist Jacques Charles observed that all gases contract in the same way when cooled. A table of the volume of a gas as a function of its temperature is shown below.

Temperature in °C, t	50	−30	−110	−135	−220
Volume of gas, v	120	90	60	50	20

8. Draw a pair of axes as shown at the top of the next page and plot the five points in this table.

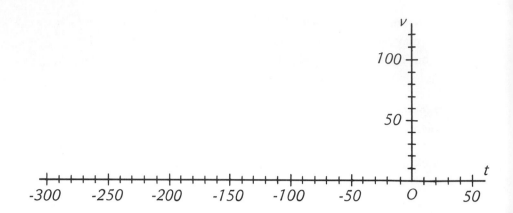

9. What do you notice about the five points?

Draw a line through them, extending it until it meets the t-axis. Refer to the graph to answer the following questions.

10. What is the approximate volume of the gas at $-60°C$?

11. Did you *interpolate* or *extrapolate* to obtain this answer?

12. At approximately what temperature does the volume of the gas appear to become 0?

13. Did you *interpolate* or *extrapolate* to obtain this answer?

Set II

This figure shows a set of weights suspended by a rubber band. The length to which the rubber band is stretched is a function of the weight supported. A table for this function is shown here.

Weight in kilograms, w	0.2	0.3	0.5	1.0	1.2
Length in centimeters, ℓ	15	16	18	23	25

1. Draw a pair of axes as shown at the top of the next page and graph this function.

Use your graph to estimate the length of the rubber band if

2. a weight of 0.8 kilograms is suspended from it.

3. no weight is suspended from it.

4. a weight of 2.0 kilograms is suspended from it.

5. Which one of your three estimates do you think could be entirely wrong? Explain.

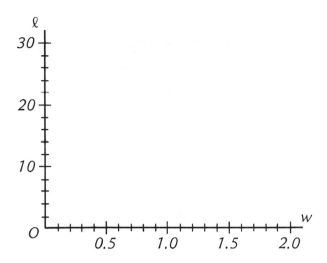

The earth's population is increasing at a rapidly growing rate. A table expressing it as a function of time is shown here.

Time (the year), t	1600	1700	1800	1900	1950	1975
Population in billions, p	0.5	0.6	0.9	1.6	2.5	3.9

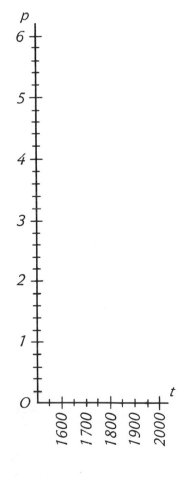

6. Draw a pair of axes as shown at the right and plot the six points in this table. Connect them with a smooth curve.

7. What happens to the steepness of the curve as you look at it from left to right?

8. Use your graph to estimate the earth's population in 1500.

9. Use your graph to estimate the earth's population in 2000.

10. Which of your estimates did you think was easier to make?

11. Why?

The temperature of the water in a lake is a function of the depth. A typical table for this function for a lake 20 meters deep at its center is shown here.

Depth in meters, d	0	4	8	12	16	20
Temperature in °C, t	15	14	13	6	5	4

12. Graph this function by plotting the points in this table and connecting them with a smooth curve. Let 2 units on the d-axis represent 4 meters.

13. If you saw only the part of the table shown below, graphed it, and were asked to guess the temperature at 20 meters by extrapolation, what would be a reasonable answer?

Depth in meters, d	0	4	8
Temperature in °C, t	15	14	13

14. Between what depths does it seem the most difficult to determine the temperature by interpolation?

15. Why is it meaningless to extrapolate the temperature beyond 20 meters?

Set III

The Indianapolis 500 automobile race is run each year on Memorial Day.

Photograph by Ron McQueeney. © Indy 500 Photos. Indianapolis Motor Speedway Corp.

The table below shows some of the winning speeds in the race.

Year	Winning speed in miles per hour	Year	Winning speed in miles per hour
1920	89	1960	139
1930	100	1970	156
1940	114	1980	143
1950	124	1990	186

1. In general, how has the winning speed changed over the years?

2. For which year does the winning speed seem the most surprising?

3. Graph this function, labeling your axes as shown here. Draw a line through the points for 1920 and 1960.

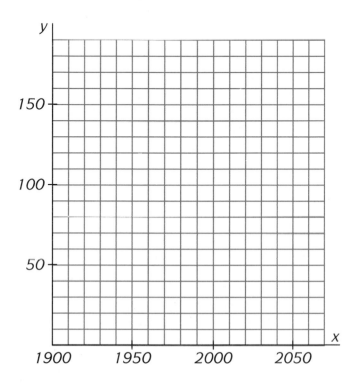

4. In 1960, what winning speeds do you suppose people might have predicted for 1970, 1980, and 1990?

5. The Indianapolis 500 was first held in 1911. Judging from your graph, guess the winning speed for that year.

6. Judging from your graph, when was the winning speed 0 miles per hour?

7. Why doesn't this answer make any sense?

8. Does it seem reasonable that you could correctly guess from your graph what the winning speed in the year 2050 will be?

CHAPTER

3

Summary and Review

In this chapter we have become acquainted with:

The idea of a function *(Lesson 1).* A function is a pairing of two sets of numbers so that to each number in the first set there corresponds exactly one number in the second set.

A function can be represented by a table of numbers, a formula, or a graph.

The coordinate graph *(Lesson 2).* The coordinate graph was invented by the French mathematician and philosopher René Descartes.

Each point on a coordinate graph is located by a pair of numbers, called its coordinates, which are the distances of the point from the x- and y-axes.

Functions with line graphs *(Lesson 3).* To graph a function for which a formula is known, it is first necessary to create a table for that function. The pairs of numbers in the table are coordinates of points of the graph.

Functions with curved graphs *(Lessons 4 and 5).* If the graph of a function is a curved line, it is often useful to include points with negative x-coordinates to get a complete picture.

Typical formulas and graphs of functions that we have studied are shown here.

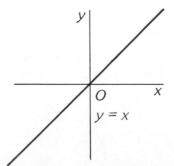

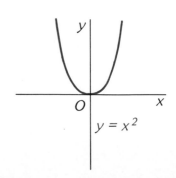

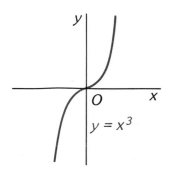

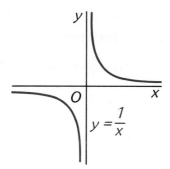

Interpolation and extrapolation *(Lesson 6).* Interpolation is guessing another value of a variable *between* values that are known. Extrapolation is guessing another value *beyond* those that are known.

EXERCISES

SET I

This figure shows a square drawn on a coordinate graph. Find the coordinates of each of its corners.

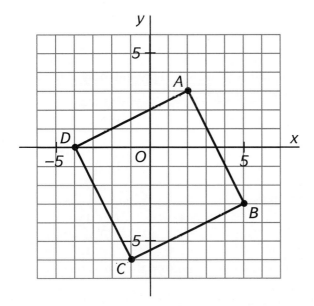

1. A.

2. B.

3. C.

4. D.

A car's air conditioner should produce air that is 30 degrees below the outside temperature.

5. Copy and complete the following table for this function.

Temperature of outside air in °F, x	80	90	100	110
Temperature of cooled air in °F, y	50	▨	▨	▨

6. What is a formula for this function? Begin your formula with $y =$.

Scuba divers use a belt with lead weights to control their buoyancy in the water. The amount of weight needed is a function of the diver's weight.

Diver's weight in pounds, x	100	120	140	160	180
Weight of lead in pounds, y	10	12	14	16	▨

7. What is the missing number in the table?

8. How can this number be found from the previous y-number?

9. How can the y-numbers in the table be found from the corresponding x-numbers?

10. Which one of the following is a formula for this function?

$$y = x + 10 \qquad y = 10x \qquad y = \frac{x}{10}$$

Guess a formula for the function represented by each of these tables. Begin each formula with $y =$.

11. Function A

x	0	1	2	3	4	5
y	0	4	8	12	16	20

12. Function B

x	0	1	2	3	4	5
y	0	1	4	9	16	25

Draw a pair of axes, letting 5 units on the y-axis represent 10.

13. Use the tables above to graph both functions on your pair of axes.

14. In how many points do the line and curve intersect each other?

15. What are the coordinates of these points?

SET II

The louder a portable radio is played, the faster its batteries will run down. A typical formula for this function is

$$y = \frac{120}{x}$$

in which x is the loudness at which the radio is played and y is the number of hours the batteries last.

1. Copy and complete the following table for this function.

Loudness, x	1	2	3	4	5	6
Number of hours, y						

2. What happens to the number of hours the radio will play when the loudness is doubled?

3. What happens to the number of hours the radio will play when the loudness is tripled?

4. Graph the function, letting 2 units on the x-axis represent 1 and 5 units on the y-axis represent 50.

The length of an ocean wave is a function of its speed. It is approximated by the formula

$$y = 0.6x^2$$

in which x represents the speed of the wave in knots and y represents its length in feet.

5. Copy and complete the following table for this function.

Speed in knots, x	0	10	20	30	40	50
Length in feet, y		60				

Warren Bolster/Tony Stone Worldwide

6. Graph the function. Let 2 units on the x-axis represent 10 and 5 units on the y-axis represent 500.

7. Use your graph to estimate the length of a wave whose speed is 45 knots.

8. Compare your estimate to the length given by the formula when $x = 45$.

Calvin and Hobbes, 1987, Universal Press Syndicate.

Except at the equator, the length of daylight changes throughout the year. The table below is typical.

Days after September 1, x	0	20	40	60
Hours of daylight, y	12.9	12.2	11.5	10.8

9. Draw a pair of axes, letting 5 units on the x-axis represent 100 days. Plot the four points in this table and draw a line through them. Extend the line beyond the last point.

Refer to the graph to answer the following questions.

10. Estimate the number of hours of daylight on the 50th day after September 1.

11. Did you *interpolate* or *extrapolate* to obtain this answer?

12. Estimate the number of hours of daylight on the 200th day after September 1.

13. Did you *interpolate* or *extrapolate* to obtain this answer?

14. Which of your two estimates do you think is less reasonable? Why?

Set III

The weight of an astronaut is a function of his or her distance from the surface of the earth. The weight of an astronaut who weighs 160 pounds at the surface of the earth is given by the formula

$$w = \frac{2{,}560}{(d+4)^2}$$

in which d is the astronaut's distance from the earth in thousands of miles and w is the astronaut's weight in pounds.

1. Use this formula to copy and complete the following table. Round each weight to the nearest pound.

Distance in thousands of miles, d	0	1	2	3	4	5	6
Weight of astronaut in pounds, w	▓	▓	▓	▓	▓	▓	▓

2. What happens to the astronaut's weight as his or her distance from the earth increases?

3. How much would the astronaut weigh at a distance of 50 thousand miles from the earth?

If an astronaut could travel from the surface of the earth to its center, the astronaut's weight would be given by the formula

$$w = 160 + 40d$$

4. Use this formula to copy and complete the following table.

Distance in thousands of miles, d	0	-1	-2	-3	-4
Weight of astronaut in pounds, w	▓	▓	▓	▓	▓

The radius of the earth is about 4 thousand miles so, when $d = -4$, the astronaut would be at the center of the earth.

5. How much would the astronaut weigh at the center of the earth?

6. Graph the astronaut's weight as a function of distance from the surface of the earth by plotting the points from the tables you made for exercises 1 and 4. Draw the x-axis from -4 to 6. Let 5 units on the y-axis represent 100.

7. Where is the astronaut with respect to the earth when he or she is heaviest?

CHAPTER
3

Further Exploration

LESSON 1

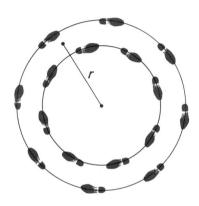

1. A person lost on a desert sometimes walks a great distance in what seems to be a straight line, only to return to the starting place without realizing it. The reason for this is that a person's legs are not exactly the same length, so that the steps taken with one foot are slightly longer than those taken with the other. An exaggerated diagram of this is shown here.

    a. If a man's left leg is longer than his right leg, would you expect him to turn to the *left* or *right*?

    The radius of the circle in which a person walks is a function of the difference between the lengths of that person's steps. A typical table for this function is shown here.

Difference in millimeters, d	5	4	3	2	1
Radius of circle in meters, r	36	45	60	90	180

    b. What happens to the radius of the circle as the difference between the lengths of the steps decreases?
    c. Write a formula for this function.
    d. If the difference between the lengths of a person's steps is 2.5 millimeters, what do you think would be the radius of the circle in which that person walked?
    e. Can the formula be used to find the radius of the circle in which someone would walk if the difference between the lengths of the steps was 0 millimeters? Explain.

2. The following pattern is based on a table for the function $y = x^2$.

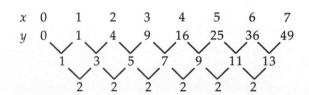

a. How can the numbers in the third and fourth rows of this pattern be obtained from the numbers above them?
b. Use the same procedure to copy and complete the following pattern based on a table for the function $y = x^3$.

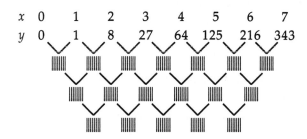

x	0	1	2	3	4	5	6	7
y	0	1	8	27	64	125	216	343

c. Make a similar pattern for the function $y = x^4$.
d. Make a pattern for the function $y = x$.
e. On the basis of the patterns for $y = x$, $y = x^2$, $y = x^3$, and $y = x^4$, what would you predict about the corresponding pattern for $y = x^5$?

LESSON 2

1. The speed of the hand during a karate strike is a function of time. A table for this function for a hammer-fist strike is given here.

Time in hundredths of a second, x	0	1	2	3	4	5	6
Speed of hand in miles per hour, y	0	4.5	6.7	6.3	5.4	5.4	6.0

(Table continued)

x	7	8	9	10	11	12	13	14
y	7.6	9.8	12.1	15.0	18.0	21.5	24.2	26.0

Photograph by Bruce Tegner; courtesy of Thor Publishing Company

a. Draw a pair of axes in which 5 units on the y-axis represent 10 miles per hour. Use the pairs of numbers in the table to plot points. Join the points with a smooth curve.

The graph shows the entire karate strike.

b. How long does it last?
c. The hand speeds up the fastest over the part of the curve that is the steepest. When is this?
d. Between what times is the hand slowing down?

e. The hand picks up speed at a steady rate over the part of the curve that is the straightest. When is this?

f. Just before the end of the strike, is the hand speeding up or slowing down?

2. In his book *On Growth and Form*, the great British zoologist D'Arcy Thompson applied Descartes' method of coordinates to the study of the shapes of living things. With a series of examples, he showed how a coordinate system could be applied to the form of an animal and the system then transformed in a certain way to obtain the form of a different animal.

One of the examples concerns a fish of the species *Argyropelecus olfersi*.

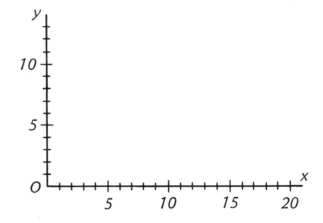

a. Draw and label a pair of axes as shown here. Plot the point (3, 7) and draw a small circle around it to represent the eye of the fish. Then connect the points in the following list in order with straight line segments to form the outline of the fish.

(0, 6) (1, 4) (3, 2) (8, 0) (8, 1) (10, 2) (10, 4) (11, 3) (12, 5) (16, 4) (15, 6) (16, 8) (12, 7) (9, 8) (11, 10) (9, 12) (8, 10) (8, 12) (6, 10) (1, 7) (2, 5) (0, 6)

b. Make another set of coordinates by changing each x-coordinate in the preceding set like this:

$$(x, y) \longrightarrow \left(x + \frac{y}{2}, y \right)$$

For example, the eye of the fish would be changed from (3, 7) to $\left(3 + \frac{7}{2}, 7 \right)$ or (6.5, 7), and the first point in the outline of the fish would be changed from (0, 6) to $\left(0 + \frac{6}{2}, 6 \right)$ or (3, 6).

c. Draw and label another pair of axes and repeat the directions given in part a of this exercise with the list of coordinates that you made in part b. The result is the outline of a fish of a species of a different genus, *Sternoptyx diaphana*.

LESSON 3

1. Metal nails were so valuable in Colonial America that it was not unusual for people to burn their houses down when they moved in order to have enough nails to build the next one!

The 16 most common nail sizes are shown in this table.

a. Draw a pair of axes in which the *x*-axis represents nail size and the *y*-axis represents length. Let 5 units on the *x*-axis represent 10 and 4 units on the *y*-axis represent 1. Use the pairs of numbers in the table to plot 16 points.

Your graph should reveal two patterns: one in the sizes of the smaller nails and the other in the sizes of the larger nails.

b. Which nails do not fit either pattern?

Draw a line through the points that fit the first pattern and extend it beyond them.

c. If this pattern continued, what would be the size of a 6-inch nail?

Draw a line through the points that fit the second pattern and extend it beyond them.

d. If this pattern continued, what would be the length of a size-5 nail?

Size	Length in inches
2	1
3	1¼
4	1½
5	1¾
6	2
7	2¼
8	2½
9	2¾
10	3
12	3¼
16	3½
20	4
30	4½
40	5
50	5½
60	6

2. In about 450 B.C., the Greek philosopher Zeno made up a puzzle about a race between Achilles and a tortoise on which the following exercise is based. Achilles runs much faster than the tortoise, and so the tortoise is given a head start.

Suppose Achilles runs at the rate of 500 meters per minute and the tortoise runs at the rate of 50 meters per minute.

a. Make a table showing the distances Achilles covers in 0, 1, 2, 3, 4, and 5 minutes.
b. Write a formula for the distances run by Achilles, *d*, as a function of time, *t*.

Suppose the tortoise is given a head start of 1,800 meters so that its distance at the beginning of the race is 1,800.

c. Make a table showing the tortoise's distances in 0, 1, 2, 3, 4, and 5 minutes, supposing that its distance at the beginning of the race is 1,800 meters.

d. Write a formula for the distances of the tortoise along the track, *d*, as a function of time, *t*.

e. Draw a pair of axes in which the *x*-axis represents time and the *y*-axis represents distance. Let 2 units on the *x*-axis represent 1 minute and 5 units on the *y*-axis represent 500 meters. Plot the points from your tables for parts a and c on the graph and connect each set of points with a straight line.

f. When and where does Achilles overtake the tortoise?

LESSON 4

1. The number of pieces into which a pancake can be cut by a series of straight cuts, each of which crosses the others, is a function of the number of cuts. One cut across a pancake divides it into two pieces; a second cut crossing the first one results in four pieces; and a third cut crossing the first two cuts results in seven pieces.

a. Draw two figures representing pancakes that have been cut with four cuts and five cuts respectively, in which each cut crosses each of the others. (To get as many pieces as possible, be careful not to have more than two cuts meet in a single point.)

b. Refer to the figures above and the figures that you have drawn to copy and complete the following table.

Number of cuts, *x*	1	2	3	4	5
Number of pieces, *y*	▓▓▓	▓▓▓	▓▓▓	▓▓▓	▓▓▓

c. Graph it, letting 2 units on the *x*-axis represent 1.

d. What do you think the value of *y* would be if $x = 0$? Does your answer make any sense? Explain.

e. Use your graph to decide the value of *y* when $x = 3.5$. Does your answer make any sense? Explain.

2. Large-screen television pictures that are produced by projection look brighter when viewed from some positions than they do from others.

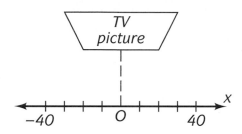

A typical formula for brightness as a function of position is

$$y = 100 - 0.03x^2$$

in which y is the brightness and x indicates the horizontal distance from the center of the picture.

a. Use this formula to copy and complete the following table.

Distance from center, x	−40	−30	−20	−10	0	10	20	30	40
Brightness, y	52	▓	▓	▓	▓	▓	▓	▓	▓

For example, if $x = -40$, $y = 100 - 0.03(-40)^2 =$ $100 - 0.03(1600) = 100 - 48 = 52$.

b. Graph this function, letting 2 units on the x-axis represent 10 and 5 units on the y-axis represent 50.

c. From what position does the picture look brightest?

d. Approximately how much does the brightness drop off to the sides?

Lesson 5

1. When a basketball player shoots for the basket, the distance of the ball above the floor is a function of the horizontal distance from the point at which it was thrown.

For a particular shot, the formula for this function is

$$y = 7.5 + 0.6x - 0.02x^2$$

in which x is the horizontal distance in feet of the ball from the point at which it was thrown and y is the distance in feet of the ball above the floor.

a. Copy and complete the following table for this function.

x	0	5	10	15	20	25
y						

The calculation for $x = 5$ is shown below.

$$y = 7.5 + 0.6(5) - 0.02(5)^2 = 7.5 + 3 - 0.02(25) = 10.5 - 0.5 = 10$$

b. Graph the function, labeling the axes as shown here.

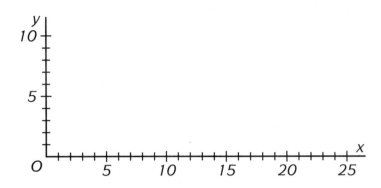

c. How high above the floor is the ball when it is shot?
d. What is the greatest distance of the ball above the floor?
e. How high is the basket above the floor? (The last point of your graph represents the position of the basket.)
f. How far back on the floor from the basket was the ball shot?

2. The brightness of the light produced by fireflies is a function of temperature. An approximate formula for this function is

$$y = 100 + 3x + 4x^2 - 0.1x^3$$

in which y is the light intensity and x is the temperature in degrees Celsius.

a. Copy and complete the following table for this function.

x	0	10	20	30	40
y					

b. Graph the function, letting 2 units on the x-axis represent 10 degrees and 5 units on the y-axis represent 500.
c. What happens to the brightness of the light produced by fireflies as the temperature increases?

Painting by Tom Prentiss. Copyright © 1976 by Scientific American, Inc. All rights reserved.

d. From the table you made for exercise a, it seems as if fireflies are at their brightest at 30°C. Can you find a temperature at which they produce an even brighter light? (*Hint:* Try some other numbers for x that are not in the table.)

LESSON 6

1. The sex ratio of a population is the number of males for every 100 females. The following table shows its value for the United States as determined by four censuses.

Year, y	1920	1940	1960	1980
Sex ratio, r	104	101	98	95

a. Graph the function, labeling the axes as shown here. (The scale on the r-axis jumps from 0 to 90 to save space.)
b. What do you notice about the graph?
c. If the sex ratio continued to change at the same rate, when would it become 0? Explain your reasoning. (Because of the gap in the graph, you will need to extrapolate from the numbers in the table.)

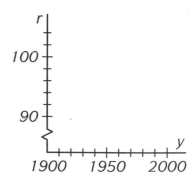

2. An application of extrapolation about which there is still much uncertainty concerns the age of the universe. That the universe is expanding seems certain, but whether it will expand forever is not known.

According to an article in *Scientific American,*

> If the motions of the galaxies are extrapolated into the past as far as possible, a state is eventually reached in which all the galaxies were crushed together at infinite density. That state represents the big bang, and it marks the origin of the universe and everything in it.*

The graph below, from the article, shows three possible ways in which the scale of the universe may be changing with the passage of time.

   a. What does line *a* suggest as one possible history of the universe?

   b. What does curve *b* suggest? (Assume that it continues to rise as time passes.)

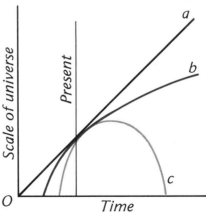

After "Will the Universe Expand Forever?" by J. Richard Gott III, James E. Gunn, David N. Schramm, and Beatrice M. Tinsley. Copyright © 1976 by Scientific American, Inc. All rights reserved.

   c. What does curve *c* suggest?

   d. Which of these three possibilities suggests the greatest present age for the universe?

   e. What is it about the line and two curves at the line marked "present" that makes it difficult to decide which one represents the actual history of the universe?

---

*"Will the Universe Expand Forever?" by J. Richard Gott III, James E. Gunn, David N. Schramm, and Beatrice M. Tinsley in *Scientific American,* March 1976.

Large Numbers and Logarithms

Calvin and Hobbes, © 1987, Universal Press Syndicate.

1

Large Numbers

The word *million* may have been invented in the fourteenth century by the Italian merchant Marco Polo, who, in describing his trips to China, said that he saw a "milli-one" (literally "many thousand") people there. Since then it has entered many other languages in basically the same form: in French, German, and English as *million* and in Spanish as *millón.*

Although the word *million* has become a part of everyone's vocabulary, it is difficult to grasp the size of a number as large as a million. It is easy to write 1,000,000, but how big is that? If you counted to 1 million and could name a number every second without stopping, it would take you nearly 12 days. And how long is a million days? A million days ago was in the eighth century B.C. How far is a million inches? Almost 16 miles!

If our idea of the size of a million is vague, our notions of larger numbers must be even more vague. How big, for example, is a billion? Although McDonald's has sold more than a billion hamburgers, if it were possible to eat a hamburger every minute, *day and night without stopping,* it would take more than 1,900 years to eat a billion hamburgers. A billion seconds is almost 32 years!*

The names for large numbers are based on numbers that are powers of 10:

Hundred:	$100 = 10 \times 10 = 10^2$
Thousand:	$1,000 = 10 \times 10 \times 10 = 10^3$
Million:	$1,000,000 = 10 \times 10 \times 10 \times 10 \times 10 \times 10 = 10^6$
Billion:	$1,000,000,000 = 10 \times 10 \times 10 \times 10 \times 10 \times 10 \times 10 \times 10 \times 10 = 10^9$

and so forth

This suggests the use of exponents as a way to write large numbers compactly. The numbers printed in smaller type, called exponents, tell how many times 10 is multiplied and, hence, how many zeros follow the 1 if the number is written in decimal form.

When a number is written as a power of 10, it is written in **exponential form.** The *exponent* indicates the number of zeros that follow the 1 when the number is written in decimal form.

This suggests that it makes sense to write

$$10 \text{ as } 10^1$$

and even

$$1 \text{ as } 10^0 \text{ (1 followed by 0 zeros).}$$

Each number in the sequence

$$1 \quad 10 \quad 100 \quad 1,000 \quad 10,000 \quad 100,000 \quad 1,000,000 \quad \ldots$$

is 10 times as large as the number preceding it. These numbers can be written in exponential form as

$$10^0 \quad 10^1 \quad 10^2 \quad 10^3 \quad 10^4 \quad 10^5 \quad 10^6 \quad \ldots$$

Comparing these two forms reveals that multiplying a number by 10 results in increasing the exponent by 1.

---

*Oddly enough, the size of a billion depends on where you live. In the United States a billion is 1,000,000,000, whereas in England it is 1,000,000,000,000. Perhaps the reason that the difference remains is that until recently there was very little need for such a large number.

Here is a list of the names and exponential forms of some large numbers.

10^6	million	10^{21}	sextillion
10^9	billion	10^{24}	septillion
10^{12}	trillion	10^{27}	octillion
10^{15}	quadrillion	10^{30}	nonillion
10^{18}	quintillion	10^{33}	decillion

Exercises

Set I

A large number such as one hundred trillion can be written as a power of 10 by observing that, since a trillion is 10^{12}, a trillion has 12 zeros. So one hundred trillion is

$$100,000,000,000,000 \text{ or } 10^{14}$$

Write each of the following numbers as a power of 10.

1. One.

2. Ten.

3. One thousand.

4. Ten billion.

5. One hundred quadrillion.

6. One thousand decillion.

Write the name of the number that is equal to each of the following powers of 10.

7. 10^2.

8. 10^7.

9. 10^{18}.

10. 10^{32}.

A tourist from England visited Yankee Stadium to see a baseball game. He did not understand the game and left when the scoreboard read

$$1 \ 0 \ 0 \ 0 \ 0 \ 0 \ 0 \ 0$$
$$1 \ 0 \ 0 \ 0 \ 0 \ 0 \ 0 \ 0$$

When a boy outside the game asked him the score, he answered: "It's up in the millions."

11. What score did the tourist think each team had?

12. What score did each team actually have?

13. How did the tourist get so confused?

The Milky Way galaxy, of which our solar system is a part, is estimated to contain approximately

© 1986 Dennis L. Mammana

$$100,000,000,000$$

stars. The universe is estimated to contain

$$1,000,000,000,000$$

galaxies. If each galaxy contains the same number of stars as our own, there are approximately

$$100,000,000,000 \times 1,000,000,000,000 = 100,000,000,000,000,000,000,000$$

stars in the universe.

14. Rewrite this multiplication problem, expressing each of the three numbers as a power of 10.

This star-filled sky above Northern Mexico shows Halley's Comet and a portion of the Milky Way (upper right).

A glass of water contains approximately

$$10,000,000,000,000,000,000,000,000$$

water molecules and the Atlantic Ocean contains approximately

$$1,000,000,000,000,000,000,000$$

glasses of water.

15. Write each of these numbers as a power of 10.

16. Approximately how many molecules of water does the Atlantic Ocean contain? (Write the number as a power of 10.)

Use what you have observed in exercises 14–16 to find answers to the following multiplication problems.

17. $10^3 \times 10^4$.

18. $10^{10} \times 10^{10}$.

19. $10^0 \times 10^9$.

20. $10^1 \times 10^{33}$.

21. To multiply two numbers that are written as powers of 10, what do you do to their exponents?

Think of each of the following numbers as a power of 10 in finding answers to these multiplication problems.

22. 10×10^{15}.

23. 100×10^{24}.

24. $1{,}000 \times 10^{100}$.

The small figure at the left below contains one hundred, or 10^2, dots.

25. Estimate the number of dots in the large figure both in words and as a power of 10.

Most arithmetic calculators can display numbers having as many as eight digits.

26. What is the largest number that can be displayed on one of these calculators?

27. What is the smallest number that is a power of 10 that cannot be displayed on one of these calculators?

Googol is the name of the number

10,000,000,000,000,000,000,000,000,000,000,000,000,000,000,000,
000,000,000,000,000,000,000,000,000,000,000,000,000,000,000,000.

The name was invented by a young nephew of the American mathematician Edward Kasner when the boy was asked to make up a name for a very large number.

28. How many digits does the number contain?

29. Write a googol as a power of 10.

Most scientific calculators can display numbers as powers of 10 in which the exponent can have as many as two digits.

30. Can a googol be displayed on one of these calculators?

Set II

The Greek mathematician Archimedes, who lived in the third century B.C., wrote a book called *The Sand Reckoner.* He addressed the book to Gelon, King of Syracuse, and wrote:

> Many people believe, King Gelon, that the grains of sand are without number. Others think that although their number is not without limit, no number can ever be named which will be greater than the number of grains of sand. But I shall try to prove to you that among the numbers which I have named there are those which exceed the number of grains in a heap of sand the size not only of the earth but even of the universe.

To do this, Archimedes invented a method for forming large numbers. He started with the largest number given a name by the Greeks, a *myriad,* which we call "ten thousand."

1. Write this number as a power of 10.

Archimedes began by thinking of a "myriad of myriads," or ten thousand times ten thousand.

Archimedes

2. What would we call this number?

He referred to the numbers from 1 to 100,000,000 as numbers of "the first order."

3. Write 100,000,000 as a power of 10.

Archimedes referred to the numbers from 100,000,000 to $100,000,000^2$ as numbers of "the second order."

4. Write $100,000,000^2$ out the long way. (Remember that $100,000,000^2$ means $100,000,000 \times 100,000,000$.)

5. Write it as a power of 10.

The numbers from $100,000,000^2$ to $100,000,000^3$ were considered to be the numbers of "the third order."

6. Write $100,000,000^3$ as a power of 10.

The numbers from $100,000,000^3$ to $100,000,000^4$ were considered to be the numbers of "the fourth order."

7. Write $100,000,000^4$ as a power of 10.

Archimedes continued in this way until he got to the numbers of "the 100,000,000th order": the numbers from $100,000,000^{99,999,999}$ to $100,000,000^{100,000,000}$.

8. Write $100,000,000^{100,000,000}$ as a power of 10.

This number is extraordinarily large.

9. If it were written out in full, how many digits would it contain?

10. Do you think someone could actually write this many digits? (Recall from this lesson that 1,000,000 seconds is nearly 12 days.)

Archimedes calculated that, in contrast with this immense number, the number of grains of sand needed to fill the entire universe was no more than 10^{63}. According to current estimates of the size of the observable universe, the number of grains of sand of the size used by Archimedes needed to fill it completely is no more than 10^{93}.

Set III

Perhaps the easiest way to appreciate the size of large numbers is in terms of time. A second is a very brief moment of time, yet a billion seconds is approximately 32 years.

Use this information to copy and complete the following tables.

1.

Seconds from now	Years ago	Date
0	0	(Fill in the present year here.)
100 million	▥	▥
1 billion	32	▥
10 billion	▥	▥
100 billion	▥	▥
1 trillion	▥	▥

2.

Seconds from now	Years from now	Date
0	0	(Fill in the present year here.)
100 million	▥	▥
1 billion	32	▥
10 billion	▥	▥
100 billion	▥	▥
1 trillion	▥	▥

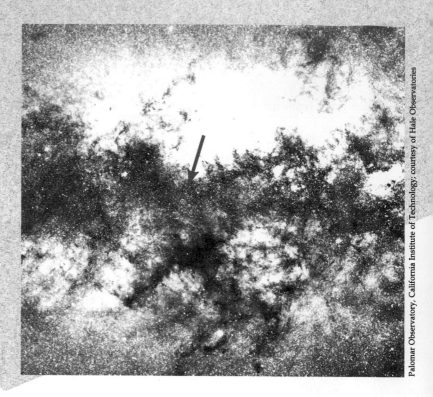

Palomar Observatory, California Institute of Technology; courtesy of Hale Observatories

Scientific Notation

Galileo once described our galaxy, the Milky Way, as "nothing else than a mass of luminous stars . . . set thick together in a wonderful way." At the center of the Milky Way is a group of stars that rapidly rotate about what is thought to be a black hole with a mass about 3,500,000 times that of our sun.*

The distance between the earth and this point is approximately

$$300,000,000,000,000,000$$

kilometers. This number can be written in a more convenient form by observing that

$$300,000,000,000,000,000 = 3 \times 100,000,000,000,000,000.$$

*A Journey into Gravity and Spacetime by John Archibald Wheeler (Scientific American Library, 1990).

Writing 100,000,000,000,000,000 as a power of 10, we have

$$3 \times 10^{17}$$

Written in this form, the number is in *scientific notation.*

A number is in **scientific notation** if it is written in the form

$$a \times 10^b$$

in which a is a number that is at least 1 but less than 10. The number a is called the *coefficient* and the number b is called the *exponent.*

Compare 300,000,000,000,000,000 with its form in scientific notation, 3×10^{17}. In the scientific form, the decimal point has been moved from the end of the number to just after its first digit. The point has been moved 17 decimal places and the 17 has become the exponent of the 10.

$$3\,0\ 0,0\ 0\ 0,0\ 0\ 0,0\ 0\ 0,0\ 0\ 0,0\ 0\ 0 = 3 \times 10^{17}$$

17 16 15 14 13 12 11 10 9 8 7 6 5 4 3 2 1

The mass of the Milky Way is thought to be about

400,000,000,000,000,000,000,000,000,000,000,000,000,000

kilograms. Following the procedure described above, we can write this number as

$$4 \times 10^{41}$$

It is easy to see that the larger the number, the greater the advantage of writing it in scientific notation.

When we look at the center of the Milky Way, we are looking into the past because it takes 3.2×10^4 years for light from it to reach us. How many years is that? We can reverse the procedure described above to find out.

$$3.2 \times 10^4 = 3\,2\ 0\ 0\ 0$$

1 2 3 4

Writing in a comma, we get 32,000. When we look at the center of our galaxy, we are seeing it as it existed 32,000 years ago.

EXERCISES

SET I

It would take 3 octillion candles to give as much light as the sun.

1. Write this number in decimal form.

2. How many digits did you write?

3. Write the number in scientific notation.

4. How many digits did you write?

5. Which is easier: writing a large number in decimal form or writing it in scientific notation?

The star Betelgeuse is losing matter at the rate of 2×10^{15} pounds per second.

6. What is the coefficient and what is the exponent in this number?

7. Write the number in decimal form.

8. Name it.

The number of different possible hands in the game of bridge is about 635 billion. One way of writing this number is

$$635 \times 10^9$$

9. This number is not in scientific notation. Refer to the definition of scientific notation given in this lesson to explain why not.

10. Write 635 billion in decimal form.

11. Write 635 billion in scientific notation.

Biologists have found that mammals can expect to live for 3.3×10^8 breaths and 1.5×10^9 heartbeats.*

12. Write each of these numbers in decimal form.

13. Name each number in words.

The number of hairs on your head varies with the color of your hair.

14. Which one of these numbers is biggest?

15. If numbers are written in scientific notation, how can you tell which one is biggest without changing them to decimal form?

Color of hair	Number of hairs
Black or Brown	1.05×10^5
Blond	1.4×10^5
Red	9×10^4

*On Size and Life by Thomas A. McMahon and John Tyler Bonner (Scientific American Library, 1983).

The brightest star in the constellation Scorpius is the red supergiant Antares. Its radius is 2×10^8 kilometers.

16. The diameter of a star is twice its radius. Write the diameter of Antares in scientific notation.

17. Check your answer by writing the radius and diameter in decimal form.

A dentist in Italy holds the record for pulling the most teeth: 2,000,744 of them.

18. Write this number in scientific notation.

Rounded off, the number of teeth is about 2,000,000.

19. Write this number in scientific notation.

20. Which kind of numbers are simplified the most by scientific notation: *exact* or *approximate*?

Set II

There are about 4×10^6 microbes on each square centimeter of your skin, and you have about 2×10^4 square centimeters of skin.

1. Write each of these numbers in decimal form.

2. Use the numbers in decimal form to find out about how many microbes are on your skin.

3. Write your answer in scientific notation.

4. On the basis of your answers to exercises 1–3, to multiply two numbers that are in scientific notation, what do you do to their coefficients and what do you do to their exponents?

Do the following multiplication problems.

5. $(2 \times 10^{10}) \times (3 \times 10^5)$.

6. $(5 \times 10^8) \times (1.6 \times 10^{12})$.

Multiplying two numbers written in scientific notation may not immediately give the result in scientific notation. If this is the case, an additional step is required, as illustrated in the example below.

$$\begin{aligned} (3 \times 10^4) \times (7 \times 10^2) &= 21 \times 10^6 \\ &= 2.1 \times 10 \times 10^6 \\ &= 2.1 \times 10^7 \end{aligned}$$

Do the following multiplication problems. Write the results in scientific notation.

7. $(5 \times 10^3) \times (6 \times 10^8)$.

8. $(8 \times 10^{10}) \times (7.5 \times 10^5)$.

9. $(9 \times 10^9) \times (9 \times 10^9)$.

The following sentence appeared on a Kellogg's cereal box:

"Did you know that if Americans recycled all of our Sunday newspapers, we could save over 500,000 trees each week or 26,000,000 every year?"

10. Write 500,000 and 26,000,000 in scientific notation.

11. There are 52 weeks in a year. Show that

$$52 \times 5 \times 10^5 = 2.6 \times 10^7$$

One of the largest swarms of locusts ever seen was estimated to contain 4×10^{10} insects. A locust is capable of consuming as much as 15 grams of grain in a week.

12. How many grams of grain could be consumed in a week by this swarm of locusts? (Write your answer in scientific notation.)

In five days, 1 gram of fresh yeast can grow to about 750,000 grams, enough to make about 50,000 loaves of bread.

13. Divide 50,000 into 750,000 to find the number of grams of yeast needed to make one loaf of bread.

14. Write each of the numbers in this division problem in scientific notation:

$$\frac{750,000}{50,000} = \text{[your answer to exercise 13]}$$

15. On the basis of your answer to exercise 14, to divide two numbers that are in scientific notation, what do you do to their coefficients and what do you do to their exponents?

A swarm of locusts in East Africa

Do the following division problems. Write the results in scientific notation.

16. $\dfrac{8 \times 10^{12}}{2 \times 10^{10}}$.

17. $\dfrac{9 \times 10^{15}}{5 \times 10^2}$.

18. $\dfrac{3 \times 10^{40}}{6 \times 10^5}$.

The neocortex of the brain contains cells called "neurons." The numbers of neurons for three different mammals are listed below.

Mammal	Number of neurons
Human	3×10^{10}
Gorilla	7.5×10^9
Cat	6.5×10^7

19. Which one of these numbers is biggest?

20. The neocortex of a human brain has how many times as many neurons as the neocortex of the brain of a gorilla?

21. The neocortex of a human brain has about how many times as many neurons as the neocortex of the brain of a cat?

The sun is approximately 93,000,000 miles from the earth. Light travels a distance of 186,000 miles each second.

22. Do the following division problem to find how many seconds it takes light from the sun to reach the earth:

$$\frac{9.3 \times 10^7}{1.86 \times 10^5}$$

23. Approximately how many *minutes* does it take light from the sun to reach the earth?

Although sound travels a distance of about 7.7×10^2 miles each hour through air, sound cannot travel from the sun to the earth because there is no air between them to carry it.

24. If there *were* air so that we could hear sound from the sun, approximately how long would it take the sound to reach us?

SET III

Suppose that it cost just one penny to travel 1,000 miles. How much money would each of the following trips cost?

1. A trip around the world. (The distance around the world is 25,000 miles.)

2. A trip to the moon. (The distance to the moon is 234,000 miles.)

3. A trip to the sun. (The distance to the sun is 93,000,000 miles.)

4. A trip to the nearest star beyond the sun. (The distance to this star is 25,000,000,000,000 miles.)

5. A trip to the nearest galaxy beyond our Milky Way. (The distance to this galaxy is 1,000,000,000,000,000,000 miles.)

An Introduction to Logarithms

An amoeba can do an unusual mathematical trick: it multiplies by dividing. After it has grown to a certain size, the amoeba's single cell divides in half to produce two amoebas. In about a day, the two amoebas have grown to the point at which they are ready to divide and form four; the day after that, there are eight amoebas, and so forth. How many amoebas will there be at the end of a week?

The number of amoebas is a function of the time that has passed. A table for this function looks like this:

Time in days	0	1	2	3	4	5	6	7	8	9	10	. . .
Number of amoebas	1	2	4	8	16	32	64	128	256	512	1,024	. . .

The answer is in the table: at the end of seven days there will be 128 amoebas.

Notice that the second line of numbers in this table is the binary sequence, a geometric sequence in which each term is twice the preceding term. The first line is an arithmetic sequence in which each term is one more than the preceding term.

Something remarkable about this table was discovered by the Scottish mathematician John Napier in the early seventeenth century. Suppose that we choose two numbers from the second sequence, say 4 and 32. If we multiply these numbers,

$$4 \times 32 = 128$$

we get another number of the sequence. The numbers in the first sequence that correspond to 4, 32, and 128 are 2, 5, and 7, and here is the remarkable part:

$$2 + 5 = 7$$

0	1	2	3	4	5	6	7	8	9	10
1	2	4	8	16	32	64	128	256	512	1,024

A *multiplication* in the second sequence corresponds to an *addition* in the first sequence.

Here is another example:

0	1	2	3	4	5	6	7	8	9	10
1	2	4	8	16	32	64	128	256	512	1,024

These examples suggest that if we do not want to *multiply*, we can *add* instead by using the table. For example, what is 16×64? Finding these numbers in the second sequence, we look above them and see 4 and 6; adding 4 and 6, we get 10; looking below 10 we see 1,024.

0	1	2	3	4	5	6	7	8	9	10
1	2	4	8	16	32	64	128	256	512	1,024

So, without doing any multiplying, we have found that $16 \times 64 = 1,024$ by simply adding 4 and 6.

The numbers in the first sequence are called the *logarithms** of the corresponding numbers of the second sequence:

Logarithms, x	0	1	2	3	4	5	6	7	8	9	10	. . .
Numbers, y		1	2	4	8	16	32	64	128	256	512	1,024 . . .

A formula for this table is

$$y = 2^x$$

in which y represents a number in the binary sequence and x represents its logarithm. This formula enables us to say exactly what a logarithm is.

> When a number is expressed as a power of 2, the exponent of 2 is the **binary logarithm** of the number.

For example, the binary logarithm of 8 is 3 because $8 = 2^3$. The binary logarithm of 16 is 4 because $16 = 2^4$, and so forth.

EXERCISES

SET I

Here is a longer version of our logarithm table. It is arranged in columns with the terms of the binary sequence listed in the first column and their logarithms listed in the second.

Number	Log	Number	Log	Number	Log
1	0	2,048	11	4,194,304	22
2	1	4,096	12	8,388,608	23
4	2	8,192	13	16,777,216	24
8	3	16,384	14	33,554,432	25
16	4	32,768	15	67,108,864	26
32	5	65,536	16	134,217,728	27
64	6	131,072	17	268,435,456	28
128	7	262,144	18	536,870,912	29
256	8	524,288	19	1,073,741,824	30
512	9	1,048,576	20	2,147,483,648	31
1,024	10	2,097,152	21	4,294,967,296	32

---

*The word *log* is a shortened form of *logarithm* and is often used instead.

Numbers in this table can be multiplied by adding their logarithms. For example, compare multiplying 64 × 2,048 without and with logarithms.

Without logs   *With logs*

	2048	Numbers		Logs
×	64	2048	→	11
	8192	64	→	+ 6
	12288	131072	←	17
	131072			

1. Find the answer to this problem without using logarithms: 32 × 128. Show your work.

2. Referring to the table, show how the answer to the same problem can be found by using logarithms.

3. Find the answer to this problem without using logarithms: 512 × 1,024. Show your work.

4. Referring to the table, show how the answer to the same problem can be found by using logarithms.

Use the table to find answers to the following multiplication problems.

5. 16 × 65,536.

6. 8,192 × 16,384.

7. 64 × 256 × 2,048.

Numbers can also be divided by using logarithms. For example, compare dividing 4,096 by 128 without and with logarithms.

Without logs   *With logs*

	Numbers		Logs
32	4096	→	12
128)4096	128	→	7
384	32	←	‖‖‖‖‖
256			
256			
0			

8. What number is missing in this example?

9. On the basis of this example, to divide two numbers, what does it seem that you do with their logarithms?

10. Find the answer to this problem without using logarithms: 32,768 ÷ 2,048. Show your work.

11. Show how the answer to the same problem can be found by using logarithms.

Use the table to find answers to the following division problems.

12. 262,144 ÷ 256.

13. 16,384 ÷ 16,384.

14. 536,870,912 ÷ 64.

Logarithms can also be used to raise numbers to powers. For example, compare raising 8 to the fourth power without and with logarithms.

	Without logs	*With logs*		
	$8^4 = 8 \times 8 \times 8 \times 8$	Numbers		Logs
	8	8	⟶	3
	× 8	8	⟶	3
	64	8	⟶	3
	× 8	8	⟶	+ 3
	512	4096	⟵	▓
	× 8		or	
	4096	8	⟶	3
				× 4
		4096	⟵	▓

15. What number is missing in this example?

16. On the basis of this example, to raise a number to a power, what does it seem that you do with the logarithm of the number?

17. Find the value of 4^5 without using logarithms. Show your work.

18. Show how to find the value of 4^5 by using logarithms.

Use logarithms to evaluate the following numbers.

19. 128^2.

20. 64^3.

21. 32^4.

22. 8^{10}.

Set II

Why the logarithm table works as it does can be understood by looking at it in the following form.

Number	Log
1	0
$2 = 2$	1
$4 = 2 \times 2$	2
$8 = 2 \times 2 \times 2$	3
$16 = 2 \times 2 \times 2 \times 2$	4
$32 = 2 \times 2 \times 2 \times 2 \times 2$	5
$64 = 2 \times 2 \times 2 \times 2 \times 2 \times 2$	6
$128 = 2 \times 2 \times 2 \times 2 \times 2 \times 2 \times 2$	7
$256 = 2 \times 2 \times 2 \times 2 \times 2 \times 2 \times 2 \times 2$	8
$512 = 2 \times 2 \times 2 \times 2 \times 2 \times 2 \times 2 \times 2 \times 2$	9
$1{,}024 = 2 \times 2 \times 2 \times 2 \times 2 \times 2 \times 2 \times 2 \times 2 \times 2$	10
$2{,}048 = 2 \times 2 \times 2 \times 2 \times 2 \times 2 \times 2 \times 2 \times 2 \times 2 \times 2$	11
$4{,}096 = 2 \times 2 \times 2 \times 2 \times 2 \times 2 \times 2 \times 2 \times 2 \times 2 \times 2 \times 2$	12

1. How many 2's must be multiplied to get the number 32?

2. What is the logarithm of 32?

3. What is the logarithm of 4,096?

4. How many 2's must be multiplied to get the number 4,096?

The figure below shows why the multiplication problem

$$8 \times 16 = 128$$

can be solved by doing the addition problem

$$3 + 4 = 7$$

$$
\underbrace{8}_{\underbrace{2 \times 2 \times 2}_{3}} \times \underbrace{16}_{\underbrace{2 \times 2 \times 2 \times 2}_{4}} = \underbrace{128}_{\underbrace{2 \times 2 \times 2 \times 2 \times 2 \times 2 \times 2}_{7}}
$$

$$3 \quad + \quad 4 \quad = \quad 7$$

5. Draw a figure like this to show why the multiplication problem $4 \times 64 = 256$ can be solved by doing the addition problem $2 + 6 = 8$.

The next figure shows why the division problem

$$512 \div 64 = 8$$

can be solved by doing the subtraction problem

$$9 - 6 = 3$$

$$\frac{\cancel{2} \times \cancel{2} \times \cancel{2} \times \cancel{2} \times \cancel{2} \times \cancel{2} \times 2 \times 2 \times 2}{\cancel{2} \times \cancel{2} \times \cancel{2} \times \cancel{2} \times \cancel{2} \times \cancel{2}} = 2 \times 2 \times 2$$

6. Draw a figure like this to show why the division problem $128 \div 4 = 32$ can be solved by doing the subtraction problem $7 - 2 = 5$.

The next figure shows why the raising-to-a-power problem

$$8^4 = 4096$$

can be solved by doing the multiplication problem

$$4 \times 3 = 12$$

$$8^4 = \quad 8 \quad \times \quad 8 \quad \times \quad 8 \quad \times \quad 8 \quad = \quad 4096$$

$$\underbrace{2 \times 2 \times 2}_{3} \times \underbrace{2 \times 2 \times 2}_{3} \times \underbrace{2 \times 2 \times 2}_{3} \times \underbrace{2 \times 2 \times 2}_{3} = \underbrace{2 \times 2 \times 2 \times 2 \times 2 \times 2 \times 2 \times 2 \times 2 \times 2 \times 2 \times 2}_{4 \times 3 = 12}$$

7. Draw a figure like this to show why the raising-to-a-power problem $32^2 = 1024$ can be solved by doing the multiplication problem $2 \times 5 = 10$.

The patterns that we have been considering can be expressed more compactly with exponents. See the table at the right.

8. Which number is the exponent in the equation $32 = 2^5$?

9. What is the logarithm of 32?

10. What is the logarithm of 4,096?

11. Which number is the exponent in the equation $4,096 = 2^{12}$?

Notice that logarithms are *exponents*.

12. Rewrite the multiplication problem $32 \times 128 = 4,096$ so that the three numbers are expressed as powers of 2.

13. What are the logarithms of 32, 128, and 4,096?

Number	Log
1	0
$2 = 2^1$	1
$4 = 2^2$	2
$8 = 2^3$	3
$16 = 2^4$	4
$32 = 2^5$	5
$64 = 2^6$	6
$128 = 2^7$	7
$256 = 2^8$	8
$512 = 2^9$	9
$1,024 = 2^{10}$	10
$2,048 = 2^{11}$	11
$4,096 = 2^{12}$	12

14. When powers of 2 are multiplied, what happens to their exponents to produce the exponent of the answer?

15. Rewrite the division problem $2,048 \div 8 = 256$ so that the three numbers are expressed as powers of 2.

16. What are the logarithms of 2,048, 8, and 256?

17. When powers of 2 are divided, what happens to their exponents to produce the exponent of the answer?

18. Rewrite the raising-to-a-power problem $16^3 = 4,096$ so that 16 and 4,096 are expressed as powers of 2.

19. What are the logarithms of 16 and 4,096?

20. When a power of 2 is raised to a power, what happens to its exponent to produce the exponent of the answer?

Set III

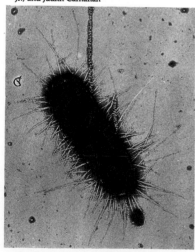

Electron micrograph courtesy of Charles C. Brinton, Jr., and Judith Carnahan

Escherichia coli

Biologists refer to the growth period of bacteria as their "log phase" because of the connection between their repeated doublings and the logarithms introduced in this lesson. *Escherichia coli*, one of the most rapidly growing bacteria, has a doubling time of about 15 minutes.

This means that, starting with 1 cell, there would be 2 cells after 15 minutes, 4 cells after 30 minutes, 8 cells after 45 minutes, and 16 cells after 1 hour.

1. Look at the table of logarithms on the next two pages to tell, if this process continued, how many cells there would be after 2 hours.

2. How many cells would there be after 3 hours?

Although it would be impossible for the cells to continue multiplying in this way for 24 hours (they would run out of space and food), it is interesting to realize what would happen if they could.

3. Approximately how many cells would there be after 24 hours?

About 10^{17} such cells weigh 1 ton.

4. Approximately how much would the cells in existence after 24 hours weigh?

Number	Log	Number	Log
1	0	34,359,738,368	35
2	1	68,719,476,736	36
4	2	137,438,953,472	37
8	3	274,877,906,944	38
16	4	549,755,813,888	39
32	5	1,099,511,627,776	40
64	6	2,199,023,255,552	41
128	7	4,398,046,511,104	42
256	8	8,796,093,022,208	43
512	9	17,592,186,044,416	44
1,024	10	35,184,372,088,832	45
2,048	11	70,368,744,177,664	46
4,096	12	140,737,488,355,328	47
8,192	13	281,474,976,710,656	48
16,384	14	562,949,953,421,312	49
32,768	15	1,125,899,906,842,624	50
65,536	16	2,251,799,813,685,248	51
131,072	17	4,503,599,627,370,496	52
262,144	18	9,007,199,254,740,992	53
524,288	19	18,014,398,509,481,984	54
1,048,576	20	36,028,797,018,963,968	55
2,097,152	21	72,057,594,037,927,936	56
4,194,304	22	144,115,188,075,855,872	57
8,388,608	23	288,230,376,151,711,744	58
16,777,216	24	576,460,752,303,423,488	59
33,554,432	25	1,152,921,504,606,846,976	60
67,108,864	26	2,305,843,009,213,693,952	61
134,217,728	27	4,611,686,018,427,387,904	62
268,435,456	28	9,223,372,036,854,775,808	63
536,870,912	29	18,446,744,073,709,551,616	64
1,073,741,824	30	36,893,488,147,419,103,232	65
2,147,483,648	31	73,786,976,294,838,206,464	66
4,294,967,296	32	147,573,952,589,676,412,928	67
8,589,934,592	33	295,147,905,179,352,825,856	68
17,179,869,184	34	590,295,810,358,705,651,712	69

(Continued on next page)

Number	Log
1,180,591,620,717,411,303,424	70
2,361,183,241,434,822,606,848	71
4,722,366,482,869,645,213,696	72
9,444,732,965,739,290,427,392	73
18,889,465,931,478,580,854,784	74
37,778,931,862,957,161,709,568	75
75,557,863,725,914,323,419,136	76
151,115,727,451,828,646,838,272	77
302,231,454,903,657,293,676,544	78
604,462,909,807,314,587,353,088	79
1,208,925,819,614,629,174,706,176	80
2,417,851,639,229,258,349,412,352	81
4,835,703,278,458,516,698,824,704	82
9,671,406,556,917,033,397,649,408	83
19,342,813,113,834,066,795,298,816	84
38,685,626,227,668,133,590,597,632	85
77,371,252,455,336,267,181,195,264	86
154,742,504,910,672,534,362,390,528	87
309,485,009,821,345,068,724,781,056	88
618,970,019,642,690,137,449,562,112	89
1,237,940,039,285,380,274,899,124,224	90
2,475,880,078,570,760,549,798,248,448	91
4,951,760,157,141,521,099,596,496,896	92
9,903,520,314,283,042,199,192,993,792	93
19,807,040,628,566,084,398,385,987,584	94
39,614,081,257,132,168,796,771,975,168	95
79,228,162,514,264,337,593,543,950,336	96
158,456,325,028,528,675,187,087,900,672	97
316,912,650,057,057,350,374,175,801,344	98
633,825,300,114,114,700,748,351,602,688	99
1,267,650,600,228,229,401,496,703,205,376	100

LESSON

4

Decimal Logarithms

Logarithms were given their name by John Napier, the Scottish mathematician credited with their invention. His book, *A Description of the Wonderful Law of Logarithms,* was published in 1614. On the 300th anniversary of its publication, Lord Moulton wrote:

> The invention of logarithms came on the world as a bolt from the blue. No previous work had led up to it, foreshadowed it or heralded its arrival. It stands isolated, breaking in upon human thought abruptly without borrowing from the work of other intellects or following known lines of mathematical thought.*

Napier was fascinated by the properties of the binary sequence, and the table below appears in his writings:

I	II	III	IV	V	VI	VII	. . .
1	2	4	8	16	32	64	128 . . .

John Napier

---

*"Inaugural Address: The Invention of Logarithms," *Napier Tercentenary Memorial Volume* (London, 1915).

By rewriting it in the following form, we recognize Napier's table as the logarithm table of the preceding lesson.

Logs	0	1	2	3	4	5	6	7	. . .
Numbers	2^0	2^1	2^2	2^3	2^4	2^5	2^6	2^7	. . .

This table is based on powers of 2. However, a logarithm table built on powers of *10* would be more convenient for us because our number system is based on the number 10. Such a table would look like this:

Logs	0	1	2	3	4	5	6	7	. . .
Numbers	10^0	10^1	10^2	10^3	10^4	10^5	10^6	10^7	. . .

These logarithms are referred to as *base 10,* or *decimal,* logarithms to distinguish them from the *base 2,* or *binary,* logarithms studied earlier.

> When a number is expressed as a power of 10, the exponent of 10 is the **decimal logarithm** of the number.

The connection between decimal logarithms and our number system is a very simple one: the logarithm of 1,000,000, for example, is 6 because $1,000,000 = 10^6$. We can write this more briefly as

$$\log 1,000,000 = 6$$

Unfortunately, in the form in which it is written above, the base 10 table is relatively useless. Consider, for example, the problem

$$100 \times 1,000$$

Logs	0	1	2	3	4	5	6	7
Numbers	1	10	100	1,000	10,000	100,000	1,000,000	10,000,000

There is little point in using logarithms to solve this problem, because 100 and 1,000 are so easy to multiply in the first place.

The value of the base 10 table becomes apparent only by filling in the gaps between the numbers on the second line. If we rewrite the beginning of the table, filling in the numbers 2 through 9 on the second line, we get

Logs	0									1
Numbers	1	2	3	4	5	6	7	8	9	10

The missing logarithms are evidently numbers between 0 and 1. The method for finding out what they are is quite complicated, and so no attempt will be made to explain it here.*

When the missing logarithms are computed and listed with their corresponding numbers, we have the table below.

Logs	0	0.30	0.48	0.60	0.70	0.78	0.85	0.90	0.95	1
Numbers	1	2	3	4	5	6	7	8	9	10

Each logarithm has been rounded to two decimal places. From it, we see that log 2 = 0.30, log 3 = 0.48, and so forth. How this table can be used will be illustrated in the exercises.

EXERCISES

SET I

From the preceding lesson we learned that *multiplying* numbers corresponds to *adding* their logarithms. This is true regardless of the number on which the logarithms are based. An example illustrating this property with decimal logarithms is shown here.

Numbers	1	2	3	4	5	6	7	8	9	10
Logarithms	0	0.30	0.48	0.60	0.70	0.78	0.85	0.90	0.95	1

Since $2 \times 3 = 6$, log 2 + log 3 = log 6. The logarithms of many numbers larger than 10 can be found in a similar way.

1. Since $2 \times 6 = 12$, log 2 + log 6 = log 12. Use the table above to find log 12.

2. Which other two numbers in the table multiply to give 12?

3. Check your answer to exercise 1 by adding the logarithms of these numbers.

4. Which two numbers in the table multiply to give 14?

---

*The easiest current method is to push the LOG button on a calculator (every scientific calculator has such a button) and let the calculator figure out the logarithms. The values had been accurately determined long before the invention of the calculator, however.

5. Find log 14.

6. Find log 15.

7. Find log 16.

8. Find log 20.

9. Make a table showing the logarithms of 10, 20, 30, 40, 50, 60, 70, 80, 90, and 100.

10. Make a table showing the logarithms of 10, 10^2, 10^3, 10^4, 10^5, 10^6, 10^7, 10^8, 10^9, and 10^{10}.

"If it's true that the world ant population is 10^{15}, then its no wonder we never run into anyone we know."

The insect population of the earth is estimated to be about 1,000,000,000,000,000,000. Of this population, 10^{15} are ants. What is the logarithm of

11. 10^{15}?

12. 1,000,000,000,000,000,000?

In the game of chess, the first 40 pairs of moves can be made in approximately 10^{95} different ways. What is the logarithm of

13. 40?

14. 10^{95}?

The name of the number whose logarithm is 12 is one trillion because 10^{12} is one trillion. What is the name of the number whose logarithm is

15. 3?

16. 10?

17. 30? (See the table on page 186.)

SET II

The two tables below can be used to find the decimal logarithms of numbers that are written in scientific notation.

Numbers	1	2	3	4	5	6	7	8	9
Logs	0	0.30	0.48	0.60	0.70	0.78	0.85	0.90	0.95

Numbers	10^1	10^2	10^3	10^4	10^5	10^6	10^7	10^8	10^9
Logs	1	2	3	4	5	6	7	8	9

For example, to find the logarithm of the number

$$2 \times 10^5$$

we find $\log 2 + \log 10^5$:

$$0.30 + 5 = 5.30$$

Use this method to find the logarithms of the following numbers written in scientific notation.

1. 3×10^6.

2. 4×10^8.

3. 7×10^5.

4. 1×10^9.

5. 6×10^{12}.

The number of wild rabbits in Australia is estimated to be about 200 million.

6. Write this number in decimal form.

7. Write it in scientific notation.

8. Write its logarithm.

The number of different ways in which a deck of 52 playing cards can be shuffled is about

$$80,000,000,000,000,000,000,000,000,000,000,$$
$$000,000,000,000,000,000,000,000,000,000,000.$$

9. Write this number in scientific notation.
10. Write its logarithm.

The numbers on the Richter scale for measuring earthquakes are logarithms. Earthquakes with readings of 0 have been detected by seismographs.

11. What number has a logarithm of 0? (Look in the tables at the beginning of Set II.)

It is unusual for an earthquake measuring 2 on the Richter scale to be felt.

12. What number has a logarithm of 2?

An earthquake measuring 7.6 would be a major disaster if it occurred near a populated area. To find the number whose logarithm is 7.6, notice that

$$7.6 = 0.60 + 7$$

The number whose logarithm is 0.60 is 4 and the number whose logarithm is 7 is 10^7. Since adding logarithms corresponds to multiplying numbers, the number is

$$4 \times 10^7$$

or 40,000,000. This number tells us that an earthquake measuring 7.6 on the Richter scale releases about 40,000,000 times as much energy as an earthquake measuring 0.

13. The San Francisco earthquake of 1906 is estimated to have had a magnitude of 8.3 on the Richter scale. Use the method shown above to find the number whose logarithm is 8.3.

The following are the logarithms of some numbers. Find the numbers both in scientific notation and in decimal form.

14. 6.48.
15. 2.85.
16. 11.7.

SET III

Although this picture seems to show a natural landscape, it is actually a fractal image produced with a computer. The appearance of a fractal image is affected by a number called its "dimension"; this number is calculated using logarithms.*

The dimension of this image is equal to

$$\frac{\log 32}{\log 4}$$

1. Use the table of decimal logarithms on page 211 to figure out log 32.

2. Divide your answer by log 4 to find the dimension of the landscape.

3. Do the same calculations that you did in exercises 1 and 2, but this time use *binary* logarithms. (See the table of binary logarithms on page 201.)

4. Does the fractal dimension seem to depend on what kind of logarithm is used?

The Science of Fractal Images, edited by Heinz-Otto Peitgen and Dietmar Saupe (Springer-Verlag, 1988).

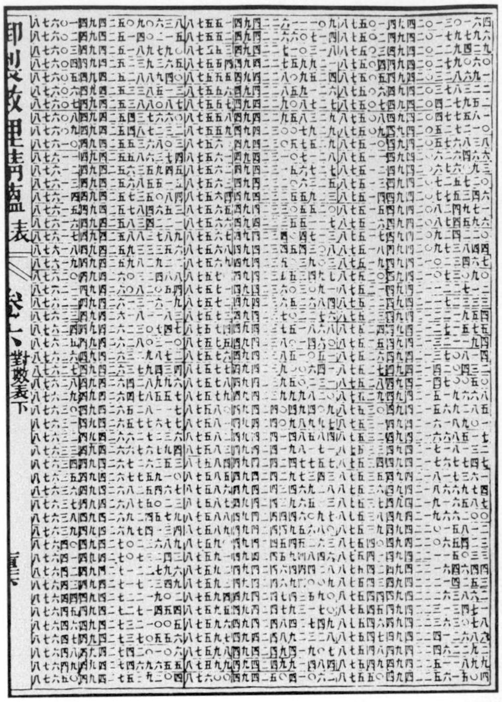

From *Science and Civilisation in China,* volume 3, by Joseph Needham, Cambridge University Press

Part of a table of logarithms published in China in 1713

LESSON 5

Logarithms and Scientific Notation

Scientists and mathematicians all over the world recognized the usefulness of logarithms immediately after their invention. This recognition was due in great part to the work of the English mathematician Henry Briggs, who wrote:

> Logarithms are numbers invented for the more easy working of questions in arithmetic and geometry. By them all troublesome multiplications are avoided and performed only by addition. . . . In a word, all questions not only in arithmetic and geometry but in astronomy also are thereby most plainly and easily answered.*

It was Briggs who suggested the number 10 as the most practical base for logarithms and Briggs who constructed the first table of decimal logarithms. Tables derived from Briggs's work were soon published in many countries. Part of one such table, shown above, is from a book

**Arithmetica Logarithmica,* 1624.

published at the request of the Chinese Emperor K'anghsi in 1713. The book, printed from wooden plates carved by hand, listed the logarithms of the numbers from 1 to 100,000, each to 10 decimal places.

The table on the next page lists the logarithms of numbers from 1 to 9.9, each to three decimal places. Although this table is brief, it can easily be used to find the logarithms of other numbers because of the relation between numbers written in scientific notation and their decimal logarithms.

Recall that a number is in scientific notation if it is written in the form

$$a \times 10^b$$

in which *a*, the *coefficient*, is a number that is at least 1 but less than 10, and *b* is the *exponent*. To find the logarithm of a number written in scientific notation, we look up the logarithm of the coefficient in the table and add it to the logarithm of 10^b, which is *b*, the exponent.

| The **logarithm** of $a \times 10^b$ is log *a* (looked up in the table) + *b*.

For example, to find the logarithm of the distance from the earth to the sun in miles.

$$93,000,000$$

we first write it in scientific notation:

$$9.3 \times 10^7$$

Because multiplying numbers corresponds to adding their logarithms, the logarithm of 9.3×10^7 is equal to

$$\log 9.3 + \log 10^7$$

The logarithm of 9.3 is in the table; it is 0.968. The logarithm of 10^7 is the exponent, 7:

$$0.968 + 7 = 7.968.$$

So the logarithm of 93,000,000 is 7.968.

More examples of how the table can be used to find the logarithms of numbers larger than 10 are shown below.

Number in decimal form	Number in scientific notation	Logarithm of number
121,000	1.21×10^5	$0.083 + 5 = 5.083$
35	3.5×10^1	$0.544 + 1 = 1.544$
	6.0×10^{23}*	$0.778 + 23 = 23.778$

*This is Avogadro's number, a large number used in chemistry, whose decimal form is 600,000,000,000,000,000,000,000.

Table of logarithms

Number	Log	Number	Log	Number	Log	Number	Log	Number	Log
1.00	.000	1.40	.146	1.80	.255	3.00	.477	6.0	.778
1.01	.004	1.41	.149	1.81	.258	3.05	.484	6.1	.785
1.02	.009	1.42	.152	1.82	.260	3.10	.491	6.2	.792
1.03	.013	1.43	.155	1.83	.262	3.15	.498	6.3	.799
1.04	.017	1.44	.158	1.84	.265	3.20	.505	6.4	.806
1.05	.021	1.45	.161	1.85	.267	3.25	.512	6.5	.813
1.06	.025	1.46	.164	1.86	.270	3.30	.519	6.6	.820
1.07	.029	1.47	.167	1.87	.272	3.35	.525	6.7	.826
1.08	.033	1.48	.170	1.88	.274	3.40	.531	6.8	.833
1.09	.037	1.49	.173	1.89	.276	3.45	.538	6.9	.839
1.10	.041	1.50	.176	1.90	.279	3.50	.544	7.0	.845
1.11	.045	1.51	.179	1.91	.281	3.55	.550	7.1	.851
1.12	.049	1.52	.182	1.92	.283	3.60	.556	7.2	.857
1.13	.053	1.53	.185	1.93	.286	3.65	.562	7.3	.863
1.14	.057	1.54	.188	1.94	.288	3.70	.568	7.4	.869
1.15	.061	1.55	.190	1.95	.290	3.75	.574	7.5	.875
1.16	.064	1.56	.193	1.96	.292	3.80	.580	7.6	.881
1.17	.068	1.57	.196	1.97	.294	3.85	.585	7.7	.886
1.18	.072	1.58	.199	1.98	.297	3.90	.591	7.8	.892
1.19	.076	1.59	.201	1.99	.299	3.95	.597	7.9	.898
1.20	.079	1.60	.204	2.00	.301	4.0	.602	8.0	.903
1.21	.083	1.61	.207	2.05	.312	4.1	.613	8.1	.908
1.22	.086	1.62	.210	2.10	.322	4.2	.623	8.2	.914
1.23	.090	1.63	.212	2.15	.332	4.3	.633	8.3	.919
1.24	.093	1.64	.215	2.20	.342	4.4	.643	8.4	.924
1.25	.097	1.65	.217	2.25	.352	4.5	.653	8.5	.929
1.26	.100	1.66	.220	2.30	.362	4.6	.663	8.6	.934
1.27	.104	1.67	.223	2.35	.371	4.7	.672	8.7	.940
1.28	.107	1.68	.225	2.40	.380	4.8	.681	8.8	.944
1.29	.111	1.69	.228	2.45	.389	4.9	.690	8.9	.949
1.30	.114	1.70	.230	2.50	.398	5.0	.699	9.0	.954
1.31	.117	1.71	.233	2.55	.407	5.1	.708	9.1	.959
1.32	.121	1.72	.236	2.60	.415	5.2	.716	9.2	.964
1.33	.124	1.73	.238	2.65	.423	5.3	.724	9.3	.968
1.34	.127	1.74	.241	2.70	.431	5.4	.732	9.4	.973
1.35	.130	1.75	.243	2.75	.439	5.5	.740	9.5	.978
1.36	.134	1.76	.246	2.80	.447	5.6	.748	9.6	.982
1.37	.137	1.77	.248	2.85	.455	5.7	.756	9.7	.987
1.38	.140	1.78	.250	2.90	.462	5.8	.763	9.8	.991
1.39	.143	1.79	.253	2.95	.470	5.9	.771	9.9	.996

EXERCISES

SET I

Find the logarithms of the following numbers by looking them up in the table on page 218.

1. 1.44.   2. 3.65.   3. 8.2.

Use your answers to exercises 1–3 in finding the logarithms of the following numbers.

4. 1.44×10^3.   5. 3.65×10^8.   6. 8.2×10^{20}.

Write the following numbers in scientific notation and then find their logarithms.

7. 22,000.   9. 45.

8. 3,150,000,000,000.   10. 1,000,000,000.

Find the numbers having the following logarithms by looking them up in the table.

11. 0.017.   12. 0.107.   13. 0.17.

Use your answers to exercises 11–13 in finding the numbers having the following logarithms. Write the numbers in both scientific notation and decimal form.

14. 5.017.   15. 2.107.   16. 10.17.

Find the numbers having the following logarithms. Write the numbers in decimal form.

17. 0.021.   19. 2.100.

18. 0.210.   20. 21.000.

The following exercises are about the decimal logarithms of the terms of a familiar number sequence.

21. Refer to the table on page 218 to copy and complete the following table.

Number	1	2	4	8	16	32	64	128
Log	0	0.301	▓	▓	▓	▓	▓	▓

22. What kind of number sequence do the numbers on the first line of this table form?

23. What is the common ratio of this sequence?

24. What kind of number sequence do the numbers on the second line of this table form?

25. What is the common difference of this sequence?

26. Judging from the patterns in this table, what numbers do you think are the logarithms of 256 and 512?

TIME chart by V. Puglisi; © 1968 Time Inc.

DECIBELS

A decibel is an arbitrary unit based on the faintest sound that a man can hear. The scale is logarithmic, so that an increase of 10 db means a tenfold increase in sound intensity; a 20-db rise a hundredfold increase, and 30 db a thousand-fold increase.

JET PLANE (100 ft. away) — 140

PNEUMATIC RIVETER — 130

ROCK MUSIC WITH AMPLIFIERS (4 to 6 ft. away) — 120
One trillion times greater than least audible sound

POWER MOWER (107) — 110

NOISY KITCHEN — 100

SUBWAY (inside) — 90

CITY TRAFFIC (inside car) — 80

FORTISSIMO SINGER (3 ft. away) — 70

ORDINARY CONVERSATION — 60

Decibel scale

SET II

Human beings can hear sounds varying over an incredibly wide range of loudness: from sounds that can barely be heard to sounds a *100 trillion* times as loud. The loudness of sounds is described by the decibel scale, explained in this chart from *Time* magazine. The least audible sound is said to be at the "threshold of hearing" and is rated as 0 decibels.

1. Refer to the explanation at the top of the chart to copy and complete the following table in which 0 decibels corresponds to a loudness of 1 unit.

Decibels	Loudness	Log of loudness
0	1	0
10	10	1
20	100	2
30	1,000	3
40	▥	▥
50	▥	▥
60	▥	▥
70	▥	▥
80	▥	▥

2. How is the decibel rating of a sound related to the logarithm of its loudness?

The sound of leaves rustling in a breeze has a rating of 20 decibels.

3. How many times as loud as the least audible sound is the sound of leaves rustling in a breeze?

According to the chart, an amplified rock group is a trillion times as loud as the least audible sound.

4. Write a trillion in decimal form.

5. What is the logarithm of a trillion?

6. How does the decibel rating of rock music compare with this logarithm?

The loudest sound that has been created in a laboratory had a rating of 210 decibels.

7. How many times as loud as the least audible sound was this sound?

Logarithms are useful in making diagrams illustrating a wide range of numbers. Such diagrams, called logarithmic scales, are commonly used in books and magazines. For example, consider the following table relating seconds to other units of time.

Interval of time	Number of seconds
1 second	1
1 minute	60
1 hour	3,600
1 day	86,000*
1 year	32,000,000
1 century	3,200,000,000

There is no way to represent these numbers on a line numbered like an ordinary ruler so that you can clearly see all six of them.

8. Copy the first column of this table. Change each number in the second column to scientific notation.

9. Add a third column to your table by using the table of logarithms on page 218 to write the logarithm of each number. Round each logarithm to one decimal place.

10. Draw and label a logarithmic scale as shown below. Let 1 centimeter represent 1 unit.

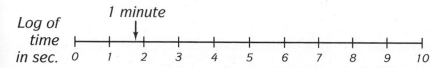

---

*The numbers beginning with this entry are approximate.

Use your answers to exercise 9 to mark each interval of time on your scale with an arrow. (The arrow for 1 minute is shown as an example.)

After you have marked the six arrows, change the label on the scale and turn the numbers into exponents as shown below.*

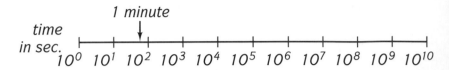

Notice that the logarithmic scale spreads out the smaller numbers.

11. What does it do to the larger numbers?

Light is a form of electromagnetic radiation. So are television signals, microwaves, and x-rays. Here is a table of some electromagnetic waves and their frequencies.

Wave	Frequency in hertz†
AC current	60
AM radio station at 980	980,000
TV channel 2	57,000,000
Microwave in oven	2,450,000,000
Red light	400,000,000,000,000
X-rays	30,000,000,000,000,000

12. Copy the first column of this table. Change each number in the second column to scientific notation.

13. Add a third column to your table by writing the logarithm of each number. Round each logarithm to one decimal place.

14. Draw and label a logarithmic scale similar to the one you drew for exercise 10, but longer. Label it "log of frequency." Mark each wave on your scale with an arrow.

After marking the arrows, change the label to "frequency" and turn the numbers into exponents as you did in exercise 10.

---

*Remember that logarithms *are* exponents.

†One hertz is one cycle per second.

SET III

Benjamin Franklin bequeathed some money to the cities of Boston and Philadelphia with the condition that it could not be spent for 200 years. He died in 1790 and the money became available in 1990.*

The amount that Franklin's gift had become was approximately equivalent to the result of putting $2,550 into a bank account drawing 4% annual compound interest for 200 years.

The following table shows how such an account would grow over the first 50 years.

Art Resource, New York

Year	Amount	Log of amount
1790	$2,550.00	3.41
1800	$3,774.62	3.58
1810	$5,587.36	3.75
1820	$8,270.66	3.92
1830	$12,242.60	4.09
1840	$18,122.04	4.26

Benjamin Franklin

Although the amount increases by larger and larger numbers, the logarithm of the amount always changes by the same number.

1. What is it?

2. Use the fact that there are 20 increases equal to this amount from 1790 to 1990 to find the logarithm of the amount in 1990.

3. Use the table on page 218 to figure out what number is closest to having this logarithm and, hence, the approximate amount of money Franklin's bequest had become in 1990.

*Reported in the *New York Times,* April 21, 1990.

Most Living Ascendants

Megan Sue Austin (b May 16, 1982) of Bar Harbor, Maine, had a full set of grandparents and great-grandparents and five great-great-grandparents making 19 direct ascendants.

LESSON

6 Exponential Functions

Megan Sue Austin of Bar Harbor, Maine, is listed in the *Guinness Book of World Records* because she had so many living ancestors when she was born. At the time of her birth in 1982, all four of her grandparents, all eight of her great-grandparents, and five of her great-great-grandparents were still alive!

A table showing the numbers of a person's ancestors going back in time is shown below.

Generations back, x	1	2	3	4	5	6	7	8	9	10
Number of ancestors, y	2	4	8	16	32	64	128	256	512	1,024

A formula for this function is

$$y = 2^x$$

Such a function is called *exponential* because one of the variables, *x*, appears as an exponent. Exponential functions are sometimes called growth functions because they often show how certain things grow, such as a population or money in an account.

The graph of an exponential function is frequently hard to draw because of the tremendous range of values of the *y*-variable. In the graph of the ancestors function shown here, the first five values of *y* are so small in comparison to the last value, 1,024, that they are squeezed close to the *x*-axis. Notice that the curve rises slowly at first and then more and more steeply as it continues to the right.

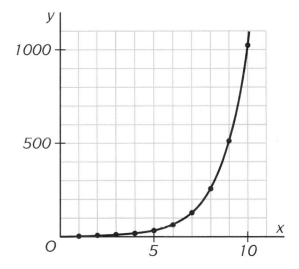

As you may recall from the previous lesson, logarithms are useful in making diagrams illustrating a wide range of numbers. Unlike a scale numbered like an ordinary ruler, a logarithmic scale spreads out the smaller numbers and squeezes the larger numbers together.

Here is the ancestors table again, but with the second line rewritten as the *logarithms* of the numbers of ancestors.

Generations back, *x*	1	2	3	4	5	6	7	8	9	10
Logarithm of number of ancestors, *y*	0.3	0.6	0.9	1.2	1.5	1.8	2.1	2.4	2.7	3.0

This table is easier to graph and the result, shown on the next page, is surprisingly simple. It is a straight line.

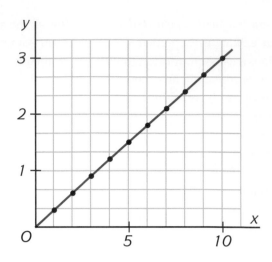

In general, this always turns out to be true. If a function is *exponential,* a graph of it on axes with ordinary scales is a curve, but a graph of it in which a logarithmic scale is chosen for *y* is a straight line.

EXERCISES

SET I

The Richter scale used to measure earthquakes has the formula

$$y = 10^x$$

in which *x* is the earthquake's magnitude and *y* is the relative amount of energy released.

1. What is a function with this type of formula called?

2. Copy and complete the following table for it.

Magnitude, x	0	1	2	3	4	5	6
Energy released, y	1	10	▊▊▊▊	▊▊▊▊	▊▊▊▊	▊▊▊▊	▊▊▊▊

3. Graph the function. Let 10 units on the *y*-axis represent 1,000,000. Join the points with a smooth curve.

4. What happens to the steepness of the curve as you look at the graph from left to right?

Your graph is probably about 2.5 inches tall.

5. If you didn't change the scale on the *y*-axis, how tall would the graph have to be in order to show the energy released by an earthquake with a magnitude of 7?

6. How tall would it have to be in order to show the energy released by an earthquake with a magnitude of 8?

HOW RICHTER SCALE GAUGES EARTHQUAKES

The Richter scale, used by seismologists to measure the magnitude of earthquakes, operates on a logarithmic basis so there is a 10-fold increase from one unit or number to the next.

A magnitude of 6, which was about the strength of Monday night's earthquake, would be 10 times greater than a magnitude 5 earthquake and 100 times a magnitude 4 earthquake.

Thus the Tehachapi earthquake in 1952 with a magnitude of 7.5 was more than 30 times stronger than Monday night's earthquake and the 1906 San Francisco earthquake with a magnitude of 8.25 was more than 110 times stronger.

Courtesy of the Los Angeles Times

It is hard to see from your graph how y changes at the beginning. The following table can be used to show this change more clearly.

x	0	0.30	0.48	0.60	0.70	0.78	0.85	0.90	0.95	1
y	1	2	3	4	5	6	7	8	9	10

7. How are the x-numbers in this table related to the y-numbers below them?

8. What is the x in the formula $y = 10^x$ called?

9. Graph the part of the function shown in the table above. Let 5 units on the x-axis represent 0.5.

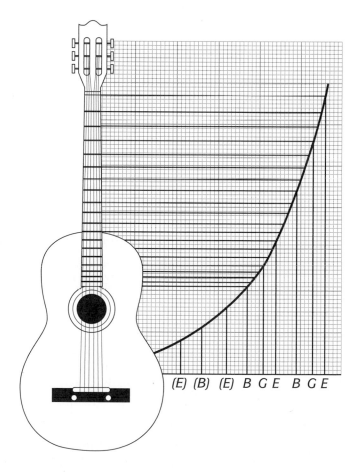

(E) (B) (E) B G E B G E

On a guitar, the distance between the frets (the ridges to guide the fingers) is greater for two lower notes than it is for two higher notes. The way in which the distances change is shown by the graph.

10. What kind of mathematical scale does the spacing of the frets remind you of?

11. What kind of function does the graph look like?

Courtesy of Pentax Corporation

The amount of light that enters a camera is determined by the *f*-stop setting of the lens. The *f*-stop numbers on the lens in this photograph are listed in the table below.

f-stop, *x*	1	2	3	4	5	6	7	8	9
f-stop number, *y*	1.4	2	2.8	4	5.6	8	11	16	22

12. Plot the nine points in this table on a graph. Let 5 units on the *y*-axis represent 10. Join the points with a smooth curve.

13. Copy and complete the table below by replacing the *f*-stop numbers (the *y*-values) in the table above with their logarithms. Refer to the table on page 218 and round each logarithm to 2 decimal places.

f-stop, *x*	1	2	3	4	5	6	7	8	9
Log of *f*-stop number, *y*	0.15	0.30	▥	▥	▥	▥	▥	▥	▥

14. Plot the nine points in this table on a graph. Let 5 units on the *y*-axis represent 0.50.

15. Do the points lie on a curve or on a line? Draw it.

SET II

The first oil well was drilled in 1859. Since then, oil has become the most important source of energy in the world.

The table at the left shows how the annual world crude oil production increased from 1880 to 1970. The table stops with 1970 because a war in the Middle East in 1973 resulted in a vast change in the amount of oil produced annually.*

Year	Millions of barrels
1880	30
1890	77
1900	150
1910	330
1920	690
1930	1,400
1940	2,200
1950	3,800
1960	7,700
1970	17,000

1. Graph this function. Start the *x*-axis with 1880 and let each unit represent 10 years. Let 5 units on the *y*-axis represent 5,000 million barrels.

2. Make a new table by replacing the numbers in the second column of the table above with their logarithms. Refer to the table of logarithms on page 218 and round each logarithm to one decimal place.

3. Draw another graph of the oil production function, this time using your new table. Let 2 units on the *y*-axis represent 1.

4. What do you notice about the points in your second graph?

---

*Based on an example in *Introduction to the Practice of Statistics*, by David S. Moore and George P. McCabe (W. H. Freeman, 1989).

The half-life of a radioactive element is the length of time it takes for half of its atoms to change into atoms of another element. Radium-230, for example, has a half-life of 1 hour. The table below shows what would happen if you started with 64 grams of it.

Time in hours, x	0	1	2	3	4	5	6
Amount left in grams, y	64	32	16	8	4	2	1

5. Graph this function. Let 5 units on the y-axis represent 50.

6. Make a new table by replacing the numbers in the second row of the table above with their logarithms. Round each logarithm to one decimal place.

7. Draw another graph of the function, this time using your new table. Let 5 units on the y-axis represent 10.

8. How does the change to a logarithmic scale on the y-axis change the shape of the graph?

9. Use your second graph to estimate the logarithm of the amount left after 1.5 hours.

10. Estimate the actual amount left after 1.5 hours by finding the number whose logarithm is your answer to exercise 9.

Set III

In an experiment to discover how quickly people forget, a psychologist gave a list of words to several people to memorize. The people were then tested at different time intervals to see what they remembered. Some of the results are shown in the table at the right.

Time until test in hours, x	Average score on test, y
0.25	73
0.5	67
1	60
2	53
8	40
24	30
48	23
96	17
192	10

1. Graph this function. Let 5 units on each axis represent 50. Join the points with a smooth curve.

2. What happens to the rate of forgetting as time goes by?

In this function, the x-numbers vary far more than the y-numbers.

3. Make a new table by replacing the x-numbers with their logarithms, each to 1 decimal place. Skip 0.25 and 0.5.

4. Draw another graph of the function, this time using your new table. Let 5 units on the x-axis represent 1.0 and 5 units on the y-axis represent 50.

5. What do you notice about the points?

6. Use your graph to guess the logarithms of 0.5 and 0.25. (*Hint:* Find the x-numbers corresponding to 67 and 73.)

CHAPTER

4

Summary and Review

In this chapter we have become acquainted with:

Large numbers and exponents *(Lesson 1).* When a large number is written as a power of 10, it is written in *exponential form.* The *exponent* indicates the number of zeros that follow the 1 when the number is written in decimal form. $10^1 = 10$; $10^0 = 1$.

The names and exponential forms of some large numbers are:

10^6	million	10^{21}	sextillion
10^9	billion	10^{24}	septillion
10^{12}	trillion	10^{27}	octillion
10^{15}	quadrillion	10^{30}	nonillion
10^{18}	quintillion	10^{33}	decillion

Scientific notation *(Lesson 2).* A number is in *scientific notation* if it is written in the form

$$a \times 10^b$$

in which *a* is a number that is at least 1 but less than 10. The number *a* is called the *coefficient* and the number *b* is called the *exponent.*

Logarithms *(Lessons 3, 4, and 5).* When a number is expressed as a power of 2, the exponent is the *binary logarithm* of the number. For example, the binary logarithm of 8 is 3 because $8 = 2^3$.

When a number is expressed as a power of 10, the exponent is the *decimal logarithm* of the number. For example, the decimal logarithm of 1,000,000 is 6 because $1,000,000 = 10^6$. The decimal logarithm of $a \times 10^b$ is $(\log a) + b$.

Numbers can be multiplied (or divided) by adding (or subtracting) their logarithms to get the logarithm of the answer.

Logarithms are useful in making diagrams that illustrate a wide range of numbers.

Exponential functions *(Lesson 6).* A function is *exponential* if one of the variables, *x*, appears as an exponent. If a function is exponential, a graph of it on axes with ordinary scales is a curve, but when a logarithmic scale is chosen for the *y*-axis, the graph is a straight line.

EXERCISES

SET I

A family in Utica, New York, spent 20 years collecting pennies in a glass jar. After reaching their goal of 1 million pennies, they deposited them in a bank and bought a new car!*

    1. How many dollars are 1 million pennies worth?

One thousand pennies weigh about 7 pounds.

    2. About how many pounds did the million pennies weigh?

There are approximately 10^{14} grains of sand in Malibu Beach, California.

    3. Write this number in decimal form.

    4. Write it in words.

Black holes are regions in space so tremendously massive that light cannot escape from them. Although such holes eventually evaporate, one having the mass of our sun would last for about a million million million million million million million million million million million years.†

    5. Write this number in decimal form.

    6. Write it as a power of 10.

    7. What advantages does writing a very large number as a power of 10 have over writing it in decimal form?

The Hindus consider "one day in the life of God" to be 4,320,000,000 years.

    8. Write this number in scientific notation.

"One day in the life of God" is approximately equal to 1.6×10^{12} human days.

    9. Write this number in decimal form.

    10. Write it in words.

---

*Time, January 22, 1979.

†A Brief History of Time, by Stephen W. Hawking (Bantam Books, 1988).

Number	Log
1	0
3	1
9	2
27	3
81	4
243	5
729	6
2,187	7
6,561	8
19,683	9
59,049	10

You have worked with binary logarithms, based on 2, and decimal logarithms based on 10. Here is a table of logarithms based on another number.

11. On what number is this logarithm table based?

12. Find the answer to this multiplication problem without using logarithms: 81×729. Show your work.

13. Show how the answer to the same problem can be found by using logarithms from this table.

14. Find the answer to this division problem without using logarithms: $2,187 \div 9$. Show your work.

15. Show how the answer to the same problem can be found by using logarithms from this table.

SET II

In the book *How to Draw Charts and Diagrams*, the author points out that logarithmic scales are good for showing a wide range of numbers but have the disadvantage that they can be understood only by readers familiar with logarithms.*

Show that you understand logarithms and logarithmic scales by doing the following.

1. Copy the first column of this table. Change each number in the second column to scientific notation.

Creature	Mass in milligrams
Spider	80
Hummingbird	2,000
Chicken	3,150,000
Human	70,000,000
Elephant	6,300,000,000
Whale	138,000,000,000

2. Add a third column to your table by using the table of logarithms on page 218 to write the logarithm of each number. Round each logarithm to one decimal place.

How to Draw Charts and Diagrams, by Bruce Robertson (North Light Books, 1988).

3. Draw and label a logarithmic scale as shown below. Let 1 centimeter represent 1 unit.

Use your answers to exercise 2 to mark each mass on your scale with an arrow. The spider is shown as an example. After you have marked the six arrows, change the label on the scale and turn the numbers into exponents, as shown below.

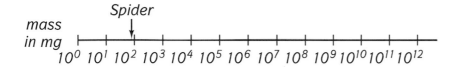

The logarithmic scale spreads out the smaller numbers.

4. What does it do to the larger numbers?

5. Would it be possible to represent these numbers on a line numbered like an ordinary ruler so that you could clearly see all six of them?

The first notes tuned on a piano are the A's. Some of their frequencies in hertz are shown in the figure below.

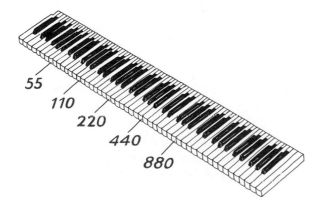

6. What kind of number sequence do the frequencies form?

7. Copy and complete the following table including the frequencies of all eight A's on the keyboard.

Position of A, x	1	2	3	4	5	6	7	8
Frequency, y	▦	55	110	220	440	880	▦	▦

Position of A, x	1	2	3	4	5	6	7	8
Frequency, y	‖‖‖‖	55	110	220	440	880	‖‖‖‖	‖‖‖‖

8. Graph this function. Let 2 units on the y-axis represent 1,000. Connect the points with a smooth curve.

9. Make a new table by replacing the numbers on the second row of the table above with their logarithms. Round each logarithm to one decimal place.

10. What kind of number sequence do the logarithms of the frequencies form?

11. Draw another graph of the function, this time using your new table. Let 2 units on the y-axis represent 1.

12. What do you notice about the points?

Set III

The deeper a scuba diver goes into the ocean, the darker it becomes. The comparative amount of sunlight that penetrates various depths is shown in the table below.

Depth in meters, x	0	50	100	150	200	250
Amount of sunlight, y	10^5	10^4	10^3	10^2	10	1

1. Graph this function. Let 2 units on the x-axis represent 50 and 10 units on the y-axis represent 100,000.

2. Make a new table by replacing the numbers in the second row of the table above with their logarithms.

3. Draw another graph of the function, this time using your new table. Let 2 units on the x-axis represent 50 and 2 units on the y-axis represent 1.

The world record for scuba diving is a depth of about 135 meters. Mark the point on the line in your second graph that shows that the logarithm of the amount of sunlight at a depth of 135 meters is about 2.3.

4. Estimate the actual amount of sunlight at this depth by finding the number whose logarithm is 2.3.

5. Find the fraction of the sunlight hitting the surface of the ocean that penetrates to this depth by dividing your answer to the previous exercise by 10^5.

Further Exploration

LESSON 1

1.   The names for large numbers beyond a million first appeared in a book written by the French mathematician Nicolas Chuquet in 1484. The list below shows the names used by Chuquet and the Latin names of the numbers from 1 through 9.

Word used by Chuquet	Meaning	Latin words for numbers	Meaning
Million	10^6	Unum	1
Byllion (billion)	10^9	Duo	2
Tryllion (trillion)	10^{12}	Tria	3
Quadrillion	10^{15}	Quattour	4
Quyllion (quintillion)	10^{18}	Quinque	5
Sixlion (sextillion)	10^{21}	Sex	6
Septyllion (septillion)	10^{24}	Septem	7
Ottyllion (octillion)	10^{27}	Octo	8
Nonyllion (nonillion)	10^{30}	Novem	9

   a. At what place do the names in the two lists first seem to be re-
       lated?

Since Chuquet made his list, other names have been added to it, including *centillion*. The related Latin word is *centum*, meaning "100."

   b. What number do you think *centillion* names? Explain your rea-
       soning.

According to *The Random House Dictionary of the English Language, zillion*
is an informal term meaning "an extremely large, indeterminate number." In other words, it does not name a specific number.

   c. How do you suppose the name *zillion* may have originated?

2.   *Writing* a large number in decimal form and *counting* to that number
are very different things. Suppose that, ignoring the commas, you wrote

at the rate of one digit per second. How long would it take you to write

    a. *one million* in decimal form?   b. *one billion* in decimal form?

Suppose that you counted at the rate of one number per second. (In other words, counting to 10 would take 10 seconds.) How long would it take you, in *days*, to *count* to

    c. one million?   d. one billion?

The name *googol* refers to the number 10^{100}. Do you think a computer could

    e. print this number in decimal form? Explain.
    f. "count" to this number by printing a list of all the numbers from one to a googol? Explain. (The number of atoms in the universe is thought to be no more than 10^{80}.)

At the same time that the word *googol* was invented, the word *googolplex* was suggested as the name of an even larger number. At first, a googolplex was considered to be "1, followed by writing zeros until you got tired."* Then it was decided that this was too vague, so a googolplex was defined to be 1 followed by a googol of zeros.

    g. Do you think that a computer could print a googolplex in decimal form? Explain.

LESSON 2

1.   Scientific notation is useful in representing not only numbers that are very large but also numbers that are very small. Nuclear physicists, for example, have discovered that particles having the shortest lifetimes exist for only

$$0.000000000000000000000000002$$

second. To discover how this incredibly small number can be written in scientific notation, look at the following table of powers of 10.

10,000	1,000	100	10	1	|||||	|||||	|||||	|||||
10^4	10^3	10^2	10^1	10^0	|||||	|||||	|||||	|||||

    a. In what kind of number sequence are the numbers in the first line of this table?
    b. In what kind of number sequence are the *exponents* of the 10's in the second line?

Mathematics and the Imagination, by Edward Kasner and James Newman (Simon & Schuster, 1940).

c. What do you think the next four numbers of the first line of the table should be? Explain your reasoning.

d. What do you think the next four powers of 10 in the second line of the table should be? Explain your reasoning.

The average lifetime of a tau-meson is 0.000000005 second.

e. How do you think this number would be written in scientific notation? (Observe that $0.000000005 = 5 \times 0.000000001$.)

f. How do you think the lifetime of the shortest lived particles, 0.0000000000000000000002 second, would be written in scientific notation?

2. The universe seems to be expanding. In the 1920s, the astronomer Edwin P. Hubble discovered that the other galaxies in space seem to be moving away from us. This suggests that the universe began with a "big bang," an explosion in which the outward motion of the galaxies began.

The farther another galaxy is from ours, the faster it seems to be moving. Galaxies 10,000,000 light-years away are estimated to be traveling 170 kilometers per second; galaxies twice as far away are estimated to be traveling twice as fast, and so on. A light-year is about 9.46×10^{12} kilometers.

a. How many kilometers from us is a galaxy that is 10,000,000 light-years away?

According to the big bang theory, the age of the universe is the time that it took this galaxy to get that far away.

b. On the assumption that it has been moving away from us at the steady rate of 170 kilometers per second, determine the approximate age of the universe. (One year is approximately 3.2×10^7 seconds.) Show your work.

Lesson 3

1. The "zone system" is a scale ranging from black to white through eight shades of gray. Invented by the landscape photographer Ansel Adams, it is illustrated in the figure below.

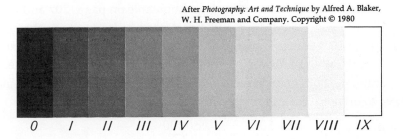

After *Photography: Art and Technique* by Alfred A. Blaker, W. H. Freeman and Company. Copyright © 1980

0   I   II   III   IV   V   VI   VII   VIII   IX

The numbers of the zones are keyed to units of exposure time, as shown in the table below.

Number of zone	0	I	II	III	IV	V	VI	VII	VIII	IX
Exposure time	0.5	1	2	4	8	16	32	64	128	256

   a. In what kind of number sequence are the exposure times of the 10 zones?

   b. Copy and complete the following version of the table, writing the exposure times as powers of 2.

Number of zone	0	I	II	III	IV	V	VI	VII	VIII	IX
Exposure time	—	—	2^1	2^2	▥	▥	▥	▥	▥	▥

Dashes were written in the spaces below 0 and I.

   c. What powers of 2 could replace them to fit the rest of the pattern?

Recall that when a number is expressed as a power of 2, the exponent is the *binary logarithm* of the number.

   d. Copy and complete the table below by replacing the exposure times in the table above with their binary logarithms.

Number of zone	0	I	II	III	IV	V	VI	VII	VIII	IX
Logarithm of exposure time	▥	▥	▥	▥	▥	▥	▥	▥	▥	▥

   e. How is the number of each zone related to the corresponding logarithm of the exposure time?

2.   Imagine that you are driving on a road along which signs are posted indicating the logarithms of the distances that you have driven in meters. The first 10 meters of the road are shown in the figure below.

Compare this figure with the logarithm table on pages 207 and 208. Notice that you would drive 1,024 meters in order to reach the sign numbered 10. Because 1 kilometer is equal to 1,000 meters, you would arrive at the sign numbered 10 after roughly 1 kilometer, or 0.6 mile.

   Estimate, both in kilometers and miles, the distance that you would have to drive to reach the sign numbered

   a. 20.   b. 50.   c. 100.

The distances to the last three posts are astonishingly great. By way of comparison, the most remote object in the entire universe is thought to be about 70,000,000,000,000,000,000,000 miles away.

LESSON 4

1. Counting on our fingers may have led to our use of 10 as a base for our number system and our decimal logarithms. Because mathematics is a universal language, it is likely that, if intelligent beings exist on a planet somewhere else in space, they also might have developed the idea of logarithms even though their number system might be very different from our own.

Suppose that we established communication with another civilization, translated its numbers into our own, and learned some of its logarithms as shown in the following table:

Number	1	2	3	4	5	6	7	8	9	10
Logarithm	0	0.39	0.61	▓	0.90	▓	1.09	▓	▓	▓

a. On the assumption that the logarithms of this other civilization behave just as ours do, determine the missing logarithms in this table.
b. What number do you think is the base of this number system?
c. What number would have 2 for its logarithm in this system?
d. If the creatures in the other civilization have two hands as we do, how many fingers do you think they might have on each hand?

2. It has been suggested that, in performing age-related experiments with animals, the ages of the animals should be in geometric sequence.

Suppose that an experiment is performed with five groups of mice, the youngest group being 16 days old and the next older group being 24 days old.

a. Copy and complete the second line of the following table so that the logarithms of the ages form an *arithmetic* sequence.

Age	16	24	▓	▓	▓
Log of age	1.204	1.380	▓	▓	▓

Then copy and complete the first line of the table by finding the numbers corresponding to the logarithms on the second line. Refer to the table on page 218 to do this.
b. Are the numbers in the first line in geometric sequence? Explain.

LESSON 5

1.   Experiment: *A Slide Rule*

The slide rule was invented by the English mathematician William Oughtred in 1622, soon after the invention of logarithms. In the following experiment, you will construct a simple slide rule and discover how it worked.

From *Science and Civilisation in China*, volume 3, by Joseph Needham, Cambridge University Press

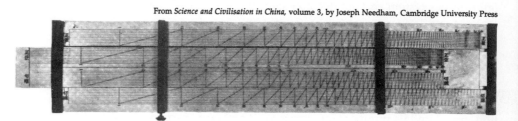

A Chinese slide rule dated at 1660

First, draw a scale 6 inches long on graph paper that has 10 squares to the inch. Divide the scale into 0.5-inch lengths and label it with numbers as shown in the figure below.

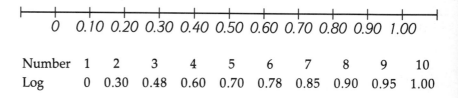

Number	1	2	3	4	5	6	7	8	9	10
Log	0	0.30	0.48	0.60	0.70	0.78	0.85	0.90	0.95	1.00

Next, use the table of logarithms above to add a logarithmic scale above the scale you have just made. Do this by locating each of the numbers from 1 through 10 directly above its logarithm. The first three numbers are shown in the figure below.

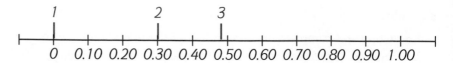

Next, draw a line on a 4-by-6-inch card to divide it in half as shown at the left, and cut it into two pieces along the line.

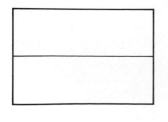

Take the lower piece of your card and place it on your graph paper so that the upper edge falls along the scale. Carefully make marks on the upper edge of the card corresponding to the numbers from 1 through 10. Your finished card should look like the one shown in the figure at the top of the next page.

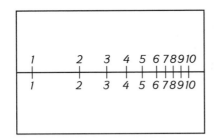

Now put it below the other half of the card and carefully copy the scale and the numbers from the finished card onto the lower edge of the other card, as shown in the figure below. When you have done this, your slide rule is ready to use.

a. Using a simple example like 2×3, explain how the slide rule can be used to multiply.

b. Explain how the slide rule can be used to divide.

2. The largest number that can be written using only two digits is

$$9^9$$

Its approximate value in decimal form can be found by using logarithms as shown here.

$$
\begin{array}{rcl}
9 & \longrightarrow & 0.954 \\
\times 9 & \longrightarrow & +0.954 \\
\times 9 & \longrightarrow & +0.954 \\
\times 9 & \longrightarrow & +0.954 \\
\times 9 & \longrightarrow & +0.954 \\
\times 9 & \longrightarrow & +0.954 \\
\times 9 & \longrightarrow & +0.954 \\
\times 9 & \longrightarrow & +0.954 \\
\times 9 & \longrightarrow & +0.954 \\
& & \overline{9(0.954) =} \\
3.85 \times 10^8 & \longleftarrow & 8.586
\end{array}
$$

So $9^9 \approx 385,000,000$, a number nine digits long.

The largest number that can be written using three digits is

$$9^{9^9}$$

When two exponents are used in this way, they are interpreted as representing

$$9^{(9^9)}$$

a. Use logarithms to find the approximate value of this number in scientific notation.
b. If this number were written in decimal form, approximately how many digits would it contain?

LESSON 6

1. The number of words in large English dictionaries has steadily increased over the past three centuries.* Here is a list of some dictionaries.

Title	Year	Number of words
New English Dictionary	1702	28,000
Johnson's Dictionary	1755	40,000
Webster's American Dictionary	1828	70,000
Century Dictionary	1891	200,000
Webster's Third New International	1961	450,000

a. Graph the numbers of words in these dictionaries as a function of the year. Start the x-axis with 1700 and let 4 units represent 100 years. Let 2 units on the y-axis represent 100,000 words. Draw a smooth curve through the points.
b. Find the logarithms of the numbers in the third column of the table above. Refer to the table on page 218 and round off each logarithm to one decimal place.
c. Draw another graph of the dictionary function, this time using the logarithms of the numbers of words. Label the x-axis as in the previous graph and let 2 units on the y-axis represent 1. Plot the points as accurately as you can.
d. One of the points on your graphs does not quite fit the pattern of the other four. On which of your two graphs is it easier to see this?
e. Which one of the five dictionaries has fewer words than you might have expected from the trend in the other four?

---

*This exercise is based on information in *The Science of Words*, by George A. Miller (Scientific American Library, 1991).

2.   The great Greek mathematician and astronomer Hipparchus, who lived in the second century B.C., classified the stars that he could see according to six classes of brightness. The brightest stars were assigned to the first class and the faintest stars that he could see were assigned to the sixth class.

In the nineteenth century, his system was refined as shown in the following table.

Magnitude	1	2	3	4	5	6
Brightness	100	40	16	6.3	2.5	1

As the table shows, a star of the first magnitude is 100 times as bright as a star of the sixth magnitude.

a. Refer to the table of logarithms on page 218 to copy and complete the following table, which is based on the table above. Round each logarithm to one decimal place.

Magnitude	6	5	4	3	2	1
Brightness	1	2.5	6.3	16	40	100
Log of brightness	▒	▒	▒	▒	▒	▒

b. What do you notice about the logarithms of the brightnesses?
c. According to this pattern, what would be the logarithm of the brightness of a star with magnitude 0?

The full moon has a magnitude of -12.

d. Use the pattern in your table to figure out how many times as bright the full moon is as a star of the sixth magnitude. Show your reasoning.

The sun has a magnitude of about -27.

e. About how many times as bright as the sun is a star of the sixth magnitude? Show your reasoning.

Symmetry and Regular Figures

1

Symmetry

The ink drawing shown here was created by the psychologist Roger Shepard. Shepard titled it "Reflecting Prince" and wrote that it "can be seen either as a frontal view of a strangely wide-faced (perhaps extra-terrestrial) being or, alternatively, as a profile view of a normally proportioned human prince with his face (and hand) . . . pressed against a mirror."*

If you place a mirror perpendicular to the page so that the edge of the mirror lies on the vertical line down the center of this figure, you will see that one side of the picture is indeed a reflection of the other. This center line is called a mirror line, and it is the line of symmetry of the figure. Shepard's drawing can be seen as a single face because the human face itself has *line symmetry*.

_____

*Mind Sights, by Roger N. Shepard (W. H. Freeman, 1990).

A figure has **line symmetry** if there is a line along which a mirror can be placed to reflect either half of the figure so that it reproduces the other half.

Another way to see if a figure has line symmetry is to fold the figure along the supposed line to see if the two halves coincide (fit together exactly). Look, for example, at the flag of Turkey shown here. It is symmetrical with respect to a horizontal line through its center. If the flag is folded along this line, the pattern on one side of the fold coincides with the pattern on the other side.

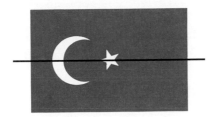

A figure can be symmetrical with respect to more than one line. A square, for example, has four lines of symmetry. It is easy to see that the horizontal and vertical lines in the figure below are lines of symmetry; the two lines through the opposite corners of the square are less obvious.

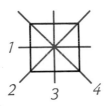

The paper windmill shown here is symmetrical, yet it does not have any lines of symmetry. It is impossible to fold it along a line so that any two halves coincide. The windmill can be *rotated*, however, into other positions that look exactly the same. For example, it can be rotated 90° (through one-quarter of a circle) and look exactly the same. Therefore, it has *rotational symmetry*.

A figure has **rotational symmetry** if it can be rotated through an angle of less than 360° so that it coincides with its original position.

The point about which the figure is rotated is called the *axis* of symmetry. The axis of symmetry of the paper windmill is the pin about which it turns.

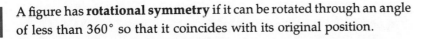

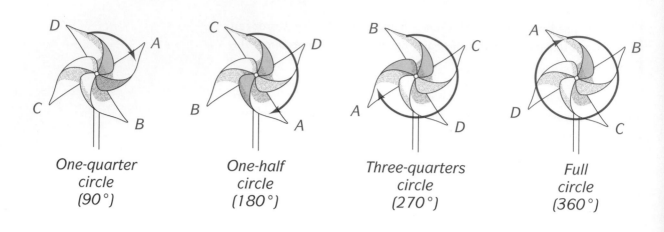

One-quarter
circle
(90°)

One-half
circle
(180°)

Three-quarters
circle
(270°)

Full
circle
(360°)

Notice that the windmill can be turned through four different angles up to and including 360° so that it looks exactly the same: 90°, 180°, 270°, and 360°.*

For this reason, the windmill is said to have "4-fold" rotational symmetry.

EXERCISES

SET I

Someone was a bit careless in painting the word ONLY on this freeway off-ramp in Los Angeles. The stencil must have accidentally been turned over so that left and right were reversed when the N was painted.

1. For which of the letters in ONLY would this have not mattered?

*Angles and their measures are discussed on pages 656–657.

The letter Y has symmetry with respect to a vertical line, as this figure shows. If a mirror were placed on the line, which point would be the reflection of

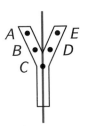

2. point A?   3. point D?   4. point C?

In general, if a figure has line symmetry, exactly where is the reflection of a point that

5. is on the line?   6. is not on the line?

The letter N has rotational symmetry because it looks the same if it is rotated 180°. If the letter N is rotated in this way, to which point does

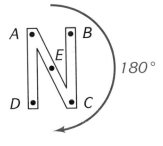

7. point A move?   8. point D move?   9. point E move?

10. In general, if a figure has rotational symmetry, which point of the figure does not move?

11. Which letter in ONLY has both line symmetry and rotational symmetry?

A reflection of the horse and its rider is clearly visible in this photograph.

12. What is acting as a mirror?

13. Does the part of the picture showing the horse and rider and their reflection appear to have line symmetry? (Ignore the rest of the picture.)

14. Does it have rotational symmetry? Explain why or why not.

Photograph by Joe Vitti

Many playing cards are symmetrical. The two cards shown here are examples.

15. What kind of symmetry do they have?

16. Why do you suppose they have this symmetry?

The star in a circle has been the symbol of Mercedes-Benz cars since 1909.

Mercedes Benz symbol

17. How many lines of symmetry does this symbol have?

The symbol has 3-fold rotational symmetry.

18. What does this mean?

19. To have rotational symmetry, must a figure look the same upside down?

A starfish has many symmetries. Trace the star in the drawing to the right of the photograph on tracing paper. Also copy the shaded part on your drawing.

20. In how many ways can you place your tracing on this figure so that the whole star fits the whole star but the shaded part

Courtesy of Barbara Ferenstein

is in a different place? (Count the original position of the tracing as one of the ways.)

21. How many lines of symmetry does the star have?

22. What kind of rotational symmetry does it have?

This "running-legs" symbol has 7-fold rotational symmetry but no lines of symmetry.

Describe, as completely as you can, the symmetry, if any, possessed by each of the following figures.

23.

©Scott Kim, 1989

24.

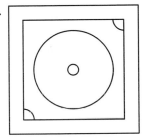

Magnet with iron filings

25.

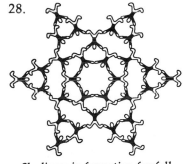

Floor plan of wrestling ring

26.

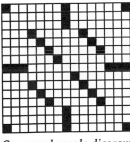

Crossword puzzle diagram

27.

© 1991 Robert Petrick

28.

Skydivers in formation freefall

29.

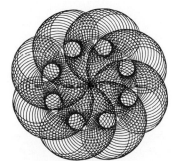

Figure by Y. Chernikhov

30.

Sea urchin
From *Art Forms in Nature*, by Ernst Haeckel, Dover Publications, Inc., New York, 1974.

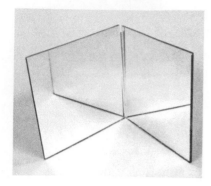

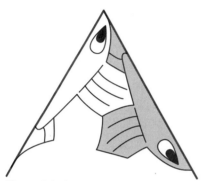

After a periodic drawing by M. C. Escher, *Baarn IV-'64*, in *Visions of Symmetry*, by Doris Schattschneider, W. H. Freeman and Company, 1990, p. 216

SET II
EXPERIMENT: SYMMETRIES WITH A KALEIDOSCOPE

The kaleidoscope, invented in 1816, originally consisted of two mirrors inside a tube. Continually changing symmetrical patterns are produced by colored pieces of glass that tumble between the mirrors as the tube is turned.

A simple kaleidoscope can be made by hinging two mirrors together with tape, as shown in the photograph at the left.

1. Stand the mirrors on the figure at the left so that they form the angle shown. What does the figure formed in the kaleidoscope look like?

Notice that it has 3-fold rotational symmetry as well as three lines of symmetry.

Stand the mirrors on the figures for exercises 2–8 as indicated. Describe the figures formed by the mirrors and their symmetries as specifically as you can.

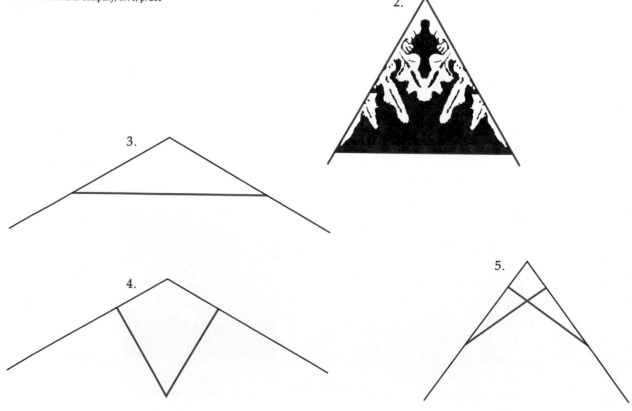

2.

3.

4.

5.

6.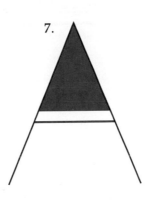

7.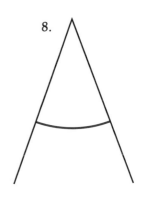

8.

The type of figure formed by the kaleidoscope is related to the angle between the mirrors. Stand the mirrors on figure A, in which the angle is 120°.

A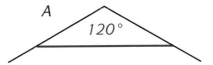

9. How many sides does the figure formed by the kaleidoscope have?

10. What number do the three angles surrounding the center of the figure add up to?

Stand the mirrors on figure B, in which the angle is 90°.

B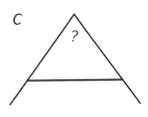

11. How many sides does the figure formed by the kaleidoscope have?

12. What number do the four angles surrounding the center of the figure add up to?

Stand the mirrors on figures C and D.

C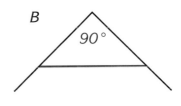

13. Use what you observe, as well as your answers to exercises 9–12, to copy and complete the following table.

Figure	A	B	C	D
Number of sides, n	▥	▥	▥	▥
Angle of mirrors, m	120°	90°	▥	▥

14. What happens to the number of sides of the figure produced as the angle between the mirrors gets smaller?

15. Write a formula for m, the angle of the mirrors, in terms of n, the number of sides of the figure formed by the kaleidoscope.

D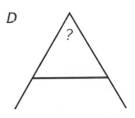

16. Copy and complete the following table.

Number of sides, n	8	9	10	12
Angle of mirrors, m	▥	▥	▥	▥

Set III

This photograph was taken as a gag.

Photograph by Peter Renz

1. How might the scene be changed to make it more convincing?

2. Explain by means of symmetry why the board would still look wrong, even with this change.

One part of the photograph is puzzling.

3. What is it that is puzzling?

4. Explain how it could look that way.

By permission of Johnny Hart and Field Enterprises, Inc.

Regular Polygons

A square is certainly not the ideal shape for a wheel, and even though a wheel in the shape of an equilateral triangle "eliminates one bump," it is clearly even less practical. The equilateral triangle and the square are the simplest geometric figures among the *regular polygons*.

Equilateral triangle   *Square*

A **regular polygon** is a figure of which all sides are the same length and all angles are equal.

The names of the rest of the regular polygons are taken from their numbers of sides. These figures were studied extensively by the early Greek mathematicians, and the names they gave them have been used ever since. The origin of the names is apparent in the table of Greek names for numbers below.

5	6	7	8	9	10	12
pente	hex	hepta	octo	ennea*	deca	dodeca

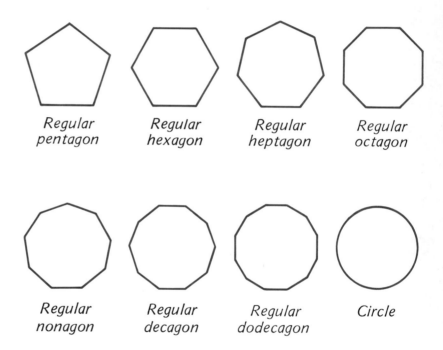

Regular pentagon	*Regular hexagon*	*Regular heptagon*	*Regular octagon*

Regular nonagon	*Regular decagon*	*Regular dodecagon*	*Circle*

The set of regular polygons, like the set of counting numbers, goes on indefinitely. The greater the number of sides that a regular polygon has, the more symmetrical it is and the more nearly circular in shape. A regular polygon with 100 sides, for example, looks so much like a circle that it might very well serve as the shape of a wheel. For this reason, some of the properties of the circle were discovered by comparing it with a regular polygon having a large number of sides.

---

*The Greek name for the regular polygon having nine sides was *enneagon.* We now use the word *nonagon,* from the Latin word *nonus,* which translates into "nine."

Regular polygons look more like a circle as their number of sides increase because of their symmetries. The figures below suggest that if a regular polygon has *n* sides, it has *n* lines of symmetry and *n*-fold rotational symmetry.

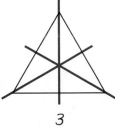 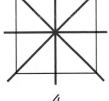 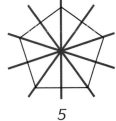

Sides:	3	4	5
Line symmetry:	3 lines	4 lines	5 lines
Rotational symmetry:	3-fold	4-fold	5-fold

Images of regular polygons can be formed in a simple kaleidoscope by adjusting the two mirrors to different angles. The mirror angle of an equilateral triangle has a measure of 120°, the mirror angle of a square, 90°, and the mirror angle of a regular pentagon, 72°.

These measures come from the fact that the number of degrees around a point is 360°.

  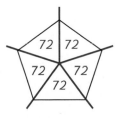

$$\frac{360°}{3} = 120°$$ $$\frac{360°}{4} = 90°$$ $$\frac{360°}{5} = 72°$$

In general, the mirror angle of a regular polygon having *n* sides has a measure of $\frac{360}{n}$ degrees.

EXERCISES

SET I

Silos used on farms to store feed for animals are usually round. The first silo to be built was square, but didn't work very well because air in the corners spoiled some of the feed.

Larry Lefever/Grant Heilman

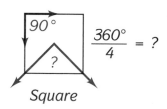

Square

The sides at each corner of a square form an angle of 90°.

1. What is the measure of the mirror angle of a square?

Here are figures illustrating angles of two more regular polygons.

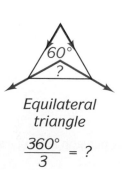

Equilateral triangle

$$\frac{360°}{3} = ?$$

Regular pentagon

$$\frac{360°}{5} = ?$$

2. Copy and complete the following table.

Number of sides, n	3	4	5
Mirror angle, m	▨	▨	▨
Angles of polygon	60°	90°	108°

As the number of sides of a regular polygon increases, what happens to

3. the size of the mirror angle?

4. the size of the angles of the polygon?

5. On the basis of your table, what seems always to be the result of adding the measures of the mirror angle and an angle of the polygon?

Perhaps the most common type of traffic sign is the one shown in this photograph.

6. What shape does it have?

7. Find the measure of the mirror angle of this polygon.

8. What do you think is the measure of one of the angles of this polygon?

The sum of the measures of the three angles in every triangle is 180°. The connection between the mirror angle and the angles of a regular polygon follows from this fact and the line symmetry of the polygon.

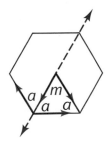

In the figure shown here, m represents the measure of the mirror angle. Notice that $2a$ represents both the sum of the other two angles of the triangle and the measure of one of the angles of the regular polygon. Since the sum of the three angles of the triangle is 180°,

$$m + 2a = 180°$$

A Wells Fargo bank in San Francisco is based on the shape of a regular polygon with 40 sides.

A building by Hertzka & Knowles and Skidmore, Owings and Merrill, 1958, in Crown Zellerbach Plaza, San Francisco. Photograph © by Morley Baer

9. Why do you suppose it is called the "round bank"?

10. What is the measure of the mirror angle of a regular polygon having 40 sides?

11. What is the measure of each angle of the polygon?

Because of the symmetries of regular polygons, one way to draw them is to locate their corners on a circle. Two tools that can be used to do this are the *straightedge*, for drawing straight lines, and the *compass*, for drawing circles and parts of circles, called arcs.

12. A regular polygon that is especially easy to draw with these
    tools is the hexagon. Use them to construct this polygon by
    doing the following.

    a. Draw a circle, label its center O, and label a point on the
       circle A as shown in the figure at right.* Put the metal
       point of the compass on A and the pencil point on O and
       draw an arc intersecting the circle at point B. Put the metal
       point on B and draw an equal arc intersecting the circle
       at C. Continue drawing equal arcs around the circle. If your
       work is accurate, the last arc should intersect the circle at A.

    b. Label the remaining points at which the arcs intersect the
       circle D, E, and F. Connect the points in order to form a
       regular hexagon.

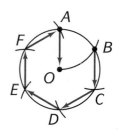

Napoleon Bonaparte had a surprising interest in mathematics. In 1797,
he told two mathematicians in Paris about new discoveries in geometry
that he had learned of in Italy. One of the mathematicians is said to have
remarked: "General, we expect everything of you, except lessons in
geometry!"

One of the problems that interested Napoleon was how to locate the
corners of a square without using a straightedge. The solution is described in the following exercise.

13. The construction begins in the same way as that in the pre-
    vious exercise.

    a. Draw a circle, mark a point on the top of it, and leaving
       the compass adjusted to the radius of the circle, draw three
       arcs like those in the first figure below. Label the points A,
       B, C, and D.

    b. Adjust the compass so that the metal point is on A and the
       pencil point is on C and draw an arc like the one shown in
       the second figure. Put the metal point on D and the pencil

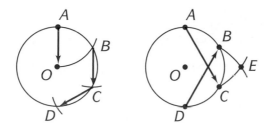

---

*In these figures, the arrows show where to put the compass on the paper. Each
arrow begins at the point at which the metal point of the compass should be
placed and points to the point at which the pencil point should be placed.

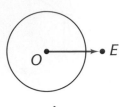

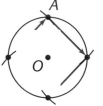

point on B and draw another arc crossing the preceding one as shown.

c. Label the point at which the arcs cross E. Adjust the compass so that the metal point is at the center of the circle, O, and the pencil point is on E. The distance between these two points is equal to the length of the side of the square.

d. Pick the compass up and, starting at A, draw four arcs around the circle, as shown in the second figure at the left. The points at which they intersect the circle are the corners of the square, and only a compass was used to find them. Of course, to draw the sides of the square you will need to use your straightedge.

SET II

The figure below, called a circular protractor, is for use in making the drawings in the exercises that follow.

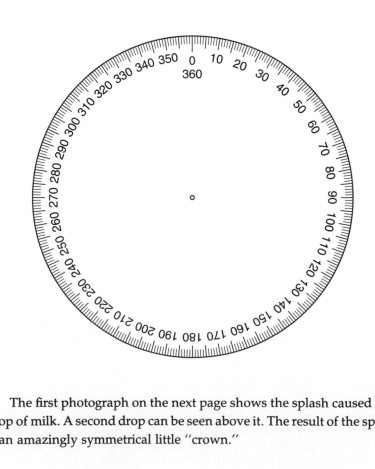

The first photograph on the next page shows the splash caused by a drop of milk. A second drop can be seen above it. The result of the splash is an amazingly symmetrical little "crown."

1. Into how many degrees is the scale of the circular protractor divided?

The milk drop "crown" has 24 spikes.

2. If these spikes are represented as 24 evenly spaced points around a circle, how many degrees apart would they be?

3. Place a sheet of tracing paper over the circular protractor and do the following.

    Mark a large dot on the circle at the point numbered 0 to represent the first spike. Then mark large dots around the circle to represent the other spikes.

    Use a ruler to draw line segments that connect the points in the order you marked them to form a regular polygon.

4. Describe the symmetries of the polygon as specifically as you can.

5. What are the measures of some of the angles about which you could rotate the polygon so that it would look exactly the same?

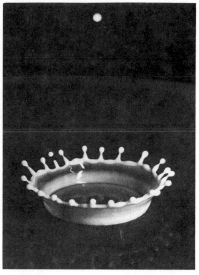

Milk drop
'Milk Drop Coronet 1957' © The Harold E. Edgerton 1992 Trust

Now use a ruler to draw an equilateral triangle by connecting three of the spikes of the crown in your drawing for exercise 3.

6. Through what angles could the equilateral triangle be rotated so that it would look exactly the same?

7. Name the other regular polygons that could be formed by connecting spikes of the crown in this way.

This figure is a computer-generated model of DNA. Except for the central part of the figure, it has 10 lines of symmetry and 10-fold rotational symmetry.

8. If 10 points are evenly spaced around a circle, how many degrees apart would they be?

9. If the first point is chosen at 0 on the circular protractor, at what numbers would the other 9 points be?

10. Place your tracing paper over the circular protractor and use these numbers to mark 10 evenly spaced points around the circle.

    Starting at the top, use a ruler to draw line segments that connect every *second* point in the order you marked them to form a regular polygon. Draw line segments that connect the corners of this polygon in *every possible way*.

© Regents, University of California; Computer Graphics Laboratory, University of California, San Francisco

The result is a figure that was used as the symbol of a secret society in ancient Greece.

11. What does the figure look like?

12. Describe its symmetries as specifically as you can.

13. Now use a pen or pencil of a different color to connect the remaining five points that you originally marked to form another polygon. Draw line segments that connect the corners of this polygon in every possible way.

Your drawing should now have the same type of symmetry as the DNA picture.

14. What polygon surrounds the center of your drawing?

The Dome of the Rock in Jerusalem has a floor plan based on regular polygons.

George Chan/Tony Stone Images

15. If 16 points are evenly spaced around a circle, how many degrees apart would they be?

16. Place your tracing paper over the circular protractor and mark 16 evenly spaced points around it.

Starting with the *second* point you marked, use a ruler to connect every *second* point to form a regular octagon. Next, starting with the point numbered 0, connect every *fourth* point to form a square.

The octagon represents the outer wall of the Dome. The square crosses the wall in the positions of the four doors. Shade the wall, leaving gaps for the doors, as shown in this figure.

Connect the remaining 4 dots to form another square. Add the remaining lines as shown in the figure. Make 12 heavy dots to show the locations of the 12 pillars inside the Dome.

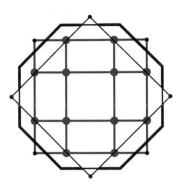

Look carefully at your drawing to answer the following questions.

17. How many regular octagons do you see?

18. Where are they?

19. How many squares are there altogether in the figure?

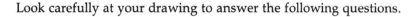

SET III

Here is an interesting puzzle. It is called a *dissection* puzzle and the problem is to cut a regular dodecagon into pieces that can be rearranged to form a square.

Make an accurate copy of the figure shown here by placing the figure you drew for exercise 3 of Set II over a 4-by-6-inch card. Use a pin or the sharp point of a compass to poke holes into the card at every *second* corner of the 24-sided polygon. Then remove the tracing paper from the card and connect the points with a ruler to form the dodecagon and the four extra line segments shown in color. Notice that the figure consists of six pieces, one of which looks like an equilateral triangle.

Cut the six pieces apart with scissors and try to rearrange them to form a square. If you solve the puzzle, either make a sketch to show what the arrangement of the pieces looks like or tape them together on your paper.

Some dissection puzzles were discovered by the Greeks and, in the tenth century, the Persian astronomer Abul Wefa wrote an entire book about them. The most complete book on them currently available was written by Harry Lindgren, an Australian patent examiner.*

Recreational Problems in Geometric Dissections and How to Solve Them, by Harry Lindgren (Dover, 1972).

Photograph from the Moody Institute of Science

3

Mathematical Mosaics

Bees are expert mathematicians when it comes to building honeycombs. The fourth-century Greek mathematician Pappus said:

> Though God has given to men the best and most perfect understanding of wisdom and mathematics, He has allotted a partial share to some of the unreasoning creatures as well. . . . This instinct is specially marked among bees. They prepare for the reception of the honey the vessels called honeycombs, with cells all equal, similar and adjacent, and hexagonal in form.*

Pappus observed that bees use the regular hexagon exclusively for the shape of the cells in the honeycomb. The photograph shows how the

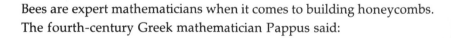

*Selections Illustrating the History of Greek Mathematics, by Ivor Thomas (Harvard University Press, 1939).

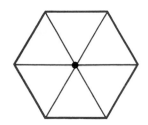

cells are arranged. Each corner point in the honeycomb is surrounded by exactly three hexagons, and there is no wasted space between cells.

Is this hexagonal arrangement of cells the most efficient one? Would other regular polygons work just as well?

The regular polygon with the least number of sides is the equilateral triangle. The angles of an equilateral triangle are smaller than those of a regular hexagon, and so more than three triangles will fit around a point. In fact, there is room for exactly six.

If bees used equilateral triangles, their honeycomb would look like the second figure at the right. This arrangement, however, would require more wax to form the cells than does the hexagonal honeycomb, if the cells in both patterns are equal in size.

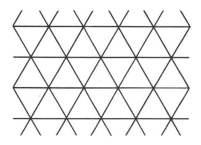

What if bees used a regular polygon having *more* sides than a hexagon rather than fewer sides? How would, say, regular octagons work? The angles of a regular octagon are larger than those of a hexagon; so there is room for only two around a point, with some space remaining unfilled. However, there is just the right amount of room for a square in

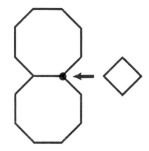

this space; so two regular octagons and one square surround a point exactly. If this pattern were repeated, the honeycomb would look like the figure at the right. This design has the disadvantage of being more complicated than the simple hexagonal pattern, and it, too, would require more wax.

Each of the designs that we have considered is a *mosaic*. Although a mosaic is any arrangement of shapes fitted together to cover a surface with no gaps or overlapping, we will consider only mosaics of the following type in this lesson.

> A **mathematical mosaic** is a set of regular polygons arranged so that the polygons share their sides and the polygons at each corner point are alike in number, kind, and order.

If all the polygons are alike, as in the case of the honeycomb mosaic, the mosaic is called *regular.* If the polygons are not all alike, the mosaic is called *semiregular.*

EXERCISES

SET I

The pattern used by bees in building a honeycomb is also used in the construction of doors. The doors, made by gluing thin sheets of wood to both sides of a paper honeycomb, have the advantage of being light in weight yet exceptionally strong.

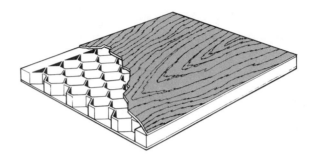

1. What regular polygon is used in the shape of the cells of the honeycomb?

2. How many of them surround each corner point of the honeycomb?

In the last lesson, you learned that the size of the angles of a regular polygon having n sides can be found by subtracting the mirror angle of the polygon, $\dfrac{360}{n}$, from 180°.

For example, each angle of a regular hexagon has a measure of

$$180 - \frac{360}{6} = 180 - 60 = 120°$$

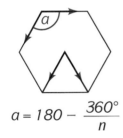

$$a = 180 - \frac{360°}{n}$$

3. Use this method to copy and complete the following table.

Number of sides, n	3	4	5	6	8	9	10	12
Mirror angle				60				
Angles of polygon				120				

A set of polygons will surround a point exactly only if the sum of the measures of the angles at the point is 360°. Three hexagons surround a point because

$$120° + 120° + 120° = 360°$$

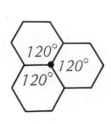

4. How many squares surround a point exactly?

5. Write an equation like the one for the hexagons to show why.

6. How many equilateral triangles surround a point exactly?

7. Write an equation to show why.

8. No number of regular pentagons will surround a point exactly. Explain why by showing that three pentagons are too few and four are too many.

Remember that if all the polygons in it are alike, a mosaic is called *regular*.

9. How many different regular mosaics seem to be possible?

EXPERIMENT: SEMIREGULAR MOSAICS

The following experiment will enable you to discover ways to surround a point exactly with regular polygons of more than one shape.

The figure below contains the five polygons that you will need. Copy it accurately on tracing paper. (All that you need to do to make the tracing

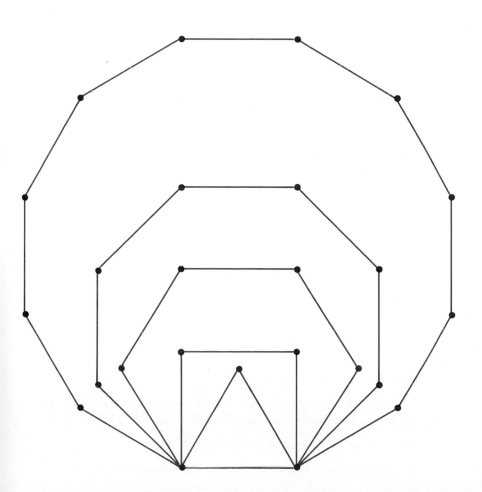

is to mark the corners of each polygon.) Put a sheet of stiff paper underneath the tracing paper and poke a hole with the metal point of a compass at each of the 12 corners of the dodecagon. Then remove the paper, use a straightedge to draw the sides of the polygon on the stiff paper, and carefully cut it out. Repeat this procedure until you have:

> two regular dodecagons
> two regular octagons
> two regular hexagons
> three squares
> four equilateral triangles

On another sheet of paper, draw a point and surround it with a square and two octagons as shown below.

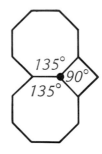

10. From the table you made for exercise 3, what is the measure of an angle of a regular octagon?

11. Write an equation to show why angles of two octagons and one square surround the point exactly.

Two dodecagons together with another polygon will exactly surround a point.

12. What is the polygon? (Use your polygons to find out which one it is.)

13. Write an equation to show why the point is surrounded exactly.

Use your polygons to discover whether a point can be surrounded by the following combinations of shapes. If it is possible to surround a point, *indicate how many of each polygon are used.* If it is not possible, say so.

14. One or more triangles and squares.

15. One or more triangles and hexagons. (Two different combinations are possible in this case. Tell what both of them are.)

16. One or more squares and hexagons.

17. One or more triangles, squares, and hexagons.

18. One or more triangles, squares, and octagons.

19. A dodecagon and polygons of two other shapes. (Two different combinations are possible in this case. Tell what both of them are.)

SET II

The pattern in this film of soap bubbles looks very much like a mosaic of regular polygons. The mosaic can be represented by the symbol 3-12-12 to show that one triangle and two dodecagons surround each point.

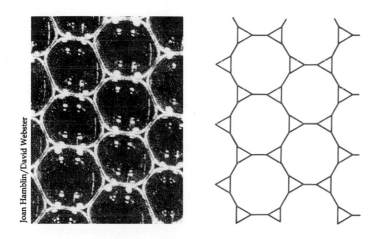

Write symbols to indicate the polygons that surround each point in the mosaics illustrated in the following figures.

1.   2.   3.   4.

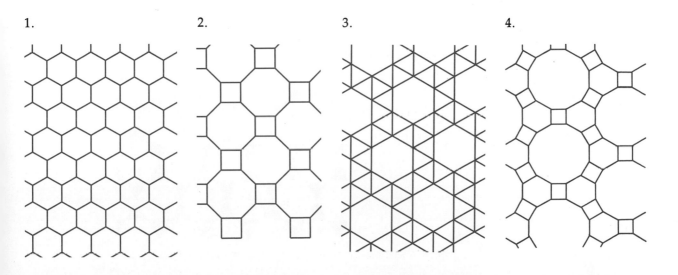

This figure suggests that a 5-6-8 mosaic might be possible.

5. Refer to the table you made for exercise 3 of Set I to find out if a 5-6-8 mosaic might exist.

This figure suggests that a 5-5-10 mosaic might be possible.

6. Refer to the table you made for exercise 3 of Set I to find out if a 5-5-10 mosaic might exist.

The figure at the left shows what happens in trying to make a 5-5-10 mosaic.

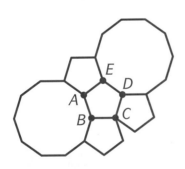

7. Which of the lettered points A, B, C, D, and E are surrounded by two pentagons and a decagon?

8. What prevents the mosaic from continuing?

9. If a particular set of regular polygons exactly surround a point, is it necessarily possible to make a mosaic from them?

This figure shows a pattern of ceramic tiles used for a kitchen countertop.

The figures below show how the tiles might be put into place in this pattern. First, point A is surrounded by two triangles and two hexagons. Next, point B is surrounded in the same way. Next, point C, and so on.

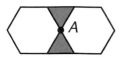 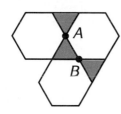 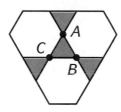

Now, suppose instead that the tiles were placed with the triangles next to each other instead of opposite. The first three steps are shown in the figures below.

 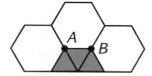 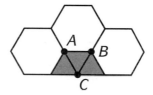

10. Is point B surrounded in the same way as point A?

11. Is it possible to surround point C in the same way as points A and B? Explain why or why not.

12. Can the order in which the polygons appear around each point affect whether or not a mosaic is possible?

13. Which of these symbols seems more appropriate to represent the ceramic tile pattern: 3-3-6-6 or 3-6-3-6?

A triangle, two squares, and a hexagon can be used to surround a point in two different orders, as shown in the figures at the right. Only one of these orders can be used to make a mosaic, however. It is shown in the figure below.

3-4-4-6   *3-4-6-4*

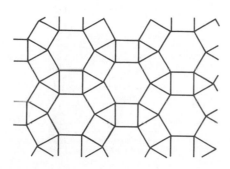

14. Which order is it?

Three different patterns consisting of equilateral triangles and squares are shown below. One of them is not a mathematical mosaic because every corner point is not surrounded by the same regular polygons in the same order.

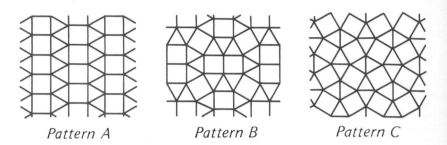

Pattern A *Pattern B* *Pattern C*

15. Which one?

16. Write appropriate symbols for the other two patterns.

Set III

We have defined a mathematical mosaic as an arrangement of *regular polygons* that completely cover a surface. It is possible to fill a surface with a repeating pattern of other shapes as well.

The Dutch artist Maurits Escher created many mosaics using picture shapes.* A remarkable example of these drawings is the one of knights on horseback shown below.

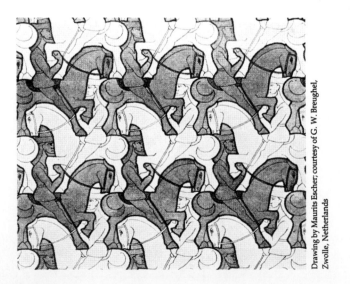

Drawing by Maurits Escher; courtesy of G. W. Breughel, Zwolle, Netherlands

*About 150 of them are included in the book *Visions of Symmetry: Notebooks, Periodic Drawings, and Related Work of M. C. Escher,* by Doris Schattschneider (W. H. Freeman, 1990).

Escher's notebooks reveal that he started with a mosaic of kite-shaped figures. Below is a diagram based upon Escher's early sketch of the horseman in relation to the kite figure.

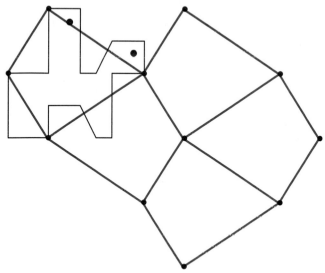

After M. C. Escher's sketch for Horseman; in *Visions of Symmetry,* by Doris Schattschneider, W. H. Freeman and Company, New York, 1990, p. 111.

1. To get an idea of how the mosaic "works," place a sheet of tracing paper over this figure and make an accurate copy of it (all five kites as well as the horseman).

2. What do you notice about the four parts of the horseman figure that stick out of the kite on which it is drawn?

3. How does the space filled by the horseman compare to the space filled by the kite?

Complete your sketch by tracing the horseman from this page onto the other four kites on your paper. (You will have to turn your paper over to draw the horseman in the kites that point to the left.)

Escher wrote of the final version of this mosaic:

At the end [the horseman] stands out clear as crystal, obeying unshakable laws, clenched on all sides by his own mirror image. Note how each contour has a double function: a single line borders the mane as well as the belly of the horse; also the face of the rider as well as the neck of the animal are defined by one and the same line. See him riding through a landscape that is himself. He is alone by himself and yet he completely fills the entire two-dimensional world in which he lives.

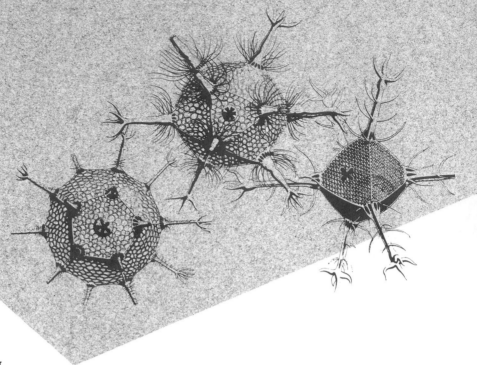

LESSON

4

Regular Polyhedra: The Platonic Solids

The skeletons of sea creatures so small that their structure can be seen only under a microscope are shown above. Called radiolarians, these one-celled animals are found floating near the surface of warm ocean water. The skeletons of radiolarians appear to be built of equilateral triangles and regular pentagons; more than that, each one has the shape of a *regular polyhedron.**

> A **regular polyhedron** is a solid having faces (surfaces) in the shape of a regular polygon. All of its faces, edges, and corners are identical.

_____

*The plural of *polyhedron* is *polyhedra.*

The simplest regular polyhedron is the regular **tetrahedron,** a solid with four equilateral triangles for its faces. Three triangles meet at each corner of a tetrahedron. The figures below show a tetrahedron and the three triangles that meet at one of its corners.

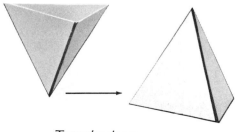

Tetrahedron

If there were four triangles at each corner instead of three, we would have a solid with eight faces, called a regular **octahedron.**

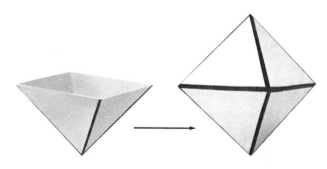

Octahedron

Five triangles at each corner result in a solid called a regular **icosahedron.** An icosahedron has twenty faces in all.

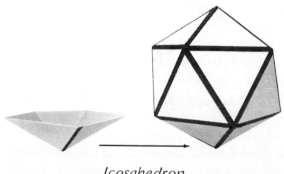

Icosahedron

The most familiar regular polyhedron is the one whose faces are squares: the hexahedron, or **cube.** Three squares meet at each corner of a cube and a cube has six faces altogether.

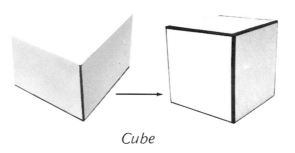

Cube

You may recall that when three regular pentagons are put together around a point, they leave a gap. If the pentagons are folded upward to close the gap, we get a corner of a regular **dodecahedron,** a solid having twelve faces in all.

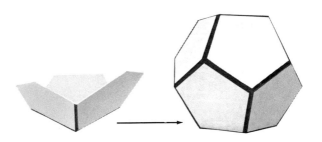

Dodecahedron

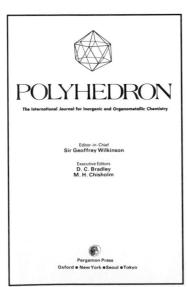

POLYHEDRON

The International Journal for Inorganic and Organometallic Chemistry

Editor-in-Chief
Sir Geoffrey Wilkinson

Executive Editors
D. C. Bradley
M. H. Chisholm

Pergamon Press
Oxford ● New York ● Seoul ● Tokyo

The regular polyhedra were first studied by a group of Greek mathematicians under the leadership of Pythagoras in the sixth century B.C. They are called the **Platonic solids,** after Plato, who gave instructions for making models of them. Plato claimed that the atoms of the four elements of ancient science had the shape of regular polyhedra. Atoms of earth were supposed to have the shape of cubes; atoms of air, octahedrons; atoms of fire, tetrahedrons; and atoms of water, icosahedrons. The universe itself was thought to be in the shape of a dodecahedron.

Although these ideas now seem naive, it is true that many crystals grow in the shape of regular polyhedra as a result of the arrangements of the atoms in them. Modern chemists have found polyhedron models so useful, in fact, that a major journal of chemistry is titled *Polyhedron*!

Exercises

Set I

A different regular polyhedron is illustrated in each of these figures. Tell which polyhedron in each case.

1.

Sodium chlorate crystal

From *Crystals and and Crystal Growing* by Alan Holden and Phylis Singer. Copyright © 1960 by Educational Services Inc. Reprinted by the permission of Doubleday & Compny, Inc., and Heinemann Educational Books Ltd.

2.

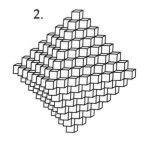

A Japanese interlocking puzzle

3.

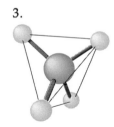

Model of methane molecule

4.

Box with Shells and Starfish
© 1963 M. C. Escher Foundation—Baarn-Holland. All rights reserved.

5.

Loudspeaker

Experiment: Models of the Regular Polyhedra

To do this experiment, you will need seven 4-by-6-inch file cards (preferably unlined), a sheet of tracing paper, a ruler, a compass (or a straight pin), and a pair of scissors.* You will also need the patterns on this and the following page.

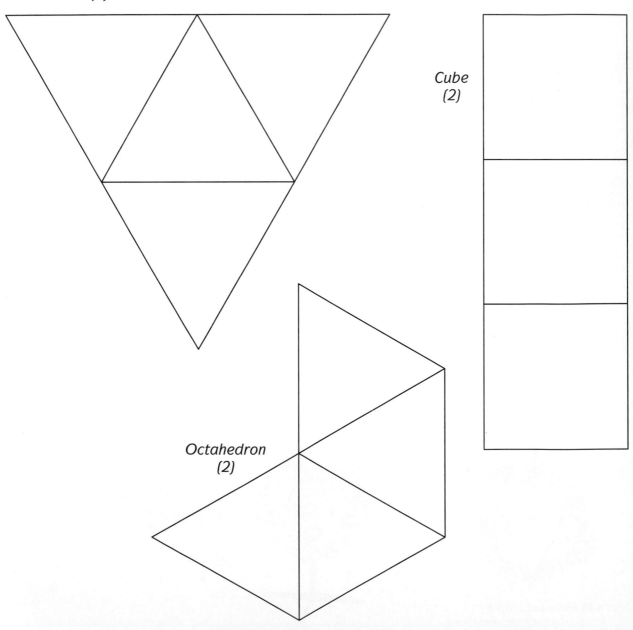

Tetrahedron
(1)

Cube
(2)

Octahedron
(2)

*Optional: A size-19 rubber band and a sheet of stiff card (such as from a file folder).

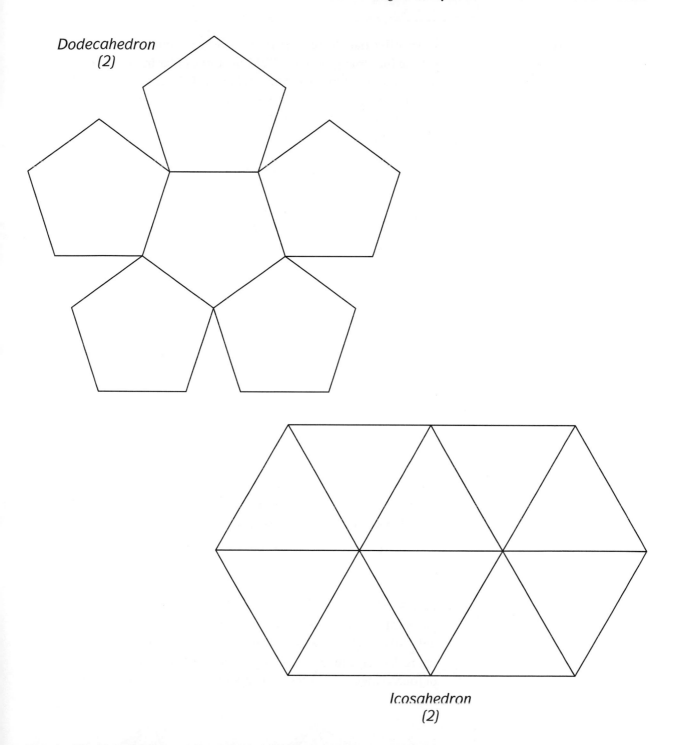

*Dodecahedron
(2)*

*Icosahedron
(2)*

Part 1. The Tetrahedron   Copy the tetrahedron pattern in one corner
of the sheet of tracing paper. (All that you need to do is to mark the
corners of the triangles.) Put a file card underneath the tracing paper and
use the metal point of a compass (or a straight pin) to poke a hole at each
of the six points of your tracing. Remove the tracing paper and use a ruler

to draw the triangles on the card. Cut the figure out of the card (but do not cut the four triangles apart.) Fold the three outer triangles one at a time onto the triangle in the center, making a sharp crease along each fold.

Use some short pieces of tape to tape the edges of the outer triangles together to form the tetrahedron. The result should look like the figure shown here.

Look at your model to answer the following questions.

6. How many faces (triangles) does a tetrahedron have?

7. How many corners does it have?

8. How many edges does it have?

9. How many faces meet at each corner?

10. How many faces meet at each edge?

Part 2. The Cube   Copy the cube pattern on the tracing paper. (Again, all that you need to do is to mark the corners of the squares.) Using the method described in Part 1, make two copies of the pattern on a file card. Cut the two figures out of the card. Fold the two outer squares one at a time onto the square in the middle, making a sharp crease along each fold. Tape the two strips together to form the cube.

Look at your model to answer the following questions.

11. How many faces (squares) does a cube have?

12. How many corners does it have?

13. How many edges does it have?

14. How many faces meet at each corner?

15. How many faces meet at each edge?

Part 3. The Octahedron   Copy the octahedron pattern on tracing paper. Make two copies of the pattern on a file card. Cut the two figures out of the card. Make sharp creases along all three inside edges of each figure. Use tape to make each figure into a three-dimensional "pyramid" of four triangles. Tape the two "pyramids" together to form the octahedron.

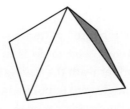
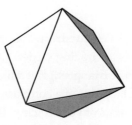

Look at your model to answer the following questions.

16. How many faces does an octahedron have?

17. How many corners?

18. How many edges?

19. What regular polygon other than equilateral triangles do the edges form?

20. How many of these polygons can you find in the octahedron?

21. How many faces meet at each corner?

22. How many faces meet at each edge?

Part 4. The Dodecahedron

Method 1.   Copy the dodecahedron pattern on tracing paper. Put two file cards underneath the tracing paper and poke holes through all the points of the tracing far enough so that they go through both cards. Draw the pattern on each card and cut the figures out of the cards. Fold the five outer pentagons in each figure one at a time onto the pentagon in the center, making a sharp crease along each fold.

Use tape to make each figure into a three-dimensional "bowl." Tape the two "bowls" together to form the dodecahedron.

Method 2.   Follow the directions for Method 1 but with the following changes. Cut two 4-by-6-inch pieces from a sheet of stiff card (such as a file folder) and use them in place of the ordinary file cards. You may have to poke the holes through one card at a time.

After making the sharp creases along the folds, do not use tape to form the "bowls." Instead, hold the "bowls" so that they face each other and press them together so that they flatten out. Take a rubber band (size 19) and weave it alternately above and below the corners as shown in the figure at the right, while holding the two pieces flat. If you carefully let go, a dodecahedron will pop into shape.

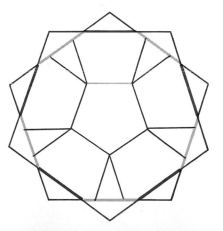

Look at your model to answer the following questions.

23. How many faces does a dodecahedron have?

24. How many corners?

25. How many edges?

26. How many faces meet at each corner?

27. How many faces meet at each edge?

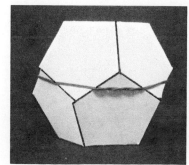

Part 5. The Icosahedron   Copy the icosahedron pattern on tracing paper. Make two copies of the pattern (use two file cards). After cutting

the figures out of the cards, cut one of the figures into two pieces as shown in figure A below and the other into two pieces as shown in figure B.

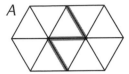 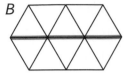

Make sharp creases along all four inside edges of each piece. Use tape to make the two halves of figure A into two separate "bowls." Turn one of the two halves of figure B over and tape it to the other half to make a "ring" of 10 triangles. Tape one "bowl" to each side of the "ring" to form the icosahedron.

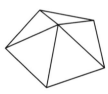 

Look at your model to answer the following questions.

28. How many faces does an icosahedron have?

29. How many corners?

30. How many edges?

31. What regular polygon other than equilateral triangles do the edges form?

32. How many of these polygons can you find in the icosahedron?

33. How many faces meet at each corner of an icosahedron?

34. How many faces meet at each edge?

Set II

The regular polyhedra are comparable to mathematical mosaics in that the regular polygons that meet at each corner point are alike in number, kind, and order. In a cube, for example, three squares meet at each corner.

In a mathematical mosaic consisting of squares, four squares meet at each corner point. The appropriate symbols for these patterns are shown in the adjoining figures.

Write the appropriate symbols for the following mosaics and regular polyhedra.

4-4-4   *4-4-4-4*

1.

2.

3.

4.

5.

6.

For every mosaic or regular polyhedron, two polygons always meet at each edge. The number of polygons that meet at each corner point, however, depends on the mosaic or polyhedron. For example, in a cube, three squares meet at each corner, but in a square mosaic, four squares meet at each corner point.

7. Copy and complete the following table for the number of polygons that meet at each corner point of each of the figures in exercises 1–6.

Exercise	1	2	3	4	5	6
Number of polygons						

8. Judging from your answer to exercise 7, what is the *fewest* number of polygons that can meet at one corner point of a mosaic or regular polyhedron?

Look at the measures of the angles of the regular polygons shown below.

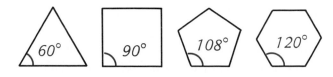

Because each corner of a cube is surrounded by three squares, the sum of the measures of the angles at each corner is

$$90° + 90° + 90° = 270°$$

What is the sum of the measures of the angles that surround each corner of

9. a tetrahedron? (See exercise 1.)

10. an octahedron? (See exercise 2.)

11. an icosahedron? (See exercise 3.)

12. a dodecahedron? (See exercise 5.)

In general, what is true about the sum of the measures of the angles that surround each corner point

13. of a mosaic?

14. of a regular polyhedron?

Lewis Carroll once referred to the regular polyhedra as being "provokingly few in number." The answer to exercise 14 is part of the reason for the fact that there are only five of them.

A regular tetrahedron has four triangular faces. One way to figure out how many corners and edges it has without counting them is to imagine putting it together from the four triangles.

Four separate triangles

Four triangles have
$4 \times 3 = 12$ corners and
$4 \times 3 = 12$ sides

Three triangles meet at each corner; so a tetrahedron has

$$\frac{12}{3} = 4 \text{ corners}$$

Two triangles meet at each edge; so a tetrahedron has

$$\frac{12}{2} = 6 \text{ edges}$$

A tetrahedron

Show how the same method of reasoning can be used to figure out how many corners and edges each of the following regular polyhedra has.

15. A cube.

Six separate squares

A cube

16. An octahedron.

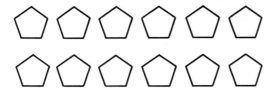

Eight separate triangles

An octahedron

17. A dodecahedron.

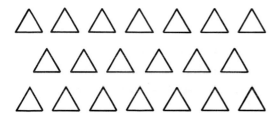

Twelve separate pentagons

A dodecahedron

18. An icosahedron.

Twenty separate triangles

An icosahedron

19. Copy and complete the following table. Use your results for exercises 15–18 to fill in the missing numbers.

Regular polyhedron	Number of faces	Number of corners	Number of edges
Tetrahedron	4	4	6
Cube	▨	▨	▨
Octahedron	▨	▨	▨
Dodecahedron	▨	▨	▨
Isocahedron	▨	▨	▨

20. How is the number of edges of each polyhedron related to its number of faces and corners?

In the model at the left, each of the four corners of a small tetrahedron touches the center of one of the four faces of a larger tetrahedron.

21. What does the model below show?

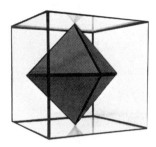

22. What does the model at the left show?

Look again at your table for exercise 19.

23. Could any other models like these be built? If so, what would they look like?

Set III

A clever puzzle called "Back in the Box" consists of a cubical box that is packed with 17 tetrahedral pieces. The idea is to dump the pieces out of the box and then figure out how to fit all of them back into it again.

To figure out how to put just the largest piece, a regular tetrahedron, back into the box is something of a puzzle in itself. Make a model of it by constructing an open box from five squares measuring 5 centimeters on each side and a tetrahedron from four equilateral triangles measuring 7 centimeters on each side. Convenient arrangements for doing this are shown at the left.

When you have figured out how to fit the tetrahedron completely inside the box, make a drawing to illustrate the solution.

OCTOCEDRON ABSCISVS VACVVS.

EXACEDRON ABSCISVS VACVVS.

VCOCEDRON·ABSCISVS VACVVS.

Courtesy of the Moffitt Library, University of California at Berkeley

Semiregular Polyhedra

Leonardo da Vinci, the great Renaissance artist, scientist, and inventor, began one of his books with these words: "Let no one who is not a mathematician read my works." Da Vinci was so obsessed with mathematics for a time that he neglected his painting. He built a set of wooden models of geometric solids and from them created a series of beautiful illustrations for a book titled *The Divine Proportion* by Luca Pacioli, published in 1509. Three of these drawings appear at the top of this page. Each drawing is of a *semiregular polyhedron.*

> A **semiregular polyhedron** is a solid that has faces in the shape of more than one kind of regular polygon, yet every corner is surrounded by the same kinds of polygons in the same (or reverse) order.

Among the semiregular polyhedra are 13 solids called the Archimedean solids.* Named for Archimedes, who wrote a book about them, they have interested mathematicians and designers ever since.

---

*Recall that there are only five *regular polyhedra,* or *Platonic solids.*

Courtesy of AMF Voit, Inc.

Compare the polyhedron illustrated in the third of da Vinci's drawings with the pattern on the soccerball shown in this photograph. Both consist of regular pentagons and regular hexagons with one pentagon and two hexagons meeting at each corner. The symbol for this pattern is 5-6-6.

This polyhedron, called a *truncated icosahedron,** is almost as round as the ball. It is round enough, in fact, that it could be put inside a sphere so that each corner would touch the surface.

The other solids pictured in da Vinci's drawings at the beginning of this lesson are a *truncated octahedron,* in which each corner is surrounded by a square and two hexagons, and a *cuboctahedron,* in which each corner is alternately surrounded by two equilateral triangles and two squares. The symbols for these solids are 4-6-6 and 3-4-3-4, respectively.

The cuboctahedron is an especially popular shape with designers. It has been used, for example, in the design of lamps, paperweights, and soap. The semiregular polyhedra, like so many other creations of mathematics, have been found to be useful in a wide variety of applications.

EXERCISES

SET I

Photographs of models of the Archimedean solids are shown below. Look at the regular polygons that surround each corner and write the appropriate symbol for each solid.

1. Truncated tetrahedron.

2. Truncated cube.

3. Truncated octahedron.

4. Truncated dodecahedron.

---

*This name, like those of the other Archimedean solids, is rather complicated. Do not be concerned about trying to remember the names of any of these solids.

5. Truncated icosahedron.

6. Cuboctahedron.

7. Small rhombi-cuboctahedron.

8. Great rhombi-cuboctahedron.

9. Icosidodeca-hedron.

10. Small rhombicosi-dodecahedron.

11. Great rhombicosi-dodecahedron.

12. Snub cube.

13. Snub dodecahedron.

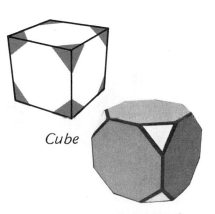

Cube

Truncated cube

The *truncated cube* is a cube whose corners have been cut off.

14. How many square faces meet at each corner of a cube?

15. What regular polygon is formed in place of each corner of the cube when the corners are cut off?

16. What regular polygon does each square face of the cube become when the corners are cut off?

The corners of the other regular polyhedra also can be cut off to form Archimedean solids. The *truncated dodecahedron* is shown here.

Dodecahedron

Truncated dodecahedron

17. How many pentagonal faces meet at each corner of a dodecahedron?

18. What regular polygon is formed in place of each corner of the dodecahedron when the corners are cut off?

19. What regular polygon does each pentagonal face of the dodecahedron become when the corners are cut off?

The other three truncated polyhedra are shown here.

Tetrahedron

Truncated tetrahedron

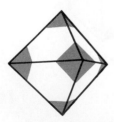

Octahedron

Truncated octahedron

Icosahedron

Truncated icosahedron

20. What do the tetrahedron, octahedron, and icosahedron have in common?

21. What regular polygon does each face of these polyhedra become when the corners are cut off?

How many faces meet at each corner of

22. the tetrahedron?   24. the icosahedron?

23. the octahedron?

When the corners are cut off, what regular polygons are formed in place of each corner of

25. the tetrahedron?   27. the icosahedron?

26. the octahedron?

Another Archimedean solid, the *cuboctahedron*, can be obtained from either of two regular polyhedra as shown below.

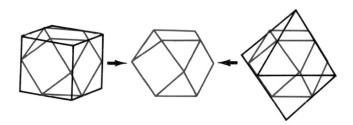

28. What are the two regular polyhedra?

29. How many square faces does a cuboctahedron have?

30. How many triangular faces does it have?

The *icosidodecahedron* also can be obtained from either of two regular polyhedra.

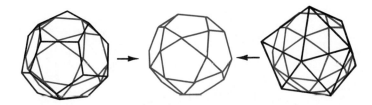

31. What are the two regular polyhedra?

32. How many pentagonal faces does an icosidodecahedron have?

33. How many triangular faces does it have?

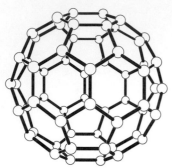

From *Fractal Music, Hypercards, and More,* by Martin Gardener, W. H. Freeman and Company, New York, 1992

Set II

In 1989 chemists created a new carbon molecule called the "bucky ball."[*] Its structure is shown in this figure. The white spheres at its vertices represent carbon atoms and the black sticks along its edges represent bonds between the atoms.

    1. Which Archimedean solid does a bucky ball look like?

To find out how many carbon atoms and bonds a bucky ball contains without trying to count them, we can reason in the following way.

Altogether, the solid has 12 faces that are pentagons and 20 faces that are hexagons.

12 separate pentagons contain $12 \times 5 = 60$ corners and sides.

20 separate hexagons contain $20 \times 6 = 120$ corners and sides.

Altogether, this makes 180 separate corners and sides.

3 of these corners meet at each corner of the solid, so it has $\dfrac{180}{3} = 60$ corners.

2 of these sides meet at each edge of the solid, so it has $\dfrac{180}{2} = 90$ edges.

Therefore, a bucky ball contains 60 carbon atoms with 90 bonds between them.

This drawing is from a book published in 1525 by the great German artist, Albrecht Dürer. It is a pattern for one of the Archimedean solids.

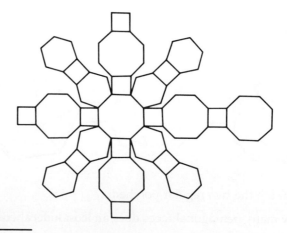

[*]Its full name is *buckminsterfullerene,* named for the engineer and architect, Buckminster Fuller, who invented the geodesic dome.

2. Compare it with the pictures of the solids shown on pages 290–291. Which Archimedean solid is it?

3. What regular polygons are its faces?

4. How many faces of each type does it have?

5. How many faces does it have altogether?

6. Use the method shown on the preceding page to figure out how many corners and edges it has. Write your ideas in words, including the appropriate equations.

Use the following information and the same method of reasoning to find the number of corners and number of edges in each of these solids. Write your ideas in words, including the appropriate equations.

7. Truncated cube. Symbol: 3-8-8. Faces: 8 triangles and 6 octagons.

8. Small rhombicosidodecahedron. Symbol: 3-4-5-4. Faces: 20 triangles, 30 squares, and 12 pentagons.

9. Snub cube. Symbol: 3-3-3-3-4. Faces: 32 triangles and 6 squares.

Here is a table for the other Archimedean solids.

Polyhedron	Number of faces	Number of corners	Number of edges
Truncated tetrahedron	8	12	18
Truncated octahedron	14	24	36
Truncated dodecahedron	32	60	90
Cuboctahedron	14	12	24
Small rhombicuboctahedron	26	24	48
Icosidodecahedron	32	30	60
Great rhombicosidodecahedron	62	120	180
Snub dodecahedron	92	60	150

10. How is the number of edges of each polyhedron in this table related to its numbers of faces and corners?

11. Does the same relation hold for the solid whose pattern by Dürer is shown on the preceding page? (Look at your answers to exercises 5 and 6.)

12. Does it hold for each of the solids in exercises 7 through 9?

Set III

It is easy for a child playing with blocks to discover that the cube can be used to fill space. Is it possible to fill space with any other regular or semiregular polyhedron?

The answer is "yes": truncated octahedra can be fitted together to fill space. The photograph below shows a set of them.

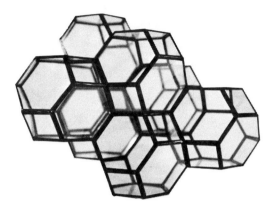

On tracing paper, make a copy of the pattern below. (All that you need to do is to mark the corners of the hexagons.) Use your tracing to make ten copies of the pattern on 4-by-6-inch file cards. Cut each pattern

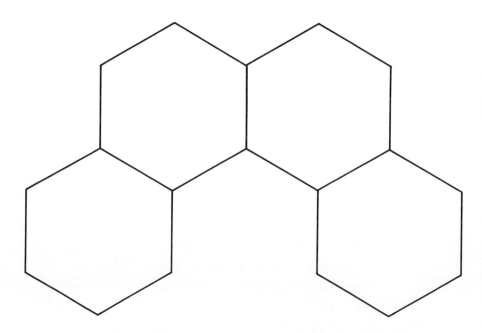

out and fold along each of the lines. Use tape to make two of the patterns into a truncated octahedron as shown in this photograph.

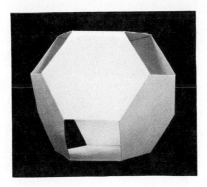

For simplicity, the pattern is designed so that each of the square faces of the solid is open. Make the eight remaining patterns into four more truncated octahedra.

After you have made the five polyhedra, see if you can fit them together. Use tape to hold them in place. (Even though you can refer to the photograph on the preceding page, you may find that fitting them together is not as easy to do as it might seem.)

Alinari/Art Resource, New York

LESSON

6

Pyramids and Prisms

The largest of all man-made geometric solids was built in about 2600 B.C. It is the Great Pyramid in Egypt, the only one of the "seven wonders of the world" still in existence. This pyramid, one of about 80 such structures built by the ancient Egyptians, was put together from more than 2 million stone blocks, weighing between 2 and 150 tons each.

The Great Pyramid is one member of an unlimited set of geometric solids called *regular pyramids.*

| *Triangular pyramid* | *Square pyramid* | *Pentagonal pyramid* | *Hexagonal pyramid* |

A **regular pyramid** is a solid that has a *regular polygon* for its base and *isosceles triangles* that are identical in size and shape for the rest of its faces.

An *isosceles triangle* is a triangle that has at least two sides of the same length. Examples of isosceles triangles are shown below.

Notice from these examples that the equilateral triangle, shown in the middle, is isosceles but that many isosceles triangles are not equilateral.

A pyramid is named according to the shape of its base. The pyramids of Egypt, for example, are square pyramids because their bases are squares.

There is another unlimited set of geometric solids called the *regular prisms.*

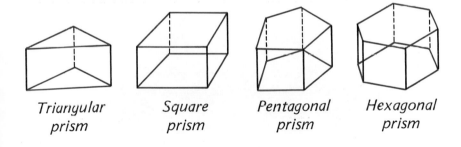

| Triangular prism | Square prism | Pentagonal prism | Hexagonal prism |

A **regular prism** is a solid that has a pair of *regular polygons* for its bases and *rectangles* that are identical in size and shape for the rest of its faces.

Examples of *rectangles* are shown below.

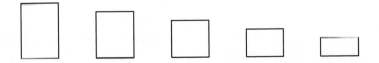

Notice from these examples that the square, shown in the middle, is a rectangle but that many rectangles are not squares.

The two bases of a prism are always parallel to each other, and a prism, like a pyramid, is named according to the shape of its bases.

Prisms are frequently found in nature. An awesome display of them is the Devil's Postpile in the Sierra Nevada range of California. The Devil's Postpile is a group of tall columns of rock, some of which are 60 feet high. Many of these columns have the shape of regular prisms, most of them hexagonal and others pentagonal. So even geologists find mathematics in their study of the earth.

Courtesy of the National Park Service

Devil's Postpile

EXERCISES

SET I

The word *prism* is often used to mean a block of glass by which light can be broken up into a spectrum of colors.

1. What shape do the bases of the glass prism shown in the photograph at the left have?

2. What shape do the rest of the faces of this prism have?

3. What is the name of a prism with this shape?

One of the oldest games known was played in the biblical city of Ur of the Chaldees in 2500 B.C. Dice like those shown in the photograph below were used in the game.

4. How many faces does each die have?

5. What type of regular polyhedron do the dice appear to be?

6. What type of pyramid do they appear to be?

Snow crystals grow around atmospheric dusts having particles of several different shapes, two of which are shown here.

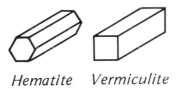

Hematite   Vermiculite

What shapes do the faces of particles

7. of hematite have?

8. of vermiculite have?

What type of geometric solid do particles

9. of hematite appear to be?

10. of vermiculite appear to be?

EXPERIMENT: PYRAMIDS WITH REGULAR FACES

In this experiment, we will consider some pyramids whose bases are regular polygons and whose other faces are also *regular*.

Copy the patterns on the next page on tracing paper. Use the tracing paper to copy pattern A on a 4-by-6-inch file card. Cut the pattern from the card and then fold and tape it to form a pyramid.

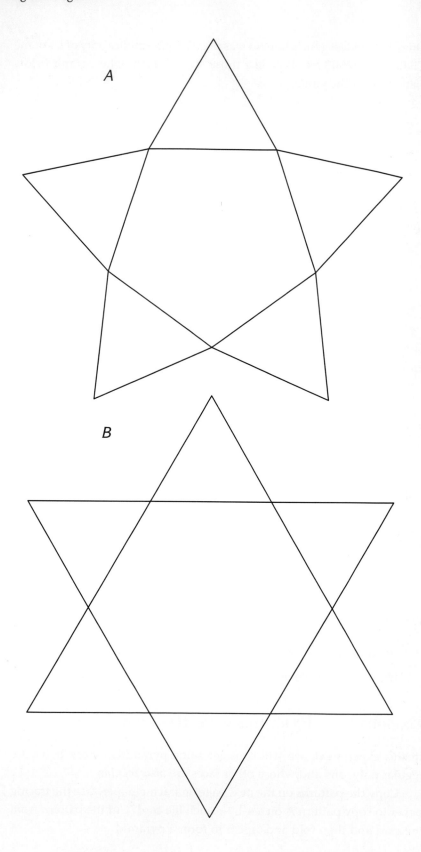

11. How many faces, and of what type, meet at the top of the pyramid?

12. How many faces, and of what type, meet at each corner of its base?

13. Including its base, how many faces does the pyramid have altogether?

14. How many corners?

15. How many edges?

Copy pattern B on another file card. Cut the pattern from the card and fold it.

16. What happens when you try to make a pyramid from it?

Use what you have discovered to figure out an answer to the following question.

17. Given that the base of a regular pyramid is a regular polygon, how many different pyramids are possible if the other faces are equilateral triangles?

If the pattern shown at the right were cut out, it could be folded to form a pentagonal prism. What solids could be formed from the following patterns?

18.   19.   20.   21.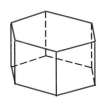

These figures illustrate a series of prisms all of whose faces are regular polygons.

The first prism can be represented with the symbol 3-4-4.

22. What does this symbol indicate?

23. Write symbols for the other three prisms.

24. Identify the prism that is a regular polyhedron by giving its name.

25. Do you think that there are an unlimited number of prisms whose faces are all regular polygons?

The figures below show a series of regular polygons in which each successive polygon has twice as many sides as the one preceding it.

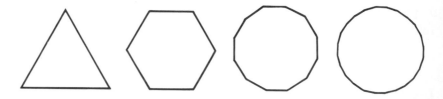

26. What geometric figure does a regular polygon become more and more like as the number of its sides becomes larger and larger?

The figures below show a series of regular pyramids having the polygons above as their bases.

27. What geometric solid does a regular pyramid become more and more like as the number of sides in its base becomes larger and larger? (*Hint:* The answer contains four letters and starts with the letter C.)

The figures below show a series of regular prisms having for their bases the same polygons as those of the pyramids in exercise 27.

28. What geometric solid does a regular prism become more and more like as the number of sides of its bases becomes larger and larger? (*Hint:* The answer contains eight letters and starts with the letter C.)

Set II

Photograph courtesy of National Museum of the American Indian, Smithsonian Institution

The tepees of the Sioux Indians, built by stretching skins around 18 poles, are somewhat like regular pyramids. An overhead view of a regular pyramid having a base of 18 sides is shown here.

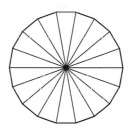

1. Including its base, how many faces does the pyramid have al together?

2. How many corners?

3. How many edges?

In 1752, the Swiss mathematician Leonhard Euler* stated a formula relating the numbers of faces, corners, and edges of certain geometric solids. The formula is

$$F + C = E + 2$$

in which F represents the number of faces, C represents the number of corners, and E represents the number of edges. You have already found that this formula applies to both the regular polyhedra and the Archimedean solids.

---

*Euler is pronounced "*oi*ler."

4. Does Euler's formula apply to the Sioux tepees? Show why or why not.

| *Triangular pyramid* | *Square pyramid* | *Pentagonal pyramid* | *Hexagonal pyramid* |

5. Refer to the figures above to copy and complete the following table.

Pyramid	Number of sides in base	Number of faces	Number of corners	Number of edges
Triangular	▦	▦	▦	▦
Square	▦	▦	▦	▦
Pentagonal	▦	▦	▦	▦
Hexagonal	▦	▦	▦	▦

6. Does Euler's formula fit every pyramid in your table?

The number of sides in the base of the "tepee" pyramid is 18, and the pyramid has $18 + 1$, or 19, faces altogether. In general, if a pyramid has a base with n sides, its number of faces is $n + 1$.

7. What is its number of corners in terms of n?

8. What is its number of edges in terms of n?

Substituting these expressions into Euler's formula, we get

$$F + C = E + 2$$
$$n + 1 + n + 1 = 2n + 2, \text{ or}$$
$$2n + 2 = 2n + 2$$

This result shows that the formula is true regardless of what number n represents. So Euler's formula is true for all pyramids.

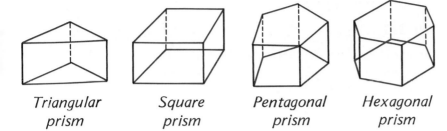

Triangular prism   *Square prism*   *Pentagonal prism*   *Hexagonal prism*

9. Refer to the figures above to copy and complete the following table.

Prism	Number of sides in base	Number of faces	Number of corners	Number of edges
Triangular	▨	▨	▨	▨
Square	▨	▨	▨	▨
Pentagonal	▨	▨	▨	▨
Hexagonal	▨	▨	▨	▨

10. Does Euler's formula fit every prism in your table?

Suppose that a prism has a base with 10 sides.

11. How many faces would it have?

12. How many corners would it have?

13. How many edges would it have?

Suppose that a prism has a base with n sides.

14. What is its number of faces in terms of n?

15. What is its number of corners in terms of n?

16. What is its number of edges in terms of n?

17. Substitute these expressions into Euler's formula to see if it is true for all prisms.

This building in Florence, Italy, was built in the eleventh century. It is in the shape of a regular octagonal prism with a pyramidal roof.

18. How many faces does this solid have? (Don't forget the floor.)

19. How many corners?

20. How many edges?

21. Does Euler's formula apply to it? Show why or why not.

Set III

The star-shaped polyhedron shown here was discovered by the German astronomer and mathematician Johann Kepler in about 1620. It can be constructed by adding pyramids to the faces of a regular polyhedron. The picture below shows five of the pyramids being added.

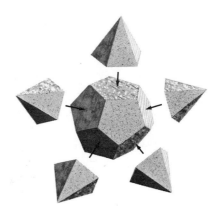

1. What kind of pyramids are they?
2. To what regular polyhedron are they added?
3. How many points does the star have altogether?

Kepler discovered another star-shaped polyhedron, shown here, at the same time. It also can be formed by adding pyramids to the faces of a regular polyhedron.

4. What kind of pyramids are they?
5. To what regular polyhedron are they added?
6. How many points does the star have altogether?

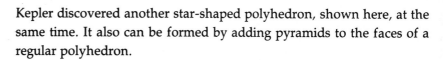

Summary and Review

*Icosapirale, a sculpture by Charles Perry at the Alcoa Golden Gate Plaza
in San Francisco*

In this chapter we have studied symmetry, regular polygons, and their
relation on a flat surface and in space.

Symmetry *(Lesson 1).* A figure has *line symmetry* if there is a line
along which a mirror can be placed to reflect either half of the figure
so that it reproduces the other half.

A figure has *rotational symmetry* if it can be rotated through an
angle of less than 360° so that it coincides with its original position.

Regular polygons *(Lesson 2).* A regular polygon is a figure of which
all sides are the same length and all angles are equal.

Regular polygons are named for their numbers of sides. The
names most frequently used are: *equilateral triangle* (three sides),
square (four sides), *pentagon* (five sides), *hexagon* (six sides), *heptagon*
(seven sides), *octagon* (eight sides), *nonagon* (nine sides), *decagon* (ten
sides), and *dodecagon* (twelve sides).

Mathematical mosaics *(Lesson 3).* A *mathematical mosaic* is a set of
regular polygons arranged so that the polygons share their sides and
the polygons at each corner point are alike in number, kind, and order.

Regular polyhedra: the Platonic solids *(Lesson 4).* A *regular polyhedron* is a solid having faces in the shape of a regular polygon. All of its faces, edges, and corners are identical.

There are only five regular polyhedra: the *tetrahedron* (four triangular faces), *cube* (six square faces), *octahedron* (eight triangular faces), *dodecahedron* (twelve pentagonal faces), and *icosahedron* (twenty triangular faces).

Semiregular polyhedra *(Lesson 5).* A *semiregular polyhedron* is a solid that has faces in the shape of more than one kind of regular polygon, yet every corner is surrounded by the same kinds of polygons in the same (or reverse) order.

Among the semiregular polyhedra are the 13 *Archimedean solids.*

Pyramids and prisms *(Lesson 6).* A *regular pyramid* is a solid that has a regular polygon for its base and isosceles triangles that are identical in size and shape for the rest of its faces.

A *regular prism* is a solid that has a pair of regular polygons for its bases and rectangles that are identical in size and shape for the rest of its faces.

Pyramids and prisms are named according to the shapes of their bases.

EXERCISES

SET I

The four symbols used on playing cards are symmetrical.

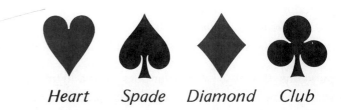

Heart   Spade   Diamond   Club

1. Which symbols have exactly one line of symmetry?

2. Which symbol has two lines of symmetry?

3. Which symbol has rotational symmetry?

If a knot is tied in a strip of paper and then pressed flat, a regular polygon is formed.

4. Which one is it?

These photographs below are of some stained glass windows created by Sheryl Cotleur of San Rafael, California. Write the symbol for the mosaic illustrated by each.

7.

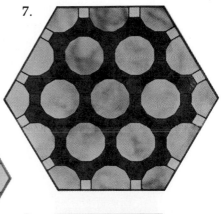

5.

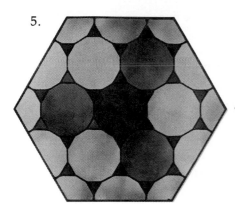

6.

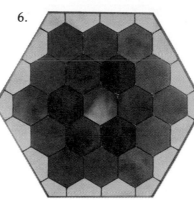

8.

Designed by Sheryl Cotleur. Stained Glass Tessela-tions Posters, © 1977, Creative Publications, P.O. Box 10328, Palo Alto, California 94303.

9. What is the sum of the measures of the angles that surround each corner point of a mosaic?

These diagrams are parts of floor plans created by the Swiss architect Le Corbusier.

10. What type of symmetry does the first plan illustrate?

11. What is a simple way to prove it?

12. What type of symmetry does the second plan illustrate?

13. What is a simple way to prove it?

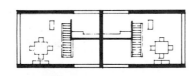

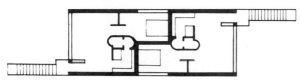

From *Handbook of Regular Patterns* by Peter S. Stevens, M.I.T. Press, 1980

The photograph at the right is of a child's ball. Although it is spherical in shape, it looks very much like a regular polyhedron.

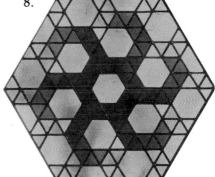

14. Which polyhedron does it look like?

15. How many faces does this polyhedron have?

This photograph shows a bunch of pencils held together with a rubber band.

16. Toward what polygonal shape is the rubber band stretched?

17. In what approximate shape is the bunch of pencils pulled?

The astronomer Johann Kepler once tried to explain the spacing of the planets in the solar system by means of the regular polyhedra. He imagined a series of six spheres, one for each planet known at the time, with the sun at their center. These spheres were separated by the five regular polyhedra, as the drawing below by Kepler shows.

18. Starting with the largest one, name the first three polyhedra as they appear in the figure. (The other two are a bit too small to see.)

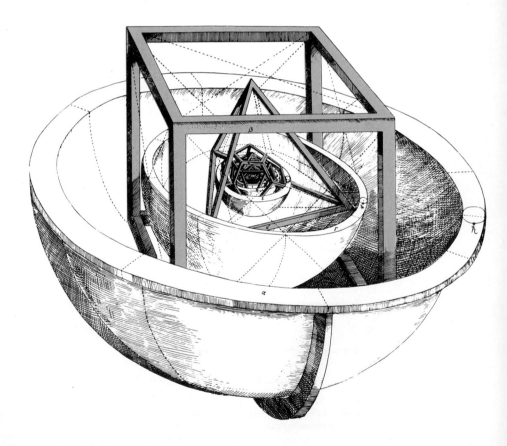

SET II

A fragment of an inscription on a Roman column of about 260 B.C.

ABCDEFGHIKLMNOPQRSTVXYZ

The Roman alphabet, which we use today, has existed since about 700 B.C. For a long time, it consisted of the 23 capital letters shown above. The letters J, U, and W were not added until the Middle Ages.

The letter A is symmetric with respect to a vertical line.

1. What other letters of the alphabet have a vertical line of symmetry?

The letter B is symmetric with respect to a horizontal line.

2. What other letters have a horizontal line of symmetry?

The letter Z has rotational symmetry.

3. What other letters have rotational symmetry?

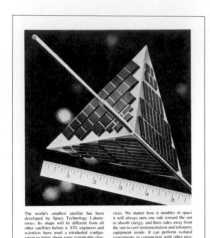

The world's smallest satellite has been developed by Space Technology Laboratories. Its shape will be different from all other satellites before it. STL engineers and scientists have used a tetrahedral configuration to bring about some remarkable characteristics in a space vehicle. There will be no need for batteries nor regulators in flight. The satellite will have no hot side, no cold side. It will require no attitude control devices. No matter how it tumbles in space it will always turn one side toward the sun to absorb energy, and three sides away from the sun to cool instrumentation and telemetry equipment inside. It can perform isolated experiments in conjunction with other projects. Or it can be put into orbit by a small rocket to make studies of its own, up to five or more separate experiments on each mission it makes.

SPACE TECHNOLOGY LABORATORIES, INC.
a subsidiary of Thompson Ramo Wooldridge Inc.

Courtesy of TRW Inc.

Small space satellites have been made in the shape shown in the advertisement at the left.

4. Give two different names for its shape.

5. How many faces, corners, and edges does it have?

6. Do these numbers fit Euler's formula: $F + C = E + 2$?

There are three simple ways of planting an orchard so that the distances between a tree and each of its closest neighbors are the same. The three ways are based on the regular polygon mosaics. One of them is shown in this figure.

7. Make sketches showing the other two.

The principal ore of lead is *galena*, a mineral that has five basic types of crystals. The crystals have the shape of polyhedra and are shown in the figures below. Tell the name and symbol of each shape. (Look at the figures on pages 290–291.)

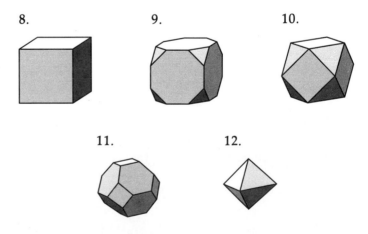

8.   9.   10.

11.   12.

This picture is of an old German top.

13. What type of geometric solid does the part of the top shown in color appear to be?

14. What shape are its bases?

15. What shape do the rest of its faces have?

The picture below, created by means of a computer program, gives a strong impression of being three-dimensional.

From *Computer Graphics, Computer Art*, by Herbert W. Franke, Verlag F. Bruckmann KG

16. On the assumption that we can see all its faces except for its base, what type of geometric solid does it appear to be?

17. Do the same number and type of faces meet at each of its corners?

18. Explain.

The nut used to turn on a fire hydrant is frequently made in the shape of a regular pentagon. It is much easier to put a wrench on this shape than on a shape that is not regular.

19. Why?

20. Which shape would be harder to turn using a normal wrench: a regular pentagon or a regular hexagon?

21. Why?

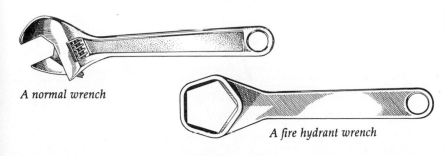

A normal wrench

A fire hydrant wrench

The geometric solid shown at the left consists of two intersecting regular polyhedra, one of which is white in this picture and the other brown.

22. What are the two regular polyhedra?

23. What kinds of pyramids appear in the solid?

In addition to the 13 Archimedean solids, which have been known for more than 2,000 years, a fourteenth solid was discovered in 1930 by J. C. P. Miller. Called a *pseudorhombicuboctahedron,* it is shown here beside a rhombicuboctahedron.

The pseudorhombicuboctahedron (the solid on the right) may have been overlooked for such a long time because of its symbol.

24. Explain why.

25. How do you think the numbers of faces, edges, and corners of the two solids compare?

Set III

Most drums make sounds with no definite pitch. The kettledrum, however, can be tuned to play notes of different pitches by adjusting the tension on the skin that is its head.

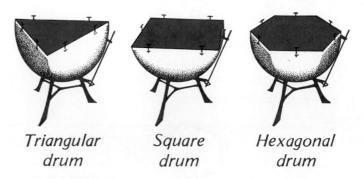

*Triangular   Square   Hexagonal
drum   drum   drum*

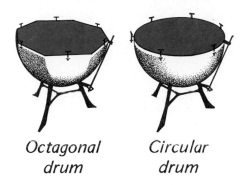

*Octagonal
drum*   *Circular
drum*

Imagine that we have a set of four special kettledrums whose heads have the shape of regular polygons and an ordinary drum whose head is circular in shape. If all the heads of these drums have the same area and the same tension, then their pitches (or frequencies) will vary according to their shape. The table below shows how the pitches of the drums might compare under these conditions.

Shape of head	Pitch (in vibrations per second)
Equilateral triangle	146
Square	136
Regular hexagon	132
Regular octagon	131
Circle	130

1. What happens to the pitch of a "regular polygon" drum as the number of sides of its head increases?

Do you think that "regular polygon" drums could be built having the same area and tension as do the drums listed in the table but having approximately the following pitches? Explain your answers.

2. 140 vibrations per second.

3. 134 vibrations per second.

4. 125 vibrations per second.

CHAPTER

5

Further Exploration

LESSON 1

1. A bill was once submitted to Congress proposing that Federal Reserve notes be printed in a manner that enables a person who is blind to determine the denomination of each note. The idea was to trim one or more corners of each note so that the values of the notes could be told from their shapes.

There are seven denominations of bills currently in circulation: $1, $2, $5, $10, $20, $50, and $100. One way in which corners of a bill might be cut is shown here.

  a. Make drawings like this to show all the other ways in which the corners of bills can be cut (or not cut) so that blind people can tell the bills apart.
  b. How many different ways are there altogether?
  c. Do you have any ideas about how each particular denomination of bill should be cut? If so, explain your reasoning.

2. In 1962, Murray Gell-Mann, a physicist at Cal Tech, used symmetry to predict the existence of an atomic particle which he named "omega minus." Of this prediction, A. Zee wrote in his book *Fearful Symmetry:*

> Even more impressive, by using symmetry considerations Gell-Mann was able to predict all the relevant properties of "omega minus.". . . A team of experimenters went out and looked. Sure enough, they found the "omega minus," with exactly the same properties as Gell-Mann predicted. For this and other fundamental contributions to physics, Gell-Mann was awarded the Nobel prize.*

Gell-Mann's prediction was based on symmetry and the existence of nine known particles. The nine particles can be represented by points on a graph in which the x-axis represents a property called "isospin" and the y-axis represents a property called "strangeness." Their coordinates are listed below:

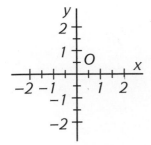

(0.5, 1)	(−0.5, −1)	(1, 0)
(−1.5, 1)	(0, 0)	(−1, 0)
(1.5, 1)	(−0.5, 1)	(0.5, −1)

Fearful Symmetry, The Search for Beauty in Modern Physics, by A. Zee (Macmillan, 1986).

a. Draw a pair of coordinate axes as shown here and plot the nine points.

A tenth point seems to be missing from the graph.

b. What do you think its coordinates should be?

The coordinates of this point enabled Gell-Mann to predict the properties of the particle "omega minus," which helped him to win the Nobel Prize.

LESSON 2

1.   The Swiss artist and designer Max Bill created a series of 16 lithographs between 1934 and 1938 titled *Fifteen Variations on a Single Theme.* Three of the lithographs, *The Theme, Variation 2,* and *Variation 5,* are shown below.

Describe, as specifically as you can, what you see in

a. *The Theme.*

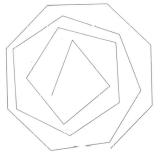

Courtesy of Max Bill

b. *Variation 2.*

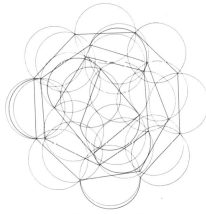

Courtesy of Max Bill

c. *Variation 5.*

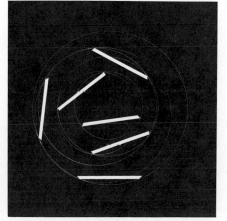

Courtesy of Max Bill

2. Sheriff's badges have traditionally been in the shape of a five-pointed star. In the nineteenth century, however, six-pointed stars were common, probably because they were simpler to make.

The six-pointed star is based on the regular hexagon, which can be easily drawn with a ruler and compass.

From the collections of the Wyoming State Museums System — Wyoming State Museum

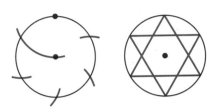

The five-pointed star, based on the regular pentagon, can also be drawn with just a ruler and compass. Several methods have been figured out for doing this, but they are all more complicated than the one for the hexagon. The one we will consider is taken from Ptolemy's great work on astronomy, the *Almagest.* Ptolemy lived and worked in Alexandria, Egypt, in about A.D. 150.

First, draw a circle with a radius of about 2 inches. Draw a diameter in it as shown in the first figure below. Then use your ruler and compass to carry out the steps illustrated. The arrows show where to put the compass on the paper. Each arrow begins at the point at which the metal point of the compass should be placed and points to the point at which the pencil point should start. To keep track of the points as you locate them, letter them as indicated in the figures.

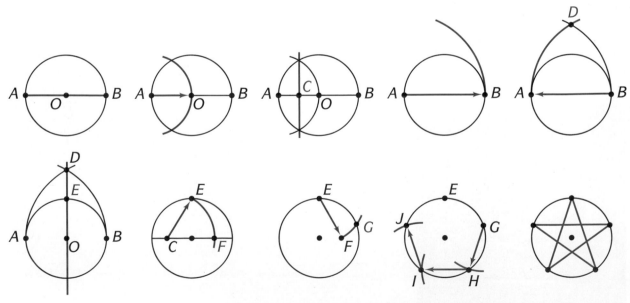

Ptolemy not only gave directions for this construction in the *Almagest,* but also explained why the method works.

Lesson 3

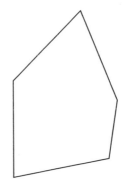

1. In the July 1975 issue of *Scientific American,* Martin Gardner's "Mathematical Games" column featured the subject of mosaics.* Although regular pentagons cannot be used to make a mosaic because they do not surround a point in a plane exactly, the article reported that pentagons of eight other types *will* work. In the December 1975 issue of the magazine, a newly discovered ninth type was reported.

This inspired one of the readers of the column, Marjorie Rice, a mother of five in San Diego, to look for more. She succeeded in discovering four more types.

One of them is based on the pentagon shown at the right. Copy the figure on tracing paper and then use the tracing paper to make at least 12 copies of the figure on stiff paper. Cut the figures out and see if you can figure out a way to place them to form part of a mosaic. Unlike the mosaics with regular polygons, you do not need to surround each corner point in the same way. Be sure, however, to place the figures so that equal sides touch each other. (You will need to turn some of the figures over.)

A possible way to begin is shown here.

When you have finished, either tape or glue the figures to your paper.

2.   The pattern shown in the figure at the right can be extended to form a mathematical mosaic having the symbol 3-4-6-4.

If we take the four numbers in this symbol, make them the denominators of unit fractions,† and add the fractions, we get

$$\frac{1}{3} + \frac{1}{4} + \frac{1}{6} + \frac{1}{4} =$$

$$\frac{4}{12} + \frac{3}{12} + \frac{2}{12} + \frac{3}{12} = \frac{12}{12} = 1$$

The result turns out to be a very simple number: 1.

---

*Reprinted as Chapter 13 of *Time Travel and Other Mathematical Bewilderments,* by Martin Gardner (W. H. Freeman, 1988).

†A unit fraction is a fraction whose numerator is 1.

There are 10 other mathematical mosaics:

3-3-3-3-3-3	3-12-12
3-3-3-3-6	4-4-4-4
3-3-3-4-4	4-6-12
3-3-4-3-4	4-8-8
3-6-3-6	6-6-6

a. Take the numbers in the symbol of each mosaic, make them the denominators of unit fractions as in the example on the preceding page, and add the fractions.

b. What do you notice about the results?

LESSON 4

1.   Jean J. Pedersen, a mathematics teacher at the University of Santa Clara, has discovered ways to weave the regular polyhedra from strips of paper. A nice model of a cube can be made from three strips of paper identical with the one shown here.

Make the strips from heavyweight construction paper, each of a different color. Fold the strips along the brown lines.

The strips can now be woven together to make a rigid cube in which each pair of opposite faces has the same color. No glue or tape is needed. See if you can figure out how to do it.

2.   Alexander Graham Bell, the inventor of the telephone, was fascinated by the regular tetrahedron. He invented and flew a number of large kites made from networks of tetrahedrons several years before the Wright brothers built their first airplane. The architecture of a museum of Bell's inventions in Nova Scotia is based on the tetrahedron shape.

Mabel and Alexander Graham Bell sharing a kiss inside the frame of his tetrahedral kite, October 16, 1903.

Prints and Photographs Division, Library of Congress

The following tetrahedron puzzle is available in some stores, but you can easily make it yourself. Make two copies of the pattern shown below on stiff paper and cut them out. Fold each one along the brown lines and tape the edges of each together to form a pair of identical solids.

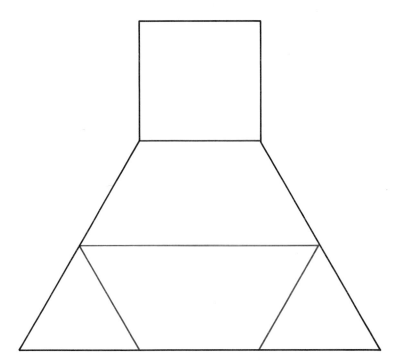

The puzzle is to put the two solids together to form a tetrahedron. Can you do it? If so, make a drawing to show the solution.

Lesson 5

1. Some of the semiregular mosaics and Archimedean polyhedra are related. For example, compare the mosaic 4-8-8 with the truncated cube, 3-8-8.

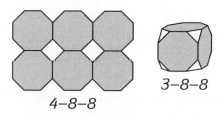

What other pairs of mosaics and polyhedra can you discover? Look at the figures on the next page. Refer to the mosaics and polyhedra by their symbols.

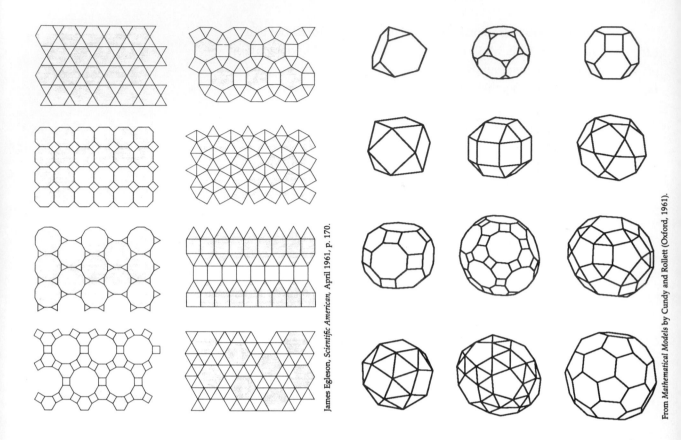

James Egleson, *Scientific American*, April 1961., p. 170.

From *Mathematical Models* by Cundy and Rollett (Oxford, 1961).

2. In 1640, Descartes made an interesting discovery about the angles in geometric solids that you may be able to discover, too.

The sum of the angles at each corner of a geometric solid is always less than 360°. Exactly how much less varies from one solid to another.

Look at the figures below and at the truncated tetrahedron pictured on page 290. An equilateral triangle and two hexagons meet at each

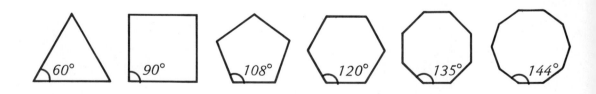

corner of the solid. The sum of the three angles surrounding each corner is

$$60° + 120° + 120° = 300°$$

which falls 60° short of 360°. We will call this difference, 60°, the *angular deficiency* of the truncated tetrahedron.

a. Use the same reasoning with each of the other Archimedean solids pictured in exercises 1–13 on pages 290–291 to copy and complete the following table.

Polyhedron	Number of corners	Sum of angles around each corner	Angular deficiency
1	12	300°	60°
2	24	▦	▦
3	24	▦	▦
4	60	▦	▦
5	60	▦	▦
6	12	▦	▦
7	24	▦	▦
8	48	▦	▦
9	30	▦	▦
10	60	▦	▦
11	120	▦	▦
12	24	▦	▦
13	60	▦	▦

b. How does the angular deficiency of a geometric solid vary with the number of corners that the solid has? Can you write a formula for the angular deficiency, a, in terms of the number of corners, n?

Lesson 6

1.   In Lesson 5, a semiregular polyhedron was defined as a solid having faces in the shape of more than one kind of regular polygon, yet having every corner the same. Among the semiregular polyhedra are the Archimedean solids, of which three have faces that are triangles and squares. They are shown at the right with their symbols.

There is another semiregular polyhedron, called a *square antiprism*, whose symbol is 3-3-3-4.

a. Cut out eight triangles and two squares from stiff paper. (The patterns on page 269 are of a convenient size.) Tape them together to make a model of a square antiprism.
b. In what way is a square antiprism like a square prism?
c. How many faces, edges, and corners does a square antiprism have?

3-4-3-4

3-4-4-4

3-3-3-3-4

In general, antiprisms are semiregular polyhedra having the symbol 3-3-3-*n*, in which *n*, the number of sides in each of the bases, is a number larger than 3.

    d. If an antiprism has a base with *n* sides, how could its numbers of faces, corners, and edges be expressed in terms of *n*?

    e. Is Euler's formula true for all antiprisms? Show why or why not.

2. This picture is a drawing of the radiolarian *Aulonia hexagona*.* From its name and appearance, you might assume that all of the cells covering its surface are hexagonal in shape. If you look at it carefully, however, you will notice some cells of other shapes.

    a. What shapes are they?

A biologist once claimed to have seen a radiolarian that was covered with only hexagonal cells. It is possible to prove, using mathematics, that the biologist must have been mistaken.

The proof is based on the fact that mathematicians have proved that the formula

$$F + C = E + 2$$

is true for every "spherical" polyhedron, even though not all of its faces are regular.

Suppose that a radiolarian has *n* hexagonal cells. Then, if the cells did not share their corners, there would be 6*n* separate corners. However, three corners come together at each corner of the solid, and so it must have $\frac{6n}{3} = 2n$ corners.

    b. Use the same type of reasoning to show how many edges the solid must have.

    c. Substitute these results into the formula

$$F + C = E + 2$$

and show that the result is false. If you can do this, you have proved that no radiolarian in the past, present, or future can ever be covered with only hexagonal cells.

The radiolarian Aulonia hexagona.
From Ernst Haeckel's *Monograph of the Challenger Radiolaria,* 1887; Science and Technology Research Center, The New York Public Library, Astor, Lenox and Tilden Foundations

---

*Other radiolarians are pictured on page 276.

Mathematical Curves

LESSON

1

The Circle and the Ellipse

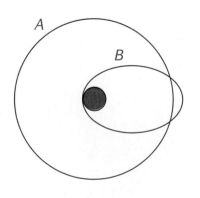

Since the launch of *Sputnik* in 1957, hundreds of satellites have been put into orbit around the earth. Many of them, including weather and communications satellites, have orbits that are circular in shape. In the figure at the left, the shaded circle represents the earth as seen from the North Pole and the curve labeled A represents the circular orbit of one of these satellites.

A circle is easily drawn with a compass. The way the compass works tells us exactly how a circle should be defined.

> A **circle** is the set of all points in a plane that are at the same distance from a fixed point in the plane. The fixed point is the *center* of the circle and the distance is its *radius.*

The figure at the top of the next page shows two points, A and B, that are 2 centimeters from point C. These two points, together with the rest of the points on the page that are 2 centimeters from point C, make up the circle shown.

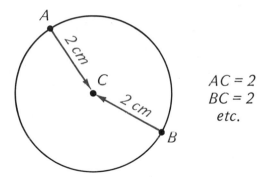

$$AC = 2$$
$$BC = 2$$
$$etc.$$

Some reconnaissance satellites have orbits in the shape of a curve called an *ellipse*. The curve labeled B in the figure on the preceding page illustrates an orbit of one of these satellites. Notice that this satellite is sometimes very close to the earth and sometimes very far away. Its orbit, an ellipse, is much harder to draw than a circle and is also harder to describe.

> An **ellipse** is the set of all points in a plane such that the sum of the two distances from each point to two fixed points is the same. The two fixed points are called the *foci** of the ellipse.

The figure below shows two points, A and B, the sums of whose distances from points F_1 and F_2 are 5 centimeters. These two points, together with the rest of the points on the page whose distances from the foci add up to 5, make up the ellipse shown.

$$AF_1 + AF_2 = 2 + 3 = 5$$
$$BF_1 + BF_2 = 4 + 1 = 5$$
$$etc.$$

All circles, regardless of their size, have the same shape. Ellipses, however, have many different shapes, ranging from nearly circular to long and narrow. For this reason, an ellipse is measured with two numbers rather than one. The two numbers are the lengths of its *major* (longer) *axis* and *minor* (shorter) *axis*. These axes are illustrated in the figures at the top of the next page.

---

*Foci (pronounced "fō-sī") is the plural of focus.

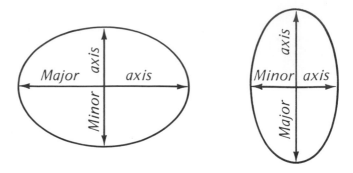

The ellipse was given its name by the Greek mathematician Apollonius in the third century B.C. Apollonius made a thorough study of the curve and discovered its properties, even though there seemed to be no practical use for such knowledge at the time. In 1609, Johann Kepler established the importance of the ellipse when he discovered that the orbits of the planets around the sun are elliptical in shape.

EXERCISES

SET I
EXPERIMENT: DRAWING ELLIPSES

Part 1. Turn a sheet of graph paper (4 units per inch or 2 units per centimeter) sideways and draw a pair of perpendicular axes with their origin at the center of the paper. Label the long axis x, the short axis y, and the origin O. Tack the graph paper on a square piece of corrugated cardboard as shown in the figure below.

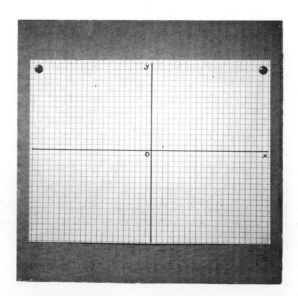

Mark the two points on the x-axis that are 6 units to the left and 6 units to the right of the origin and label them A_1 and A_2. Put a tack in each point, but do not press them all the way into the cardboard.

Take a piece of string about 10 inches long (20 centimeters if you are using metric graph paper) and, as accurately as you can, tie a knot in it to make a loop exactly 32 units around. (If you are using graph paper ruled 4 units per inch, 32 units would be 8 inches. If you are using graph paper ruled 2 units per centimeter, 32 units would be 16 centimeters.) Put the loop on the board around the tacks and, using your pencil, pull it taut to form a triangle, as shown below. Now move the pencil around the paper, keeping the string taut, to draw an ellipse. (While drawing the ellipse, you may have to hold the tacks in place with your other hand to prevent them from popping out of the board.) Label the ellipse by writing the letter A on it at some point.

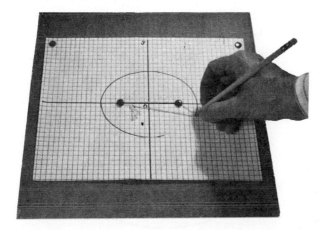

Mark the two points on the x-axis that are 4 units to the left and right of the origin and label them B_1 and B_2. Move the tacks to these points and use the same loop of string to draw another ellipse. Label it B.

Mark the two points on the x-axis that are 2 units to the left and right of the origin and label them C_1 and C_2. Move the tacks to these points and draw a third ellipse. Label it C.

Part 2. Remove the sheet of graph paper from the cardboard and tack a new sheet in place. Draw and label a pair of axes as before. Mark the two points on the x-axis that are 9 units to the left and right of the origin, label them D_1 and D_2, and put tacks in these points.

Take a longer piece of string and make, as accurately as you can, a loop that is 48 units (12 inches or 24 centimeters) around. Put the loop around the tacks, draw an ellipse, and label it D.

Mark the two points on the x-axis that are 11 units to the left and right of the origin and label them E_1 and E_2. Move the tacks to these points, draw an ellipse and label it E.

Part 3. Use your drawings of the five ellipses as a guide in answering the following questions.

1. Which ellipse looks the most like a circle?

2. Which ellipse is the most elongated; that is, the most *unlike* a circle?

3. Two of the ellipses that you have drawn have the same shape; that is, one could be enlarged to fit the other. Which two ellipses are they?

4. The two points at which you put the tacks to draw each ellipse are its foci. On which axis, the major or the minor, are the foci of an ellipse located?

5. The two foci of ellipse A are 12 units apart and the loop you used to draw ellipse A was 32 units around. What is the sum of the distances of each point on ellipse A from its foci? (See the figure below.)

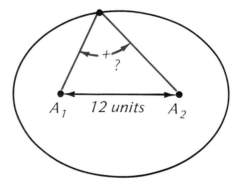

6. Copy and complete the following table.

Ellipse	Length of loop used	Distance between foci	Sum of distances of each point on ellipse from foci
A	32	12	20
B	▨	▨	▨
C	▨	▨	▨
D	▨	▨	▨
E	▨	▨	▨

7. Look at ellipses A, B, and C. What happens to the *size* of these ellipses as the sum of the distances from each point on the ellipses to the foci increases?

8. What happens to the *shape*?

9. Look at ellipses D and E. What happens to the size of these ellipses as the sum of the distances from each point on the ellipse to the foci decreases?

10. What happens to the shape?

SET II

The following exercises refer to the ellipse shown here.

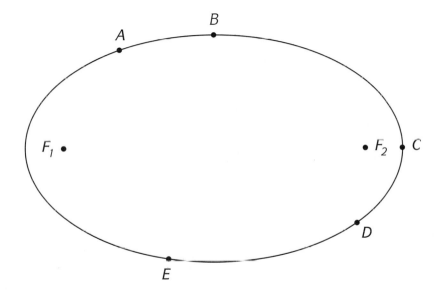

1. Use your ruler to measure, in centimeters, the distance from point A to point F_1 and the distance from point A to point F_2.

2. Make the additional measurements needed to copy and complete the following table.

Point	Distance from F_1	Distance from F_2
A	||||||	||||||
B	||||||	||||||
C	||||||	||||||
D	||||||	||||||
E	||||||	||||||

3. What is the sum of the two distances from each of these points to the points F_1 and F_2?

4. What are points F_1 and F_2 called?

5. How long is the major axis of the ellipse?

6. How long is the minor axis?

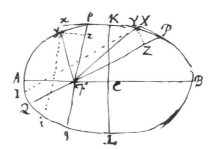

This drawing was made by the great English scientist, Sir Isaac Newton, to show the effects of gravitational forces on the orbit of a planet around the sun. To be able to use the ellipse, Newton worked with its equation.

Every mathematical curve can be associated with an equation.* Because circles and ellipses are closely related, their equations are similar. In fact, every circle and ellipse can be associated with an equation of the form

$$\frac{x^2}{a^2} + \frac{y^2}{b^2} = 1$$

in which a and b represent positive numbers.

The ellipse shown here has the equation

$$\frac{x^2}{16} + \frac{y^2}{36} = 1$$

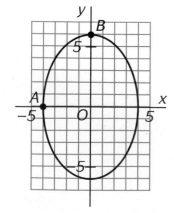

Point A is on this ellipse because its coordinates, $(-4, 0)$, make the equation true:

$$\frac{(-4)^2}{16} + \frac{0^2}{36} = 1$$

$$\frac{16}{16} + \frac{0}{36} = 1$$

$$1 + 0 = 1$$

---

*You have already studied examples of this in Chapter 3. Look at pages 168–169.

7. Show that the coordinates of point B also make the equation true.

Comparing the general equation,

$$\frac{x^2}{a^2} + \frac{y^2}{b^2} = 1$$

and the equation for the ellipse on the preceding page,

$$\frac{x^2}{16} + \frac{y^2}{36} = 1$$

reveals that, for this ellipse, $a^2 = 16$. Since a is a positive number, $a = 4$.

8. For this ellipse, what number does b^2 represent?

9. Since b is a positive number, what does b represent?

The circle shown here has the equation

$$\frac{x^2}{25} + \frac{y^2}{25} = 1$$

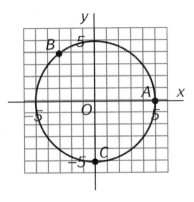

Show that the coordinates of each of the following points on the circle make the equation true.

10. Point A.

11. Point B.

12. Point C.

Compare the general equation,

$$\frac{x^2}{a^2} + \frac{y^2}{b^2} = 1$$

and the equation for the circle above,

$$\frac{x^2}{25} + \frac{y^2}{25} = 1$$

13. For this circle, what number does a^2 and b^2 represent?

14. What number does a and b represent?

15. For what kind of curve are a and b equal?

16. For what kind of curve are a and b unequal?

Look at the three curves below.

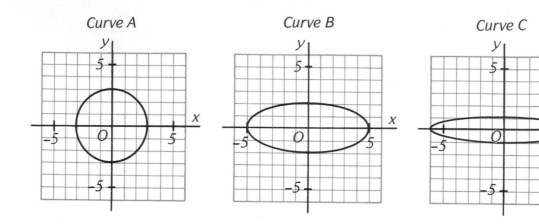

Curve A   Curve B   Curve C

17. Which curve has the equation $\dfrac{x^2}{25} + \dfrac{y^2}{4} = 1$?

18. Write equations for the other two curves.

SET III

Around 3000 B.C., the inhabitants of England, Scotland, and Ireland built large monuments of stones called "henges." The most famous of these is Stonehenge in England.

Aerofilms Ltd.

Stonehenge is circular in shape but some henges are accurately constructed ellipses. It is thought that the ellipses were constructed by using a rope stretched around two posts which served as the foci.*

*Geometry and Algebra in Ancient Civilizations, by B. L. van der Waerden (Springer-Verlag, 1983). The method was apparently similar to that used in the ellipse drawing experiment of Set I.

Drawings of three elliptical henges are shown below.

Callanish, Hebrides

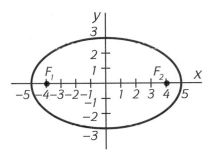

Stanton Drew, Somerset

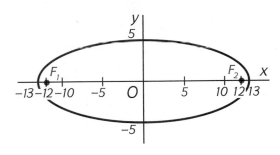

Daviot, Scotland

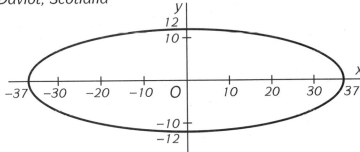

The equation for the ellipse at Callanish is

$$\frac{x^2}{25} + \frac{y^2}{9} = 1$$

1. Write equations for the other two ellipses.

In the ellipse at Callanish, the foci are 4 units from the center. Notice that

$$25 - 9 = 16 \text{ and } 16 = 4^2$$

2. Do the positions of the foci in the ellipse at Stanton Drew fit the same sort of pattern? Show why or why not.

The foci are where the posts were placed in order to trace the ellipse on the ground.

3. Exactly where do you suppose the posts for the foci were placed for the henge at Daviot?

LESSON

2

The Parabola

The water from this firehose is following a path in the shape of a curve called a *parabola*. Galileo proved that the path of an object thrown through space, such as a ball or water, is this curve. You probably recognize the parabola because you have drawn several graphs of functions with its shape.*

> A **parabola** is the set of all points in a plane such that the distance of each point from a fixed point is the same as its distance from a fixed line. The fixed point is called the *focus* of the parabola and the fixed line is called the *directrix*.

*See Chapter 3, Lesson 4.

The figure below shows three points, A, B, and C, whose distances from point F, the focus, are equal to their distances from line ℓ, the directrix. These three points, together with all other points having the same property, make up the parabola shown.

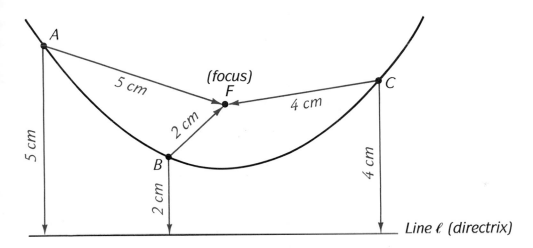

Because we can extend line ℓ indefinitely in both directions, the parabola can be extended indefinitely as well. This means that, unlike an ellipse, a parabola cannot be drawn in its entirety.

Parabolas, like ellipses and circles, can be represented by equations. An equation for the firehose parabola is

$$y = 7x - 0.7x^2$$

and its graph is shown at the right.

A parabola has one line of symmetry, which passes through its focus. It is easy to see that if the parabola is folded along this line, its two halves will coincide. The parabola can be rotated about this line to form a curved surface called a *paraboloid*, which has a very useful property. If a mirror is made in this shape and a light placed at the focus of the mirror, the light is reflected in rays parallel to the axis. This forms a straight beam of light.

Parabolic mirrors are used in automobile headlights and, on a larger scale, in searchlights. They are also used in reflecting telescopes and in

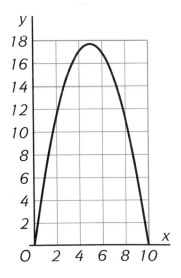

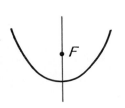

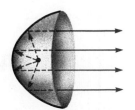

antennas to collect light and radio waves from outer space. In the latter uses, the beam comes *toward* the parabolic surface and is brought together at the focus.

Exercises

Set I

This is a photograph of the Golden Gate Bridge. The cables that are hung between the two towers from which the roadway is suspended form a curve that is very close to the shape of a parabola. In the figure below, the

Photo by Bob David, Golden Gate Bridge Highway and Transportation District Archives.

towers, cable, and roadway of the bridge are represented midway between point F and line ℓ.

• *F*

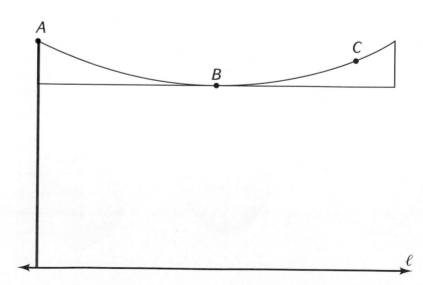

Point A, at the top of one tower, is at the end of the cable.

1. Measure its distance from F and its distance from line ℓ at the bottom, each to the nearest 0.1 centimeter. (The distance from a point to a line is measured along the *perpendicular* to the line. The perpendicular line from point A to line ℓ is shown in color in the figure.)

Point B is on the bridge midway between the two towers.

2. Measure its distance from F and its distance from line ℓ at the bottom, each to the nearest 0.1 centimeter.

Point C is another point on the cable.

3. Measure its distance from F and its distance from line ℓ at the bottom, each to the nearest 0.1 centimeter.

4. How do the distances of points A, B, and C from point F compare to their distances from line ℓ?

5. What is point F called with respect to the curve?

6. What is line ℓ called with respect to the curve?

This photograph was taken with a strobe light and shows a bouncing golf ball. An equation for the path of the outlined bounce is

$$y = 12x - x^2$$

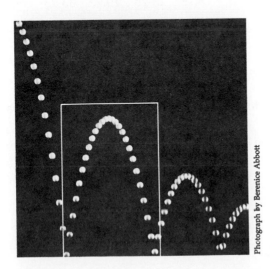

Photograph by Berenice Abbott

7. Copy and complete the following table for this equation.

x	0	1	2	3	4	5	6	7	8	9	10	11	12
y	0	11	20	▨	▨	▨	▨	▨	▨	▨	▨	▨	▨

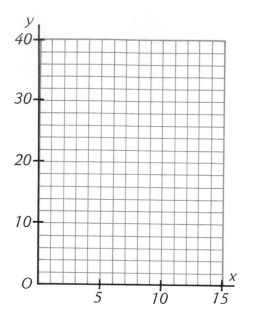

8. Draw and label a pair of axes as shown here. Then plot the points corresponding to the numbers in the table.

Your graph should look somewhat like the outlined bounce in the photograph, except that it "doesn't have the strobe flashing fast enough." The following table of numbers gives positions of the ball between the ones you plotted.*

x	0.5	1.5	2.5	3.5	4.5	5.5
y	5.75	15.75	23.75	29.75	33.75	35.75

x	6.5	7.5	8.5	9.5	10.5	11.5
y	35.75	33.75	29.75	23.75	15.75	5.75

Add the points corresponding to this table to your graph.

One way to sketch a curve is with a set of *points*. Another way is with a set of *lines*.

Points have been marked at equal intervals on each side of the angle shown below.

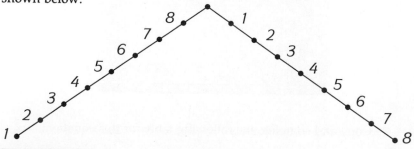

*You could also find them out from the equation, but they are somewhat tedious to calculate.

9. Trace the figure and connect each pair of points having the same number with a straight line.

Although the lines that you drew are straight, they create the illusion of a curve.

10. What kind of curve does it look like?

During a running broad jump, the path of the center of the athlete's body is along this same curve.

From Geoffrey H. G. Dyson, *The Mechanics of Athletics*, 4th ed., © 1968, Hodder & Stoughton Limited. Used with permission.

Connect the point labeled 1 at the left of your figure for exercise 9 to the point labeled 8 at the right to represent the ground. On your drawing, $\frac{1}{2}$ centimeter represents 1 foot.

Measure your drawing to estimate each of the following.

11. The length in feet of the broad jump pictured.

12. The approximate number of feet above the ground reached by the center of the jumper's body.

SET II

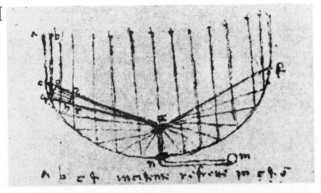

This drawing, by Leonardo da Vinci, shows how a parabolic mirror can be used to produce a beam of light. In the following exercises, you will consider the reflective property of the parabola that makes such a mirror possible.

1. Copy and complete the following table for the formula

$$y = x^2$$

x	-7	-6	-5	-4	-3	-2	-1	0	1	2	3	4	5	6	7
y	49														

2. Draw and label a pair of axes as shown here. Plot the points corresponding to the numbers in the table and connect them with a smooth curve. The focus of the parabola is at (0, 25). Label it F.

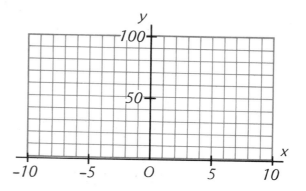

Draw lines from the focus to each plotted point on the parabola. Then draw lines upward from each of these points so that each line is parallel to the y-axis. When you have done this, your drawing should look somewhat like the drawing by da Vinci.

If the parabola were a mirror and there were a light at the focus, the lines that you drew from the focus to the parabola and from the parabola upward would represent rays of light.

3. How does the angle at which each light ray hits the parabola seem to compare in size with the angle at which it is reflected? (The two angles for one light ray are marked in this figure as an example.)

4. How does the angle at which each light ray hits the parabola change in size the farther the ray is from the y-axis?

5. The light rays leave the focus in many different directions. What happens to them when they are reflected from the parabola?

The light bulb in an automobile headlight has two filaments: one for low beam and one for high beam. One filament is located at the focus of the reflector and the other is located in a different position.

6. Based on what you learned from the drawing you made for exercise 2, which filament do you think is used for high beam? Explain.

Satellite television systems use a large "dish" antenna to collect signals.

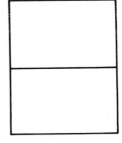

7. In what direction do you think the antenna is pointed?

8. What shape does the "dish" have?

9. What happens to the signals as they are reflected from the surface of the antenna?

Set III
Experiment: Parabolas by Folding Paper

Part 1. Fold a sheet of unlined paper in half as illustrated in the adjoining figure, and tear it along the fold into two equal pieces. Take one of the pieces of paper and mark points along the lower edge spaced 1 centimeter apart. Turn the paper over to the other side and mark a point about 3 centimeters above the center of the same lower edge; label the point A.

Fold the paper, as shown here, so that the first point that you marked on the lower edge touches point A. Make a sharp crease. Open the paper flat and fold again so that the second point on the lower edge of the paper

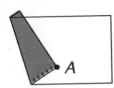

touches point A. Continue doing this, opening the paper flat each time and then folding the paper so that the remaining points on the lower edge touch point A. When you have done this, a parabola should appear. Trace the curve with your pencil.

Part 2. Do the experiment described in Part 1 again with the other piece of paper but make the following change. Mark a point in the *center* of the paper and label it B. Repeatedly fold the paper so that each point on the lower edge touches point B, and trace the curve that results.

Part 3.

1. What are the two points called with respect to the parabolas?

2. Where do you suppose the directrix of each parabola is?

3. Although the two curves seem to have different shapes, they are actually the same parabola seen from different distances away. Which parabola looks like an enlargement of part of the other?

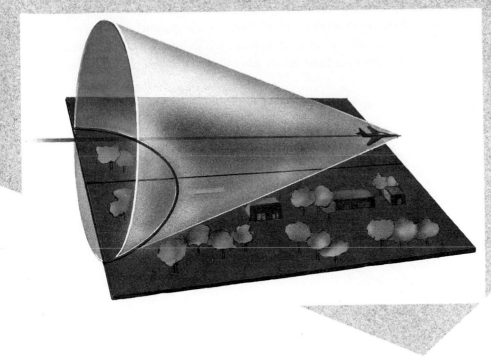

The Hyperbola

In 1953, a pilot flying faster than the speed of sound over a western air base created a shock wave that damaged almost every building on the base. This was the first evidence of the destructive power of the shock wave that is heard as a "sonic boom."

The sonic boom shock wave has the shape of a cone, and it intersects the ground in part of a curve called a *hyperbola*. It hits every point on this curve at the same time, so that people in different places along the curve on the ground hear it at the same time. Because the airplane is moving forward, the hyperbolic curve moves forward along the ground and eventually the boom can be heard by everyone in its path.

The definition of a hyperbola is very much like the definition of an ellipse.

A **hyperbola** is the set of all points in a plane such that the difference between the two distances from each point to two fixed points is the same. The two fixed points are called the *foci* of the hyperbola.

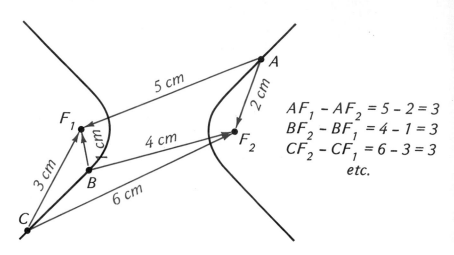

$$AF_1 - AF_2 = 5 - 2 = 3$$
$$BF_2 - BF_1 = 4 - 1 = 3$$
$$CF_2 - CF_1 = 6 - 3 = 3$$
$$etc.$$

The figure above shows three points on the hyperbola: A, B, and C. The difference between the two distances of each of these points from points F_1 and F_2 is 3 centimeters. Points A, B, and C, together with the rest of the points on the page having the same property, make up the hyperbola shown. Notice that the hyperbola consists of two separate parts, called its *branches*. The sonic-boom curve described earlier is only one branch.

The early Greek mathematicians knew all about the hyperbola. In fact, it was Apollonius* who gave the curve its name. The Greeks were equally familiar with the parabola, as well as the ellipse. How did they come to be acquainted with these curves so long ago? Their knowledge was recorded in eight books written by Apollonius on the curves that could be produced by slicing a cone in different directions. If the slice is in the same direction as the side of the cone, the curve that results is a parabola.

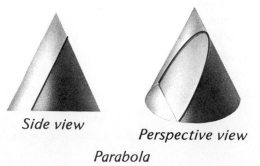

Side view   *Perspective view*

Parabola

---

*See page 330.

If the slice is tilted from the direction of the side of the cone toward the horizontal, the curve is an ellipse instead.

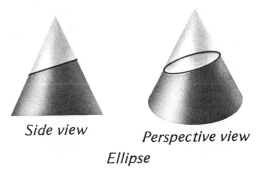

Side view   *Perspective view*

Ellipse

If the slice is tilted in the other direction — that is, away from the direction of the side of the cone and toward the vertical — part of a hyperbola is formed.

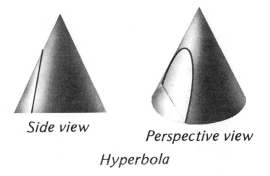

Side view   *Perspective view*

Hyperbola

To see the rest of the hyperbola, imagine a second cone balanced upside down on the point of the first one. If the slice is continued through the second cone, the other branch of the hyperbola is formed.

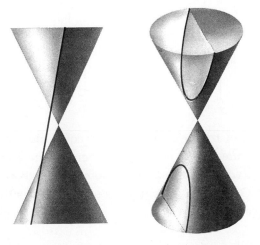

Because the parabola, ellipse, and hyperbola are formed when a cone is sliced into sections, they are called *conic sections*. Apollonius titled his work the *Conics*. Although it seemed to have no practical use at the time, many important applications of the conic sections have been discovered since.

EXERCISES

SET I
EXPERIMENT: HYPERBOLAS FROM TRACING PAPER

Part 1. One way to draw an accurate graph of a hyperbola is based on points in which lines and circles intersect. In the figure below, ten evenly spaced circles with a common center are intersected by a set of evenly spaced lines.

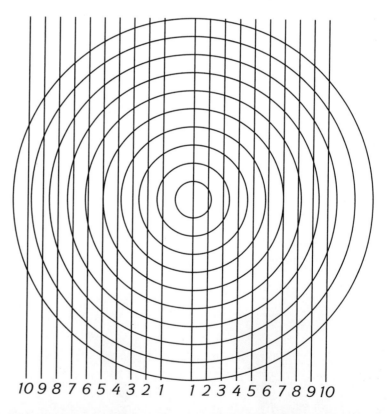

10 9 8 7 6 5 4 3 2 1   1 2 3 4 5 6 7 8 9 10

Tear a sheet of tracing paper into two equal pieces and place one piece over the figure. One of the lines numbered 1 intersects the smallest circle, which we will call circle 1, at two points. Mark these points on the tracing paper. One of the lines numbered 2 intersects the next larger circle, which we will call circle 2, at two points. Mark these points on the tracing paper.

Continue in the same way, marking the points where line 3 intersects circle 3, and so on. As the circles get larger, you will notice that they are intersected by two lines having the same number. For example, circle 5 intersects the line numbered 5 on the left at one point and the line numbered 5 on the right at two points.

After you have marked all of the points, join them with smooth curves to form the two branches of a hyperbola.

Part 2. Another way to create an accurate picture of a hyperbola is by folding paper. Draw a circle having a radius of about 4 centimeters on the other piece of tracing paper. Mark a set of points *on* the circle spaced approximately 1 centimeter apart. Mark a point about 3 centimeters *below* the circle and label it A.

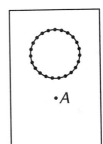

Now fold the paper so that point A touches one of the points you marked on the circle and make a sharp crease. Open the paper flat and fold again so that point A touches the next point marked on the circle. Continue in this way, repeatedly folding the paper and opening it flat again until point A has fallen on every point that you marked around the circle.

You should be able to find both branches of a hyperbola when you are finished. Trace them in on the paper.

1. Where are the two foci of the hyperbola you have just drawn located?

2. How many lines of symmetry does the hyperbola have?

SET II

Part of the two branches of a hyperbola can be seen in this photograph of the McDonnell Planetarium in St. Louis. The designer, Gyo Obata, was inspired to use the hyperbola by the fact that certain comets travel in hyperbolic orbits.

Photograph by Kiku Obata

The figure below shows the hyperbola upon which the design of the planetarium is based.

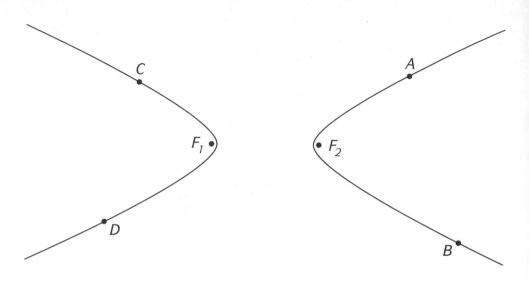

Point	Distance from F_1	Distance from F_2
A	▥	▥
B	▥	▥
C	▥	▥
D	▥	▥

1. Use your ruler to check that the distance from point A to point F_1 is 5.5 centimeters. Then measure the distance from point A to point F_2.

2. Make the additional measurements needed to copy and complete the table at the left.

3. What is the difference between the two distances from each point to points F_1 and F_2?

4. What is the shortest distance between the two branches of the curve? (Measure it with your ruler.)

5. What are points F_1 and F_2 called?

The shadows cast on a wall by a lamp with a circular shade form the branches of a hyperbola.

The two graphs below show this "lamp hyperbola" from close up and farther away.

The two dotted lines in each graph are called the *asymptotes* of the hyperbola.

6. What happens to the distances between the branches of the hyperbola and these lines as the branches move away from the center?

Just as ellipses are associated with equations of the form

$$\frac{x^2}{a^2} + \frac{y^2}{b^2} = 1$$

hyperbolas are associated with equations of the form

$$\frac{x^2}{a^2} - \frac{y^2}{b^2} = 1$$

The hyperbola shown here has the equation

$$\frac{x^2}{16} - \frac{y^2}{9} = 1$$

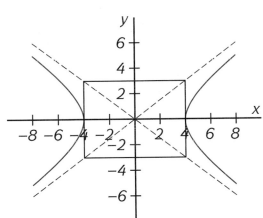

7. Show that the point with coordinates (4, 0), is on this hyperbola because the coordinates make the equation of the hyperbola true.

The dashed lines through the corners of the rectangle in the graph are asymptotes of the hyperbola. Notice that the rectangle crosses the x-axis at -4 and 4. Notice also that $(-4)^2 = 4^2 = 16$ and that 16 appears below x^2 in the equation of the hyperbola.

8. Where does the y-axis cross the rectangle?

9. How are these two numbers related to the equation of the hyperbola?

The equation of the hyperbola in the figure below is

$$\frac{x^2}{25} - \frac{y^2}{4} = 1$$

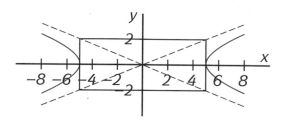

10. Where does the rectangle cross the x- and y-axes?

11. How are these numbers related to the equation of the hyperbola?

What do you think are the equations of the following hyperbolas?

12.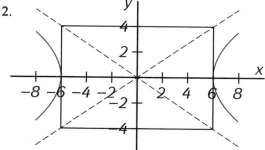

13.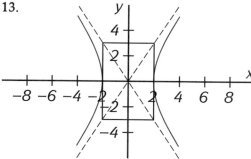

SET III

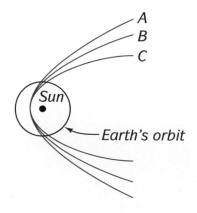

Comets travel in orbits that are either ellipses, parabolas, or hyperbolas. Unfortunately, we can only see comets when they are comparatively close to the sun and, near the sun, all three types of orbits look very much alike.

This figure shows, for each type of orbit, the part where the comet is relatively close to the earth and sun.

1. Which orbit, A, B, or C, do you suppose is part of a hyperbola?

Some comets, such as Halley's comet, travel around the sun more than once.

2. Which type of orbit do you think they have?

E. A. Janes/NHPA

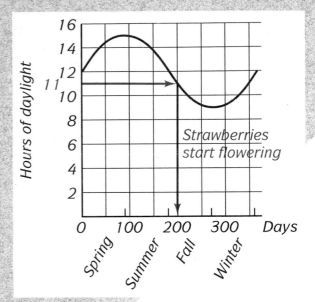

Strawberries start flowering

The Sine Curve

Many plants produce flowers as if they were aware of the calendar. Their leaves are able to detect change in the length of daylight so precisely that gardeners can predict almost to the day when some plants will bloom.

Wild strawberries, for example, flower when the length of daylight shortens to 11 hours. A typical graph showing the length of daylight as a function of the time of year is shown above.* The graph shows the number of hours of daylight each day over a period of one year, starting with the first day of spring. So many functions have graphs similar to this that the curve has been given a name. It is called a *sine curve.*

The sine curve can be understood in terms of the motion of a crank handle on a wheel that rotates at a steady rate. Imagine that a pair of

---

*This graph fits cities at a latitude of about 40°, such as Salt Lake City, Chicago, and New York.

coordinate axes is placed on the wheel as shown in the second figure. Since the number of degrees surrounding a point is 360, the various positions of the crank handle can be given in terms of numbers from 0 to 360.

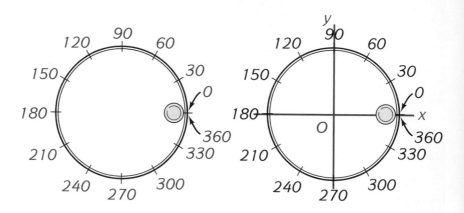

Suppose that the handle starts at 0° and that the wheel turns counterclockwise at a steady rate. The distance of the handle from the x-axis first increases as it moves from 0° to 90° and then decreases as it continues from 90° to 180°.

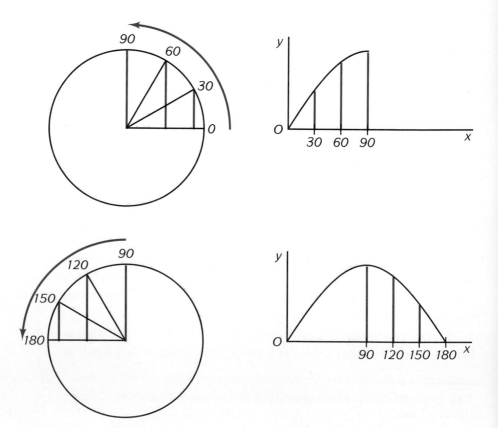

As the handle moves around the lower half of the wheel, its distance from the x-axis increases as it moves from 180° to 270° and then decreases as it continues from 270° to 360°.

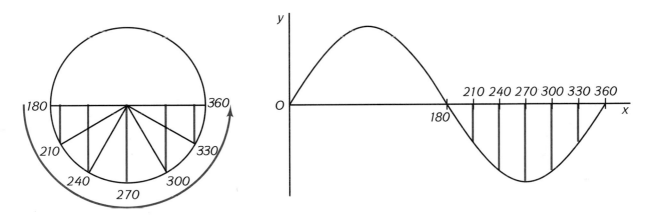

The distances can be measured in terms of the radius of the wheel, which, for convenience, is considered to be 1 unit long. The distances measured upward from the x-axis give positive y-values, and those measured downward from the x-axis give negative y-values.

The distances are called the *sines* of the angles through which the wheel has turned. The graph of these sines as a function of the angle produces the *sine curve.*

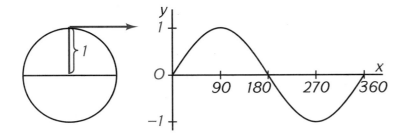

One revolution of the wheel results in one "wave" of the sine curve, shown above. If the wheel continues turning so that the graph is continued, the curve is repeated over and over. Four waves appear in the pictures of the sine curves on the postage stamp shown here.

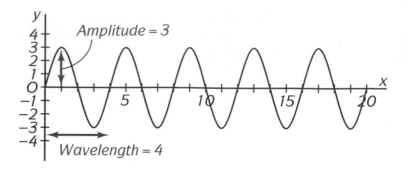

The dimensions of a sine curve are measured with two numbers: the *amplitude,* which is the maximum distance of the curve from the *x*-axis, and the *wavelength,* which is the distance along the *x*-axis required for one complete wave. In the case of the curve shown above, the amplitude is 3 and the wave length is 4.

EXERCISES

SET I

This figure can be used to estimate some sines to the nearest hundredth.

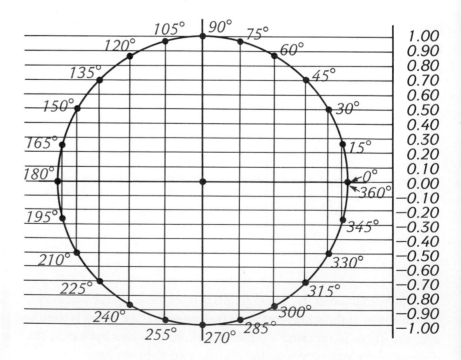

For example, the sine of 15° is approximately 0.26.

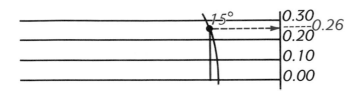

1. Refer to the diagram to copy and complete the following table.

x	0°	15°	30°	45°	60°	75°	90°
sine of x	0	0.26	▓	▓	▓	▓	▓

The figure has symmetry with respect to the vertical line through its center. As a result, the sine of 105° is equal to the sine of 75°.

2. Which angle has a sine that is equal to the sine of 45°?

3. Copy and complete the following table.

x	105°	120°	135°	150°	165°	180°
sine of x	▓	▓	▓	▓	▓	▓

The figure also has symmetry with respect to the horizontal line through its center. As a result, the sine of 195° is equal to the negative of the sine of 165°. (It is negative because it is measured downward instead of upward.)

4. Which angle has a sine that is equal to the negative of the sine of 120°?

5. Copy and complete this table.

x	195°	210°	225°	240°	255°	270°
sine of x	−0.26	▓	▓	▓	▓	▓

6. Which angle has a sine that is equal to the negative of the sine of 45°?

7. Copy and complete this table.

x	285°	300°	315°	330°	345°	360°
sine of x	▓	▓	▓	▓	▓	▓

8. Draw and label a pair of axes as shown in the figure below.

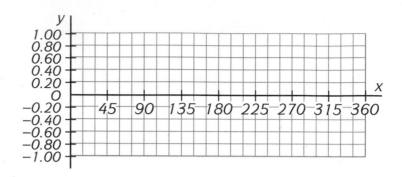

Notice that 1 unit on the x-axis represents 15° and 1 unit on the y-axis represents 0.20. Use your tables to plot points corresponding to each sine from 0° through 360° and connect the points to form one wave of a sine curve.

9. What kind of symmetry does the wave have?

Think of the curve as a roller coaster track.

10. Around what values of x is the curve the closest to being level?

11. Near what values of x is it the steepest?

The part of the sine curve that you have graphed starts at 0° and stops at 360°. The figure below shows the curve continued to the left and right.

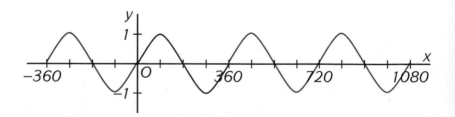

Refer to the figure to answer these questions.

12. How many waves are shown altogether?

13. What is the sine of 720°?

14. What numbers other than 90° have a sine equal to 1?

15. What numbers other than 270° have a sine equal to −1?

SET II

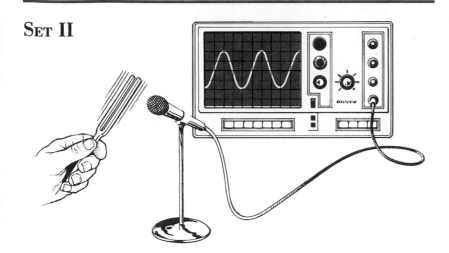

When a musical sound wave is changed into a visual image by an oscilloscope, it has a regular pattern that repeats itself many times each second. The drawing above shows the sound waves produced by a tuning fork being displayed as a sine curve on an oscilloscope.

A typical equation for such a curve is

$$y = \text{sine } 3x*,$$

the graph of which is shown below.

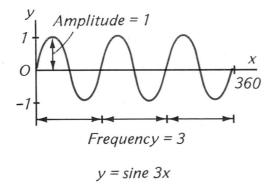

$$y = \text{sine } 3x$$

Notice from the figure that the *frequency* of a sine curve is the number of complete waves that occur when x increases by 360°.

The amplitude of the curve above is 1, its frequency is 3, and its wavelength is

$$\frac{360°}{3} = 120°$$

---

*This equation says that y is the sine of the number that is 3 times x.

1. Refer to the graphs below to copy and complete the following table.

Equation	Amplitude	Frequency	Wavelength
$y = 4\ \text{sine}\ 2x$	▓▓▓	▓▓▓	▓▓▓
$y = 2\ \text{sine}\ 5x$	▓▓▓	▓▓▓	▓▓▓
$y = 3\ \text{sine}\ \dfrac{1}{2}x$	▓▓▓	▓▓▓	▓▓▓

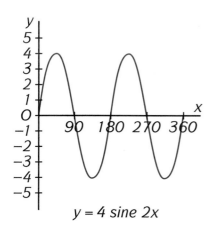

$y = 4\ sine\ 2x$

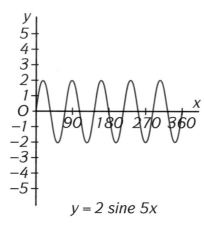

$y = 2\ sine\ 5x$

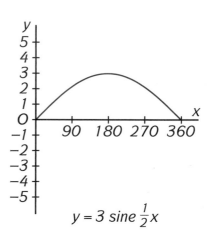

$y = 3\ sine\ \frac{1}{2}x$

The greater the amplitude of a sound wave, the louder the sound.

2. Which equation in the table in exercise 1 corresponds to the loudest sound?

3. Which equation corresponds to the softest sound?

The greater the frequency of a sound wave, the higher the pitch.

4. Which equation in the table corresponds to the sound having the highest pitch?

5. Which one corresponds to the sound having the lowest pitch?

6. What happens to the wavelength of a sound as its frequency increases?

Compare the equations of the sound waves in exercise 1 with their amplitudes and frequencies. Then write equations for the following three curves.

7.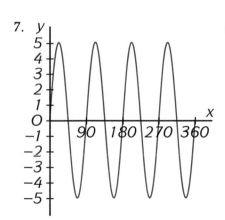

8.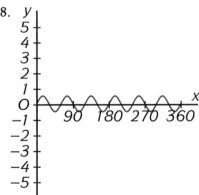

9.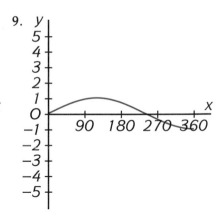

Radio stations broadcast in either AM or FM. Examples of waves sent in each form are shown below.

AM radio wave

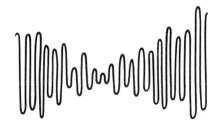

FM radio wave

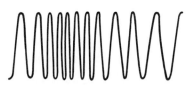

10. What is continually changing in the sine curve associated with the AM wave?

11. What stays constant?

12. What is continually changing in the sine curve associated with the FM wave?

13. What stays constant?

14. What do you think the A and F in AM and FM stand for?

SET III

The giraffe is a remarkable animal. When a giraffe 18 feet tall is standing up, its heart has to pump blood upward for 10 feet to reach its brain. As a result, the blood pressure of the giraffe is higher than that of any other animal.

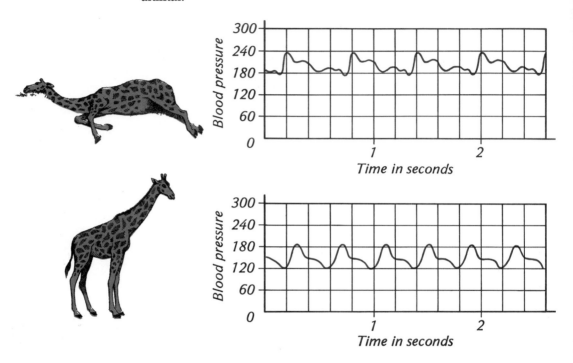

The graphs above are of the blood pressure near a giraffe's head when the animal is lying down and when it is standing up.* The changes in pressure with time can be used to measure the animal's heart rate.

1. What happens to the blood pressure near the giraffe's head when the animal stands up?

2. What is the wavelength of the graph for the giraffe when it is lying down? (Estimate it as a fraction of a second.)

3. What is the wavelength of the graph for the giraffe when it is standing up?

4. Estimate the number of times that the giraffe's heart beats each minute when it is lying down.

5. Make a comparable estimate for the giraffe's heart rate when it is standing up.

*After "The Physiology of the Giraffe," by James V. Warren. Copyright © 1974 by Scientific American, Inc. All rights reserved.

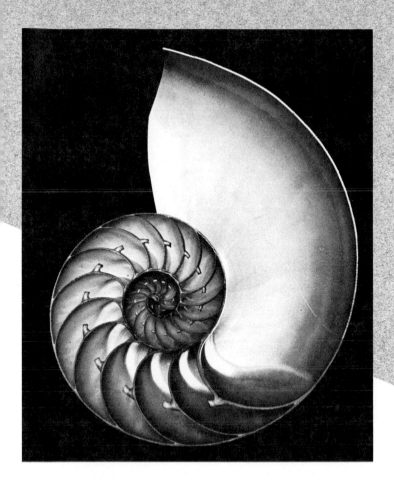

Spirals

One of the most beautiful designs in nature is the shell of the chambered nautilus, a sea creature that lives in the South Pacific. As the nautilus grows, it builds and moves through a series of ever-larger chambers. Each chamber has the same shape as the one before it. The photograph above shows the shell of a chambered nautilus that has been cut in half to reveal these chambers. It has the shape of a curve called a *spiral*.

A **spiral** is a curve traced by a point that moves around a fixed point from which it moves farther and farther away.

There are several kinds of spirals. The sound track on a compact disc is in the shape of an *Archimedean spiral.* Named after Archimedes, who wrote a book on spirals, the Archimedean spiral is one whose loops are

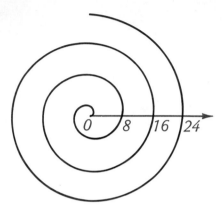

An Archimedean spiral

spaced at equal intervals. Because of this equal spacing, the successive distances of the loops from the center of the spiral form an *arithmetic sequence.* In the figure above, these distances are measured in millimeters.

The curve of the shell of the chambered nautilus is a spiral of a different kind. Called a *logarithmic spiral,* it was discovered by Descartes, the inventor of coordinate geometry. The successive distances of the loops from the center of a logarithmic spiral form a *geometric sequence.* As a result, the loops of a logarithmic spiral are spaced farther and farther apart as it winds outward.

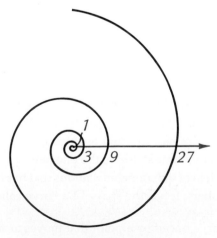

A logarithmic spiral

Photograph by Carola Gregor   Palomar Observatory, California Institute of Technology.

Spirals occur in nature in many different ways. They appear in the heads of daisies, in elephant tusks, and in the webs of certain spiders. The internal parts of the ear that sense sound are in a spiral arrangement. The most spectacular spirals in nature can be seen only through a telescope —they are the galaxies and nebulas of the universe. Most galaxies have spiral shapes, including the Milky Way, the one in which our sun is a star. Our solar system is about three-fourths of the way from the center of the spiral to the edge.

EXERCISES

SET I

If you walked from the center of a merry-go-round along a radius of the floor while the merry-go-round was turning, your path with respect to the ground would be along an Archimedean spiral.

Carousel Columbia, Paramount's Great America, Santa Clara, Ca.

The figure below represents an overhead view of a merry-go-round with a radius of 12 feet. Suppose that you walk half a foot for each 30° that the merry-go-round turns.

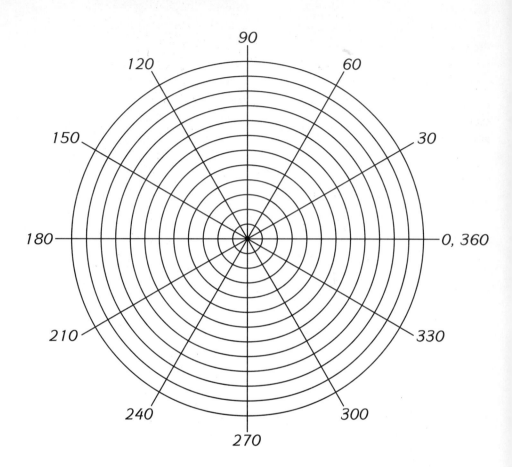

1. Copy and complete the following table relating angles and distances from the center.

Angle	0°	30°	60°	90°	120°	150°	180°
Distance	0	0.5	1				

2. Place tracing paper over the figure and mark a dot at the center to represent your starting point. Then plot a point on each of the angle lines at a distance from the center as given by your table. The points corresponding to the first three angles included in the table have been plotted in the diagram shown here as an example.

   Keep plotting points following this pattern to the outer-most circle. Connect the points in order with a smooth curve to form an Archimedean spiral.

The spiral that you have drawn represents your path on the merry-go-round with respect to the ground.

3. How far from the center are you when the merry-go-round has made one revolution?

4. How many revolutions has the merry-go-round turned when you reach its edge?

5. Copy and complete the following table for the curve that you have drawn.

Number of revolutions, x 0   1   2
Distance from center, y 0 ▊▊▊ ▊▊▊

6. What kind of number sequence do the distances form?

7. What would the distances from the center be after three and after four revolutions if the curve were continued?

The spiral of the chambered nautilus is a logarithmic spiral.

8. To draw a logarithmic spiral, place tracing paper over the figure on the facing page that you used to draw the "merry-go-round" spiral. Plot the points represented in the table below and connect them with a smooth curve.

From *Medical Radiography and Photography*, published by Radiography Markets Division, Eastman Kodak Company.

Angle	Distance from center	Angle	Distance from center
0°	1	1 rev. + 60°	3.6
30°	1.1	1 rev. + 90°	3.9
60°	1.2	1 rev. + 120°	4.3
90°	1.3	1 rev. + 150°	4.7
120°	1.4	1 rev. + 180°	5.2
150°	1.6	1 rev. + 210°	5.7
180°	1.7	1 rev. + 240°	6.2
210°	1.9	1 rev. + 270°	6.8
240°	2.1	1 rev. + 300°	7.5
270°	2.3	1 rev. + 330°	8.2
300°	2.5	2 revolutions	9
330°	2.7	2 rev. + 30°	9.9
1 revolution	3	2 rev. + 60°	10.8
1 rev. + 30°	3.3	2 rev. + 90°	11.8

9. Copy and complete the following table for this curve.

Number of revolutions, x	0	1	2
Distance from center, y	1	▓▓▓	▓▓▓

10. What kind of number sequence do the distances form?

11. What would the distances from the center be after three and after four revolutions if the curve were continued?

Set II

If the successive distances of the loops of a spiral from its center form an *arithmetic* sequence, the spiral is *Archimedean*. If they form a *geometric* sequence, the spiral is *logarithmic*.

The following questions refer to this spiral.

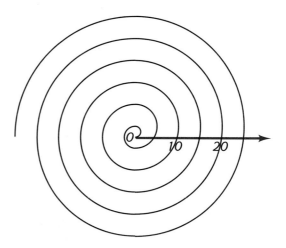

1. Copy and complete the following sequence of the distances of the successive loops from the center along the line indicated.

$$5 \quad 10 \quad ▓▓▓ \quad ▓▓▓ \quad ▓▓▓$$

2. What kind of sequence is this?

3. What kind of spiral is shown?

4. If the spiral continued, how far would the end of the 10th loop be from the center?

The following questions refer to this spiral.

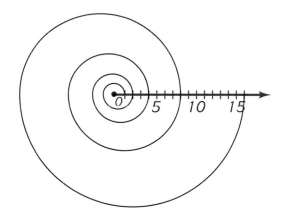

5. Copy and complete the following sequence of the distances of the successive loops from the center along the line indicated.

6. What kind of sequence is this?

7. What kind of spiral is shown?

8. If the spiral continued, how far would the end of the fifth loop be from the center?

The following questions refer to this spiral.

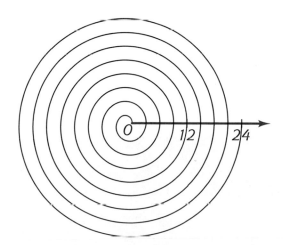

9. Copy and complete the following sequence of the distances of the successive loops from the center along the line indicated.

10. What kind of sequence is this?

11. What kind of spiral is shown?

12. If the spiral continued, how far would the end of the 20th loop be from the center?

The following questions refer to this spiral.

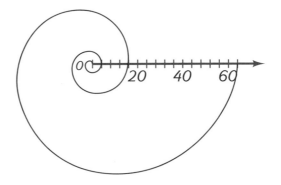

13. Copy and complete the following sequence of the distances of the loops from the center along the line indicated.

4   ▥▥▥   ▥▥▥

14. What kind of sequence is this?

15. What kind of spiral is shown?

16. If the spiral continued, how far would the end of the fifth loop be from the center?

Compare the spirals in these photographs with the spirals in exercises 1–16. Then tell what type of spiral each seems to be.

17. A spiral-mode antenna.

18. The arms of a galaxy.

Lick Observatory photograph

SET III

This figure, invented by the British psychologist J. Frazer, seems to be something other than what it really is.

From *An Introduction to Color* by Ralph M. Evans, John Wiley & Sons, 1948.

1. What does the figure seem to be?

2. What does it actually consist of?

Photograph by J. H. Lartigue; courtesy of the Association des Amis de Jacques-Henri Lartigue

6

The Cycloid

As a car moves forward, the tops of its wheels move faster with respect to the road than do the bottoms. This is shown in the picture above, taken many years ago by the French photographer Jacques-Henri Lartigue. The spokes closest to the ground can be seen distinctly, whereas those farthest from the ground are blurred.

As a wheel rolls along a straight line, its center travels at the same rate along another straight line. The diagram below shows five positions of such a wheel as it makes one revolution. The distances from A to B, from B to C, from C to D, and from D to E are equal because, if the wheel rolls forward at a steady speed, its center does also.

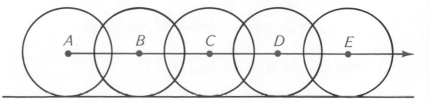

Other points on the wheel travel along more complicated paths. A point on the rim—a pebble caught in the tread, for example—follows a path called a *cycloid.*

A **cycloid** is the curve traced by a point on the rim of a wheel as the wheel rolls along a straight line.

The diagram below shows part of this curve. Again, five positions of a wheel as it makes one revolution are illustrated. In each successive position, the wheel has made a quarter turn. The first and last show the wheel with the point on the rim touching the ground; the other three show the wheel after turning 90°, 180°, and 270°.

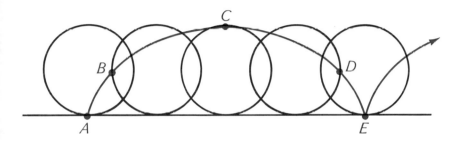

Notice that, as the wheel rolls equal distances along the ground, the point on the rim does not travel equal distances along its cycloid path. The distance from B to C is longer than the distance from A to B, yet each distance represents a quarter turn of the wheel. This shows that the point on the rim must move faster as it approaches C. Its greatest speed is at the moment that it has reached the top. As the point moves downward to E, it slows down until it stops for an instant at the bottom of the curve before beginning another arc. Points A and E of the cycloid, where the curve suddenly changes directions, are called *cusps.*

Although the early Greek mathematicians knew a lot about the conic sections (the circle, ellipse, parabola, and hyperbola) and about the spiral of Archimedes, they apparently never thought of the cycloid. In fact, no one knows who discovered this remarkable curve. Galileo, in the seventeenth century, suggested that it would be a good shape for an arch of a bridge. The great French mathematician Pascal later studied the cycloid as a result of suffering from a toothache! He decided that he needed something interesting to think about to take his mind off the pain and, as a result, made many discoveries about this "curve of a rolling wheel."

EXERCISES

SET I
EXPERIMENT: DRAWING CYCLOIDS

Part 1. The Cycloid The easiest way to draw a cycloid is to roll a wheel along a straight line and mark some of the positions taken by a point on its rim. Our procedure is based on this method.

Using a compass, draw a circle with a radius of 1 inch (2 centimeters) on a 3-by-5-inch card.* Cut the circle out and put the rest of the card aside for use in Part 3. Center the circle on the first figure below and mark 12 points on the rim as indicated by the lines. Take your circle away from the page and, using a ruler, draw diameters to connect the points that you have marked as shown in the second figure. You now have a wheel with "spokes." Number the spokes 0 through 11 as shown in the third figure, and cut a small notch in the rim of the wheel at the spoke numbered 0.

Turn sideways a sheet of graph paper ruled either 4 units per inch or 2 units per centimeter. Draw a line across it about 3 inches (7 centimeters) from the top. Mark the line with a series of 20 points that are 2 units ($\frac{1}{2}$ inch or 1 centimeter) apart. Number the points from the left 0 through 11 and 0 through 7 as shown in the diagram at the top of the next page.

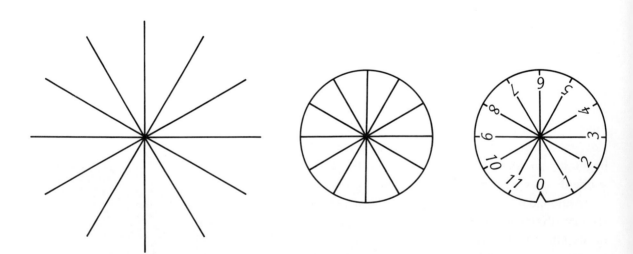

---

*The radius should be 1 inch if you are using graph paper ruled 4 units per inch. It should be 2 centimeters if you are using graph paper ruled 2 units per centimeter.

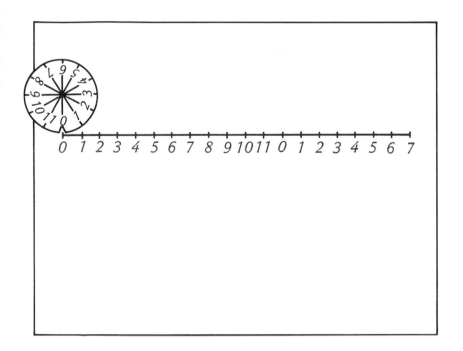

Place your wheel on the line so that the spoke numbered 0 points downward and the notch is at the point on the line numbered 0. Mark a heavy dot on the graph paper at the notch. Then roll the wheel to the right so that the spoke numbered 1 points downward and just touches the point on the line numbered 1. (You will have to adjust the position of the wheel slightly to make these numbers line up.) Mark another heavy dot on the graph paper through the notch in the wheel.

Repeat this procedure all the way across the paper so that each spoke points downward to the point on the line with its corresponding number; mark a dot in the notch each time.

Connect the dots with a smooth curve to form one full arc of a cycloid and part of a second one.

Part 2. A Curtate Cycloid The path of a point on the *inside* of a wheel that rolls along a straight line (other than the center of the wheel) is a *curtate cycloid*. To draw this curve, we will use the method used in drawing the cycloid.

Use a paper punch to punch a hole in the center of the spoke numbered 0 on your wheel. Place the wheel back on the line that you drew for Part 1 and roll it along the line, marking a dot on the paper through the center of the punched hole each time. When the dots are connected with a smooth curve, a curtate cycloid is the result.

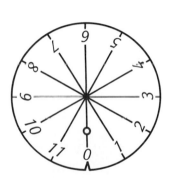

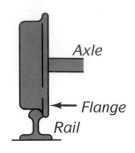

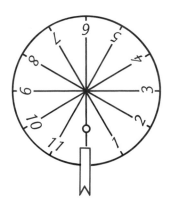

Part 3. A Prolate Cycloid   The wheels of a train have flanges that extend beyond the part of the wheel that rides on the track. Points on the flange are *outside* of the wheel and follow a *prolate cycloid*.

First, cut a small strip from the card out of which you cut the circle, cut a notch in the end of this strip, and tape it along the spoke numbered 0, as shown in the diagram at the left.

Roll the wheel along the line again, marking dots in the notch of the strip. A prolate cycloid results when the dots are connected with a smooth curve.

Part 4.   Use your drawings as a guide in answering the following questions.

A pebble caught in the rim of a tire follows a path in the shape of a cycloid.

1. The greater the spacing between successive points in your drawing, the faster the pebble is moving. Where is the pebble when it is moving the fastest?

The pebble comes to a stop for an instant when it suddenly changes direction.

2. Where is the pebble when this occurs?

A reflector attached to the middle of a spoke of a bicycle wheel follows a path in the shape of a curtate cycloid.

3. Where is the reflector when it is moving the slowest?

4. While the wheel is moving forward, does the reflector come to a stop at certain instants?

5. Is there any point inside a rolling wheel that moves forward at a steady rate?

A point on the flange of a train wheel follows a path in the shape of a prolate cycloid.

6. What is strange about the motion of such a point when it is at the bottom of its path?

7. Where is it with respect to the train track when it undergoes this motion?

Set II

The gears in an automatic transmission include wheels that roll inside and outside of other wheels. The paths of the points on the rims of such wheels are epicycloids and hypocycloids.

Experiment: Drawing Epicycloids and Hypocycloids

Part 1. An Epicycloid The path of a point on the rim of a wheel that rolls around the outside of a circle is an *epicycloid.*

Remove the small strip that was taped to the wheel made for the Set I experiment. Place the wheel on the center of a sheet of paper and trace its circumference. Mark points on the circle corresponding to the 12 points on the wheel, remove the wheel from the paper, and draw diameters connecting the points. Number the "spokes" that you have just drawn clockwise, starting with 0 at the top. The result should look like the bottom wheel in the adjoining figure.

Put the wheel at the top of the circle on your paper so that the spoke marked 0 on the wheel touches the spoke marked 0 on the circle and is in line with it. This is illustrated in the figure. Mark a heavy dot at the notch of the wheel.

Now roll the wheel to the right so that the spokes numbered 1 are in line with each other and mark another heavy dot through the notch of the wheel. Continue around the circle in the same way, being careful that the corresponding spokes are lined up each time. Finally, connect the dots with a smooth curve.

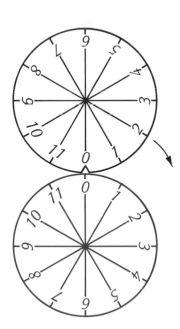

Part 2. A Hypocycloid The path of a point on the rim of a wheel that rolls around the inside of a circle is a *hypocycloid.*

Use a compass to draw a circle with a radius of 3 inches (6 centimeters) on a sheet of paper and a protractor to mark points around the circle at 10° intervals. Number the points starting from the bottom as shown in the figure below.

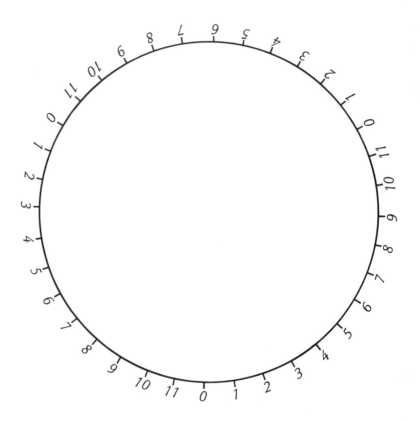

Place your wheel inside the circle so that the spoke marked 0 on the wheel touches the point marked 0 at the bottom of the circle. This is illustrated in the figure shown here. Mark a heavy dot at the notch of the wheel.

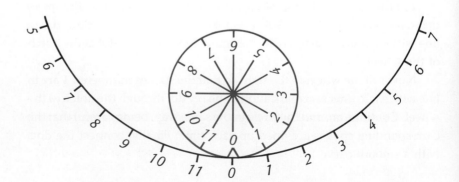

Roll the wheel counterclockwise around the circle, marking dots in the notch of the wheel as you do so. Connect the dots with a smooth curve.

Part 3. Another Hypocycloid Use a compass to draw a circle with a radius of 2 inches (4 centimeters) on a sheet of paper and a protractor to mark points around the circle at 15° intervals. Number the points starting from the bottom as shown in this figure.

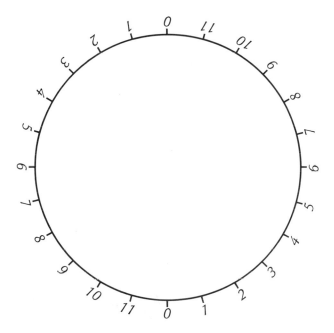

Follow the rest of the procedure described in Part 2 to finish the drawing.

Part 4. Use your drawings as a guide in answering the following questions.

The epicycloid that you drew in Part 1 is also called a *cardioid*.

1. Why? (*Hint:* The term *cardiac* has a related meaning.)

2. How many cusps does a cardioid have?

3. Describe the symmetry of a cardioid.

4. Does a point traveling in the path of a cardioid travel at a constant rate?

The hypocycloid that you drew in Part 2 is also called a *deltoid*.

5. In what way is it like the letter of the Greek alphabet named delta: Δ?

6. How many times did the wheel turn in traveling once around the circle to produce the deltoid?

7. Describe the symmetry of a deltoid.

8. Does a point traveling in the path of a deltoid travel at a constant rate?

The hypocycloid that you drew in Part 3 does not have a special name because its shape is not very interesting. If you drew it accurately, you discovered that it is not even a curve.

9. What is it?

10. How many cusps does it have?

The figures below were made by wheels rolling around other wheels.

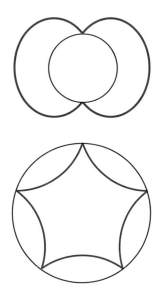

11. What kind of curve is shown in the first figure?

12. How do you think the radius of the circle in the first figure compares in length to the radius of the wheel that produced the curve?

13. Describe the symmetry of the curve.

14. What kind of curve is shown in the second figure?

15. How do you think the radius of the circle in the second figure compares in length to the radius of the wheel that produced the curve?

16. Describe the symmetry of the curve.

The figures below are of the sort produced by a toy called a "spirograph." Describe how you think each curve might have been produced.

17.

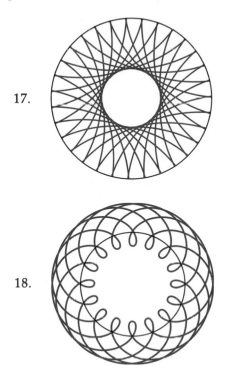

18.

SET III

The cycloid has an interesting property when it is turned upside down. If a curved ramp is built in the shape of a cycloid and a couple of marbles are released from any two points on the ramp, they will always reach the bottom at the same time!

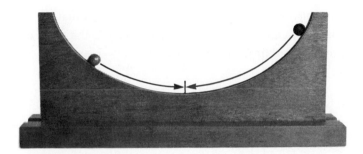

One marble will usually have farther to roll than the other; so how can this be possible?

CHAPTER

6

Summary and Review

In this chapter we have become acquainted with:

The **conic sections** *(Lessons 1, 2, and 3).* The conic sections are curves that are formed when a cone is sliced into sections. They include:

The *circle:* the set of all points in a plane that are at the same distance from a fixed point in the plane called the center.

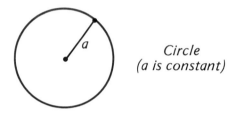

Circle
(a is constant)

The *ellipse:* the set of all points in a plane such that the sum of the two distances from each point to two fixed points, called the foci, is the same.

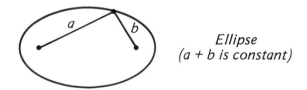

Ellipse
(a + b is constant)

The *parabola:* the set of all points in a plane such that the distance of each point from a fixed point, called the focus, is the same as its distance from a fixed line, called the directrix.

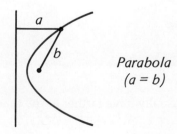

Parabola
(a = b)

The *hyperbola:* the set of all points in a plane such that the difference between the two distances from each point to two fixed points, called the foci, is the same.

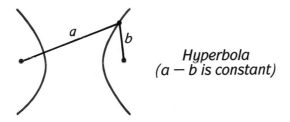

Hyperbola
(a − b is constant)

The **sine curve** *(Lesson 4)*. The *amplitude* of a sine curve is its maximum distance from the *x*-axis. The *wavelength* of a sine curve is the distance along the *x*-axis required for one complete wave. The *frequency* of a sine curve is the number of complete waves when *x* increases by 360°.

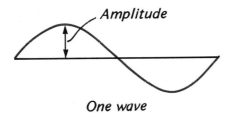

Amplitude

One wave

Spirals *(Lesson 5)*. A spiral is a curve traced by a point that moves around a fixed point from which the point moves farther and farther away.

The successive distances of the loops of an *Archimedean spiral* from its center form an arithmetic sequence.

The successive distances of the loops of a *logarithmic spiral* from its center form a geometric sequence.

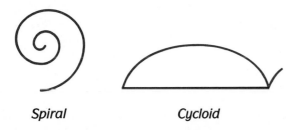

Spiral   *Cycloid*

The **cycloid** *(Lesson 6)*. A cycloid is the curve traced by a point on the rim of a wheel as the wheel rolls along a straight line.

EXERCISES

SET I

These photographs show ice cream cones that have been cut in different directions.

1

2

3

4

1. Name each curve.

2. What is the name for this set of curves?

This map shows part of the last orbit of the *Skylab* space station before it broke apart and fell to earth over the Indian Ocean and Australia.

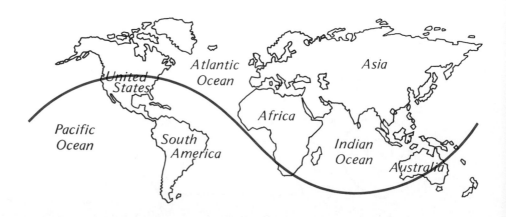

3. What kind of curve does its orbit seem to be with respect to the map?

4. How much of the curve is shown?

Gerald Ratto Photography

A model of an outdoor exhibition hall designed by the architect William Blackwell is shown in this photograph.

5. In what kind of curve are the walls of the hall arranged?

Galileo found that the curve traced by a point on the rim of a rolling wheel makes the strongest possible arch for a bridge.

Courtesy of New York State Department of Transportation

6. What curve is it?

When an alpha particle is shot toward the nucleus of an atom, the particle is repelled and changes direction. The great British scientist Ernest Rutherford showed that the particle's path is along one of the two branches of a curve that we have studied. It is illustrated on this postage stamp.

7. Which curve is it?

Here is a photograph of a fountain in which the jets of water are pointed in many directions.

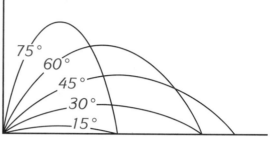

8. What curve is the shape of the path of each jet?

The figure at the right of the photograph shows how the path would change as the angle of the jet is changed. Which one of the five angles shown in this figure results in the jet going

9. the highest?

10. the farthest in a horizontal direction?

11. At which one of these angles do the jets in the photograph seem to be pointed?

12. What other angle would result in the water traveling the same horizontal distance?

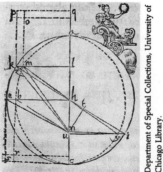

This figure appeared in a book by Johann Kepler to illustrate the orbit of the planet Mars. Before Kepler's work, it was thought that the orbit had the shape of the solid curve.

13. What kind of curve is it?

Kepler represented the actual shape of the orbit with the dotted curve.

14. What kind of curve is it?

The sun is at one focus of this curve, marked S in this figure.

15. Which axis of the curve is S on?

16. Where is Mars in this figure when it is farthest from the sun?

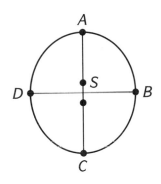

Set II

The graph of an alternating electric current is a pair of sine curves. The curve shown in dark brown in this figure represents the current and the curve shown in light brown represents the voltage.

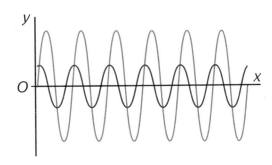

1. How do the amplitudes of the two curves compare?

2. How do the wavelengths of the two curves compare?

3. How many waves of current are shown?

Alternating current in the United States has 60 waves per second.

4. What fraction of a second is shown in the figure?

This photograph shows a round placemat made of straw.

5. What curve does it illustrate?

6. Does the curve seem to be Archimedean or logarithmic?

7. Explain the reason for your answer.

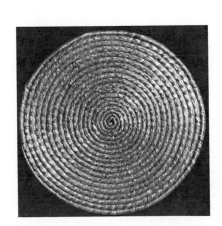

This figure from a seventeenth-century book illustrates the reflecting property of a parabolic mirror. A candle is shown at the focus of the mirror.

8. What happens to the light from the candle when it is reflected by the mirror?

9. What do you suppose would happen at the focus if the candle were taken away and the mirror were pointed toward the sun?

This figure shows the two branches of a hyperbola. The branch on the right represents a hyperbolic mirror with a source of light located at one focus.

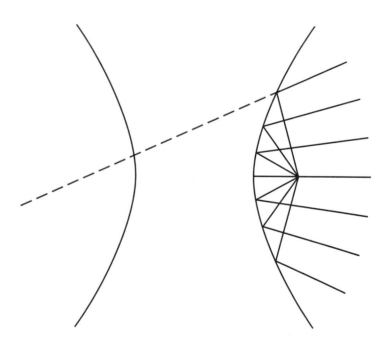

10. How does a hyperbolic mirror reflect light differently from a parabolic mirror?

11. Place a sheet of tracing paper over the figure and make an accurate tracing of it. Then use a ruler to extend the seven lines on the right across the figure to the left. (One is already drawn in as a dotted line as an example.)

12. What is interesting about the result?

There is a whispering gallery in the Capitol building in Washington, D.C. The ceiling of this room is a curved surface having the shape of an elliptical mirror. If you stand at the right spot in this large room and whisper very softly, someone standing at a certain spot many meters away can hear you clearly. The figure below represents the elliptical mirror of the gallery.

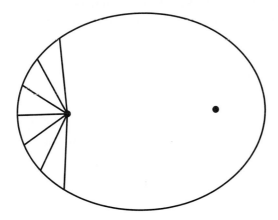

13. Where do you suppose you and the person who hears you are standing?

14. Place a sheet of tracing paper over the figure and make an accurate tracing of it. Then use a ruler to draw some lines showing the sound from you being reflected by the mirror to the person who hears you.

Sir Christopher Wren, a great English architect, discovered that an arc of a cycloid is a certain simple number of times as long as the diameter of the wheel that produces the cycloid. Look at this drawing of a cycloid and its wheel.

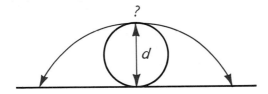

15. What do you think the number is?

The *area* under one arc of a cycloid is also a simple number of times as large as the area of the wheel that produces the cycloid.

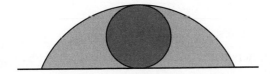

16. What do you think the number is?

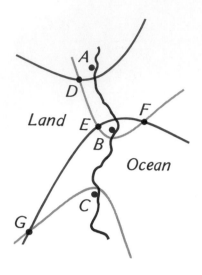

Ships at sea use the LORAN system to determine their locations by radio. On the map at the left, points A, B, and C show the positions of radio transmitters. The two curves (one shown in dark brown and the other in light brown) represent possible locations of a ship receiving certain signals from the stations.

17. What kind of curves are they?

18. In how many different points do the curves intersect?

19. From the signals received, the captain of the ship knows that the ship is at one of these points. Which one?

SET III
EXPERIMENT: CURVES FROM CUTTING A CANDLE

Cut out a strip of paper about 10 centimeters long and 3 centimeters wide. Wind it tightly around a small birthday cake candle and, with a razor blade, carefully cut through the paper and candle at a slant, as shown in this photograph. Look at the cross section of the candle that you have cut.

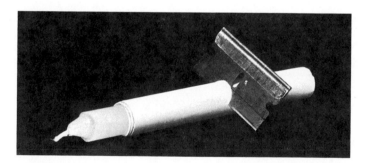

1. What curve do you see?

2. The length of one axis of this curve was determined by the diameter of the candle. Which one?

3. What determined the length of the other axis of the curve?

Unwind the paper.

4. What curve results?

5. What dimension of the curve was determined by the size of the candle?

6. What dimension of the curve was determined by the slant of the cut?

Further Exploration

LESSON 1

1. Experiment: *Curves by Folding Paper*
Use a compass to draw a circle with a radius of about 8 centimeters on a sheet of tracing paper. Mark a point approximately 3 centimeters inside the circle and label it A. Cut the circle out and fold it so that its edge falls on the point that you marked; make a sharp crease.

Unfold the paper and fold it again in a different direction so that the edge again falls on the point. Repeat this many times so that points all around the edge have been folded onto the point.

a. What kind of curve is the result?

Draw the curve on the paper. Also mark the center of the original circle and label it B.

b. Where are the foci of the curve?

Draw a radius of the circle as shown in this figure. Fold the paper so that the outer endpoint of the radius falls on point A. Notice that the fold touches the curve at one point.

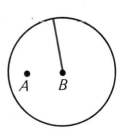

c. How is the sum of the distances from this point to points A and B related to the radius of the circle?

Open the paper flat again, draw another radius of the circle, and repeat the directions in the preceding paragraph.

d. What do you notice?
e. What do these results have to do with the definition of an ellipse?

Do the entire experiment two more times: once with a point near the center of the circle and once with a point close to the edge.

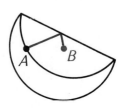

f. How does the location of the point determine the shape of the curve formed?

Photograph of Arthur P. Frigo from *Invention, Discovery, and Creativity* by A. D. Moore. Copyright © 1969 by Doubleday & Company, Inc. Reproduced by permission of the publisher.

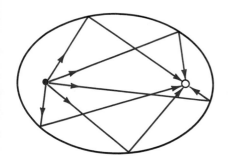

2. Elliptipool is a game played on a pool table whose shape is an ellipse. The table has only one pocket, which is located at one focus. A ball at the other focus can be hit in *any* direction and, if there are no other balls in the way, it will end up in the pocket! The drawing at the left shows some of the paths that the ball might take.

a. Suppose that you are playing Elliptipool and that you want to hit the ball at A, in the figure below, so that it goes into the pocket without hitting ball B. What will you do?

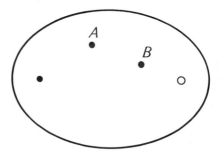

b. Trace the figure and show the path of the ball.
c. Suppose that a ball is at C on an Elliptipool table that has no pocket. You hit the ball so that it passes through the focus on

the right with enough force so that it rebounds from the cushion several times before coming to a stop. Trace the figure and show the path of the ball.

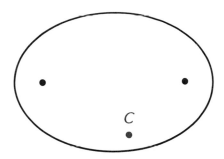

d. If the ball were to keep on going without stopping, where would its path eventually be?

LESSON 2

An accurate drawing of a parabola can be made with a ruler and a compass.

a. On a sheet of plain paper, draw nine parallel lines so that each consecutive pair of lines are 0.5 inch apart. Label the top line ℓ and then number the eight lines below it 1 through 8.

```
ℓ ─────────────────────
1 ─────────────────────
2 ──────────●P──────────
3 ─────────────────────
4 ─────────────────────
5 ─────────────────────
6 ─────────────────────
7 ─────────────────────
8 ─────────────────────
```

Mark a point at about the center of line 2 and label it P. Put the metal point of a compass into the paper at point P and adjust the radius so that it is equal to 0.5 inch. Draw an arc that touches line 1 and label the point where it touches the line, A.

Keeping the metal point of the compass at P, increase the radius of the compass to 1 inch. Draw two arcs that intersect line 2 and label the points of intersection B and C.

Keeping the metal point of the compass at P, increase the radius of the compass to 1.5 inches. Draw two arcs that intersect line 3 and label the points of intersection D and E.

What you have done so far, together with a summary of the rest of the steps, is listed below.

Radius of compass in inches	Intersections with	Names of points
0.5	line 1	A
1	line 2	B and C
1.5	line 3	D and E
2	line 4	F and G
2.5	line 5	H and I
3	line 6	J and K
3.5	line 7	L and M
4	line 8	N and O

Connect the points with a smooth curve to form a parabola.

b. Where is the focus of the parabola?

c. Choose any lettered point on the parabola and measure its distance from point P and its distance from line ℓ. What do you notice about the two distances?

d. Use the definition of a parabola to explain why this method produces one.

LESSON 3

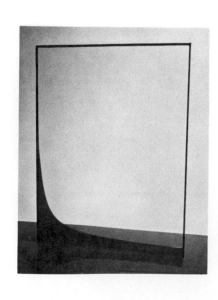

If two sheets of glass are taped together along one pair of edges, separated slightly along the opposite pair, and placed in a pan of water to which some dye has been added, one branch of a hyperbola will be formed by capillary action.

A formula for a hyperbola produced in this way is

$$y = \frac{4}{x} + 1$$

a. Copy and complete the following table for this formula.

x	$\frac{1}{3}$	$\frac{1}{2}$	1	2	3	4	6	8	10	12
y	13	9								

b. Draw an appropriate pair of axes, plot the 10 points in the table, and connect them with a smooth curve.

c. You have drawn the branch of the hyperbola formed by the water between the two sheets of glass. If the other branch of the hyperbola were added to the graph, where would it be?

d. Where are the asymptotes of the hyperbola?

LESSON 4

Everyone has good days and bad days. According to biorhythm theory, every person is in three different states of well-being throughout his or her life. These states are physical, emotional, and mental, and the degree of well-being in any given state varies to form a cyclical pattern that can be illustrated by a sine curve.

The curves are shown below.

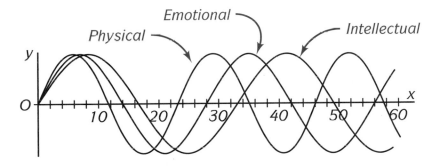

The three cycles supposedly start at birth and continue throughout life. One wave of the physical cycle takes 23 days.

a. How many days do one wave of the emotional and intellectual cycles take?

Suppose that on the day you were born, all three cycles started together as shown at the beginning of the graph above.

b. Can you figure out how many days pass until all three cycles start out together again? (*Hint:* Look at the graph below, which shows two curves with one wave taking two days and the other wave taking three days.)

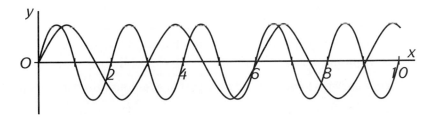

LESSON 5

From *The Language of Mathematics* by Frank Land, John Murray Ltd.

Two opposite sets of spirals appear in the head of the chrysanthemum shown in the photograph above.

    a. What kind of spirals does the head of the chrysanthemum seem to have?

    b. How many spirals are in the set that winds clockwise from the center? (Count them from the drawing below.)

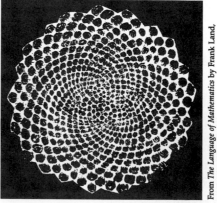

From *The Language of Mathematics* by Frank Land, John Murray Ltd.

    c. How many spirals are in the set that winds counterclockwise from the center?

Your answers to parts b and c are part of a famous number sequence.

    d. What sequence is it?

LESSON 6

Experiment: *More Hypocycloids*

With a compass, draw circles with radii of 1 centimeter and 3 centimeters on a 4-by-6-inch card. Cut the circles out, place them on the circles below, and mark points on their rims as indicated by the arrows. Take your circles away from the page and, with a ruler, draw diameters to connect the points that you have marked to form two wheels with "spokes." Number the spokes as shown, and cut a small notch in the rim of each wheel at the spoke at the bottom numbered 0.

 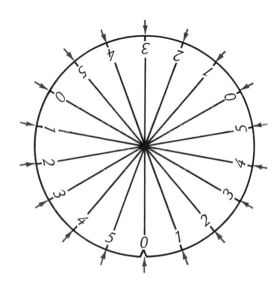

a. Use a compass to draw a circle with a radius of 4 centimeters on a sheet of paper and a protractor to mark points around the circle at 15° intervals. Number the points starting from the bottom as shown in the figure on the next page.

Place the small wheel inside the circle so that the spoke marked 0 on the wheel touches the point marked 0 at the bottom of the circle. Mark a heavy dot at the notch of the wheel.

Roll the wheel counterclockwise around the circle, marking dots in the notch of the wheel as you do so. Connect the dots with a smooth curve.

b. Now repeat the directions in part a, drawing a *new* circle with a radius of 4 centimeters and rolling the *larger* wheel around it. This time the wheel will have to be rolled around the circle more than once in order to produce the entire curve.

c. What is surprising about the results?

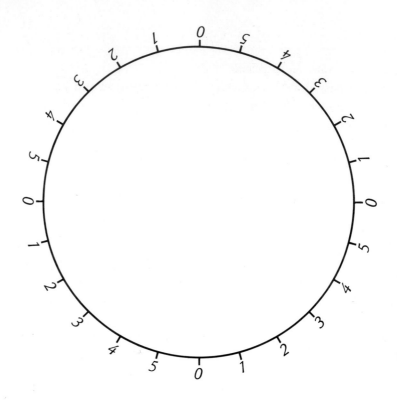

Methods of Counting

"I think you've had enough. Why don't I call you a cab?"

1

The Fundamental Counting Principle

Suppose that an ice cream store sells two drinks, sodas and milk shakes, in three sizes, small, medium, and large, and four flavors, vanilla, chocolate, strawberry, and cherry. If you order one drink, how many choices do you have?

The choices are shown in the diagram on the next page. The diagram is read from left to right, starting at the point on the left. The first choice illustrated is the size: small, medium, or large. Making this choice corresponds to choosing one of the three branches on the left and traveling along it to one of the 3 points above the first arrow. The second choice illustrated is the flavor: vanilla, chocolate, strawberry, or cherry. Making this choice corresponds to choosing one of four branches and traveling along it to one of the 12 points above the second arrow. The third choice is the type of drink: soda or milk shake. Making this choice corresponds to choosing one of two branches and traveling along it to one of the 24 points above the third arrow. This diagram, called a *tree diagram* because of its branches, reveals that there are 24 choices altogether.

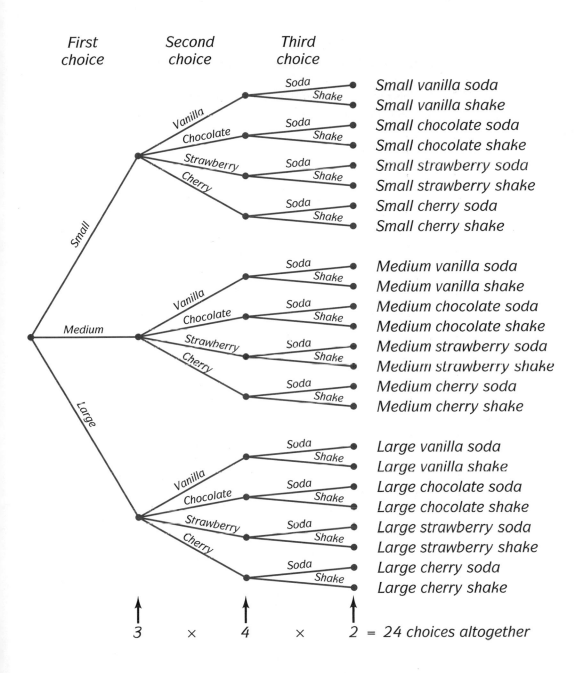

Notice that this number can be obtained by multiplying the number of choices of size, 3, by the number of choices of flavor, 4, by the number of choices of type of drink, 2:

$$3 \times 4 \times 2 = 24.$$

This method can be used to quickly count all kinds of things. We will call it the **fundamental counting principle.**

To find the number of ways in which a series of successive events can occur, *multiply* the numbers of ways in which each event can occur.

This principle is useful in counting numbers of ways for which making a tree diagram would not be practical. For example, suppose that you are at the ice cream store illustrated in the cartoon. If it sells ice cream in three forms, sodas, milk shakes and dishes, four sizes, small, medium, large, and extra large, and 150 flavors, how many choices do you have?

By the fundamental counting principle, if you have 3 choices followed by 4 choices followed by 150 choices, you have

$$3 \times 4 \times 150 = 1,800$$

choices altogether!

EXERCISES

SET I

This diagram shows all the choices of a popular brand of marker pens.

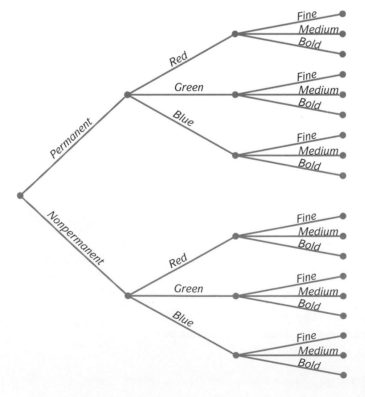

	Red	Fine	Permanent red fine
		Medium	Permanent red medium
		Bold	Permanent red bold
Permanent	Green	Fine	Permanent green fine
		Medium	Permanent green medium
		Bold	Permanent green bold
	Blue	Fine	Permanent blue fine
		Medium	Permanent blue medium
		Bold	Permanent blue bold
	Red	Fine	Nonpermanent red fine
		Medium	Nonpermanent red medium
		Bold	Nonpermanent red bold
Nonpermanent	Green	Fine	Nonpermanent green fine
		Medium	Nonpermanent green medium
		Bold	Nonpermanent green bold
	Blue	Fine	Nonpermanent blue fine
		Medium	Nonpermanent blue medium
		Bold	Nonpermanent blue bold

The pens are available in two basic types: permanent and nonpermanent.

1. How many choices of colors are available?

2. How many choices of point types are available?

3. How many different pens are available altogether?

4. How can the number of different pens be found by using the fundamental counting principle?

5. How many different *permanent* pens are available?

6. How many different *blue* pens are available?

This diagram shows all the ways in which two questions in a multiple-choice test can be answered if there are three choices for each answer.

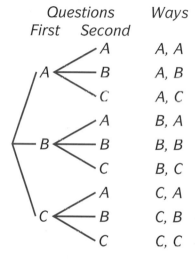

7. How many ways are there?

8. What numbers would you multiply to find this answer by using the fundamental counting principle?

In how many ways can two questions in a multiple-choice test be answered if there are

9. four choices for each answer?

10. five choices for each answer?

In how many ways can *three* questions in a multiple-choice test be answered if there are

11. three choices for each answer?

12. four choices for each answer?

13. five choices for each answer?

The discovery of the genetic code — the way in which the DNA of a living organism determines the structure of the organism's proteins — is one of the most remarkable achievements of molecular biology.

The DNA molecule is made up of four kinds of smaller molecules. These smaller molecules are often referred to by the letters A, C, G, and T.* Groups of these small molecules along the DNA molecule make up the code.

The list below shows all the different sequences of three molecules that are possible. We will refer to these sequences as "three-letter words."

AAA	CAA	GAA	TAA
AAC	CAC	GAC	TAC
AAG	CAG	GAG	TAG
AAT	CAT	GAT	TAT
ACA	CCA	GCA	TCA
ACC	CCC	GCC	TCC
ACG	CCG	GCG	TCG
ACT	CCT	GCT	TCT
AGA	CGA	GGA	TGA
AGC	CGC	GGC	TGC
AGG	CGG	GGG	TGG
AGT	CGT	GGT	TGT
ATA	CTA	GTA	TTA
ATC	CTC	GTC	TTC
ATG	CTG	GTG	TTG
ATT	CTT	GTT	TTT

14. How many three-letter words are in this list?

15. Use the fundamental counting principle and the fact that there are four choices for each letter to show that the list includes every possible three-letter word.

The fact that 16 three-letter words start with the letter C can be found by counting them in the list or by using the fundamental counting principle:

$$1 \times 4 \times 4 = 16$$

1 choice: 4 choices each:
   C   A, C, G, or T

*The molecules are nucleotides, and they contain the bases *adenine, cytosine, guanine,* and *thymine.*

Use the fundamental counting principle to show that

16. 16 words end with the letter T.
17. four words begin and end with the letter A.

Suppose the genetic code used four letters but each word was only two letters long.

18. How many different words would be possible?
19. If the genetic code used four letters and each word was four letters long, how many different words would be possible?

Suppose the genetic code used five letters but each word was two letters long.

20. How many different words would be possible?
21. If the genetic code used five letters and each word was three letters long, how many different words would be possible?

The Human Genome Project, a multimillion dollar venture currently under way, is attempting to figure out all 3,000,000,000 letters that make up human DNA.

Keys of different shapes are designed by choosing from several patterns for each of their parts. The keys of General Motors cars have six parts.

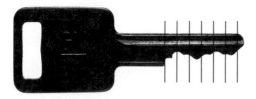

22. General Motors originally used two patterns for each part. How many different key designs were possible?
23. The number of possible key designs was later increased by changing to three patterns for each part. How many different key designs are possible in this system?
24. How many key designs would be possible if the number of patterns for each part were increased to four?

SET II

The Twenty-One Bell slot machine has three reels with 20 positions on each reel. Most of the positions have one symbol but a few positions have two. The arrangement of the symbols is summarized in the table below.

Bar

BAR

Bell

Cherry

Lemon

Melon

Orange

Plum

Seven

7

Symbol	Reel 1	Reel 2	Reel 3
Bar	3	2	1
Bell	1	5	8
Cherry	2	6	0
Lemon	0	0	4
Melon	2	2	2
Orange	5	5	4
Plum	7	3	3
7	1	1	1
(Positions)	20	20	20

The smallest payoff is for a cherry on reel 1 with no cherries on reels 2 and 3. The table shows that there are 2 cherries on reel 1, 14 spaces on reel 2 that do not have cherries, and 20 spaces on reel 3 that do not have cherries. By the fundamental counting principle, there are

$$2 \times 14 \times 20 = 560$$

ways to win this payoff.

The next smallest payoff is for cherries on the first two reels.

1. Show that there are 240 ways of winning this payoff.

One of the jackpot payoffs is for melons on reels 1 and 2 and a bar on reel 3.

2. Show that there are four ways to win this jackpot.

Use the fundamental counting principle to find how many ways there are of getting the other jackpots, which are listed below.

3. Three melons.

4. Three bars.

5. Three 7's.

6. Which one of these jackpots do you suppose is the biggest?

7. How many ways are there altogether of hitting a jackpot? (Look at your answers to exercises 2–5.)

Find how many ways there are of getting the other payoffs, which are listed below.

8. Oranges on reels 1 and 2 and a bar on reel 3.

9. Three oranges.

10. Plums on reels 1 and 2 and a bar on reel 3.

11. Three plums.

12. Bells on reels 1 and 2 and a bar on reel 3.

13. Three bells.

Look back at all the numbers of ways of getting payoffs (beginning with the 560 ways of getting a cherry on reel 1 with no cherries on reels 2 and 3.)

14. How many ways of getting a payoff are there altogether?

15. In how many different positions can the three reels of the machine stop? (Remember that each reel has 20 positions.)

16. If you played this slot machine over and over, how frequently would you expect to get a payoff? (Compare your answers to exercises 14 and 15.)

17. As you play the slot machine, what losing set of three symbols would you expect to see the most frequently?

18. What set of three symbols would you *never* see?

A manufacturer of trumpets enables its customers to "custom-build" their instruments by giving them a wide variety of choices. In describing choices for the bell of the trumpet, for example, the company writes:

36 bells. This remarkable selection includes six different flares, each of which is available in light, regular and heavy weight and your choice of yellow or gold brass.*

19. Show why these choices result in 36 different bells.

In addition to 36 choices for the bell of the trumpet, there are 8 choices for the mouthpipe and 5 choices for the bore.

20. How many different trumpets are possible from these choices?

The Vincent Bach Trumpet. The Selmer Company.

---

*Advertisement for the Bach Stradivarius trumpet made by the Selmer Company, Elkhart, Indiana.

Ten-digit telephone numbers were first planned by the Bell System in 1945. They consist of three parts: an *area code* (the first three digits), an *office code* (the next three digits), and a *line number* (the last four digits.)

In the original plan for *area codes*, the first digit could be any number from 2 through 9, the second digit was either 0 or 1, and the third digit could be any number except 0.

21. According to this plan, how many different area codes were possible altogether?

Among these possible area codes, the ones ending with 11 were reserved for other purposes.

22. How many area codes were eliminated because of this?

23. States with only one area code had a 0 as the middle digit. How many area codes having 0 as the middle digit were possible?

In the original plan for *office codes*, the first two digits could be any number from 2 through 9 and the third digit could be any number.

24. According to this plan, how many different office codes were possible altogether?

The four digits of a *line number* could be any numbers.

25. How many different line numbers were possible altogether?

26. According to the Bell System's original plan, how many different telephone numbers could have the same area code?

The following two problems are from the book *Hidden Connections, Double Meanings* by David Wells.*

27. In how many ways can a casting director choose a mother, a father, and one child, from two actresses, three actors and four children?

28. How many triangles are in this figure?

29. What is the connection between these two problems?

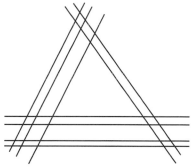

Adapted from *Hidden Connections, Double Meanings* by David Wells, Cambridge University Press, 1988, p. 109.

Set III

A book titled *Cent mille milliards de poèmes* by Raymond Queneau contains one hundred thousand billion poems.† The poems are sonnets, each one consisting of 14 lines. They are printed on the right-hand pages of the

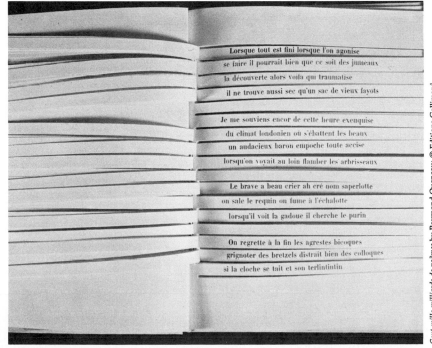

Cent mille milliards de poèmes by Raymond Queneau; © Editions Gallimard

*Cambridge University Press, 1988.

†Large numbers and their names are discussed on pages 184–186.

book only, the left-hand pages being blank. Each page is sliced into horizontal strips, with 1 line of the sonnet on a strip. The strips can be turned individually so that the lines can be selected in many different ways. Regardless of how they are chosen, however, every sonnet formed is structurally correct and makes sense.

How many pages must the book contain to produce a hundred thousand billion poems? Explain your answer.

"Blue Boy seems to be holding back a bit."

Courtesy of Joseph Zeis

Permutations

Eight horses are entered in a race in which bets are placed on which horses will finish first, second, and third. If the race is run and there are no ties, in how many orders can the first three horses come in?

Rather than trying to make a list of all the orders, we can use the fundamental counting principle. The horse that comes in first could be any one of the 8 horses. After that horse has won, the horse coming in second could be any one of the 7 horses remaining. The horse coming in third could be any one of the 6 horses remaining. By the fundamental counting principle, there are

$$8 \times 7 \times 6 = 336$$

ways in which the first three horses can come in.

The different ways in which the horses can come in are called *permutations*.

| A **permutation** is an arrangement of things in a definite order.

Another way to describe the number of ways in which three of the eight horses can finish first, second, and third is to say that we are finding "the number of permutations of eight things taken three at a time." This can be written in symbols as

$$_8P_3$$

in which P stands for "permutations" and the two numbers are written as subscripts.

The number of different orders in which all eight horses can finish the race can be written as

$$_8P_8$$

By the fundamental counting principle, this is

$$8 \times 7 \times 6 \times 5 \times 4 \times 3 \times 2 \times 1$$

or 40,320 different orders!

Mathematicians use an exclamation mark to indicate this product by writing 8!. The exclamation mark is also called a **factorial** symbol and 8! is read as "8 factorial."

| In general, $n!$ means to multiply the consecutive numbers from n all the way down to 1.

Notice that $_8P_8 = 8!$. That is, the number of permutations of "8 things taken 8 at a time" is "8 factorial."

| In general, the number of permutations of n different things taken n at a time is $n!$.

Although the symbol ! is not used in mathematics to express surprise, the numbers that it represents are, like 8! above, often surprisingly large. For example, consider 5!, the number of different orders in which the horses could come in in a race with just five horses. They are illustrated in the tree diagram shown on the next page, in which A, B, C, D, and E represent the five horses. Because

$$5! = 5 \times 4 \times 3 \times 2 \times 1 = 120$$

1st 2nd 3rd 4th 5th Order

Order	#
ABCDE	1
ABCED	2
ABDCE	3
ABDEC	4
ABECD	5
ABEDC	6
ACBDE	7
ACBED	8
ACDBE	9
ACDEB	10
ACEBD	11
ACEDB	12
ADBCE	13
ADBEC	14
ADCBE	15
ADCEB	16
ADEBC	17
ADECB	18
AEBCD	19
AEBDC	20
AECBD	21
AECDB	22
AEDBC	23
AEDCB	24
BACDE	25
BACED	26
BADCE	27
BADEC	28
BAECD	29
BAEDC	30
BCADE	31
BCAED	32
BCDAE	33
BCDEA	34
BCEAD	35
BCEDA	36
BDACE	37
BDAEC	38
BDCAE	39
BDCEA	40
BDEAC	41
BDECA	42
BEACD	43
BEADC	44
BECAD	45
BECDA	46
BEDAC	47
BEDCA	48
CABDE	49
CABED	50
CADBE	51
CADEB	52
CAEBD	53
CAEDB	54
CBADE	55
CBAED	56
CBDAE	57
CBDEA	58
CBEAD	59
CBEDA	60
CDABE	61
CDAEB	62
CDBAE	63
CDBEA	64
CDEAB	65
CDEBA	66
CEABD	67
CEADB	68
CEBAD	69
CEBDA	70
CEDAB	71
CEDBA	72
DABCE	73
DABEC	74
DACBE	75
DACEB	76
DAEBC	77
DAECB	78
DBACE	79
DBAEC	80
DBCAE	81
DBCEA	82
DBEAC	83
DBECA	84
DCABE	85
DCAEB	86
DCBAE	87
DCBEA	88
DCEAB	89
DCEBA	90
DEABC	91
DEACB	92
DEBAC	93
DEBCA	94
DECAB	95
DECBA	96
EABCD	97
EABDC	98
EACBD	99
EACDB	100
EADBC	101
EADCB	102
EBACD	103
EBADC	104
EBCAD	105
EBCDA	106
EBDAC	107
EBDCA	108
ECABD	109
ECADB	110
ECBAD	111
ECBDA	112
ECDAB	113
ECDBA	114
EDABC	115
EDACB	116
EDBAC	117
EDBCA	118
EDCAB	119
EDCBA	120

EXERCISES

SET I

In the Olympics, six gymnasts compete for the gold, silver, and bronze medals in each event. The symbol for the number of ways in which these medals can be won in a given event is

$$_6P_3$$

1. What does P stand for?

2. What do the 6 and 3 stand for?

3. Use the fundamental counting principle to find the number of ways in which these medals can be won.

At the beginning of a game of Scrabble, one of the players draws the seven letters shown here.

The word STARE is an example of one of the arrangements that can be made from five of these seven letters.

4. Write the symbol for the number of arrangements of seven different objects taken five at a time.

5. Use the fundamental counting principle to find the number of different arrangements that can be made from these seven letters taken five at a time.

The word PIRATES is an example of one of the arrangements that can be made from all seven of these letters.

6. Write the symbol for the number of arrangements of seven different objects taken seven at a time.

7. Use the fundamental counting principle to find the number of different arrangements that can be made from all seven of these letters.

8. What counting problem concerning these letters does the symbol $_7P_6$ represent?

For a Little League game, just nine players on one of the teams show up.

9. Write the symbol for the number of ways in which the pitcher and catcher can be chosen from the nine players.

10. Find the number of ways.

11. Write the symbol for the number of ways in which the first baseman, second baseman, third baseman, and shortstop can be chosen from the remaining seven players.

12. Find the number of ways.

13. Write the symbol for the number of ways in which the left fielder, center fielder, and right fielder can be chosen from the remaining three players.

14. Find the number of ways.

15. Find the number of ways in which all nine positions can be chosen from the nine players.

16. Could the team's coach try out every possible way of choosing the nine positions by trying each way in a different game? Explain.

In the game of bingo, 75 balls numbered with the numbers from 1 to 75 are mixed and drawn one at a time.

In how many ways can

17. the first number be chosen?

18. the second number be chosen?

19. the third number be chosen?

20. the first two numbers be chosen?

21. the first three numbers be chosen?

Set II

Here is a table of the values of 1! through 15!.
Use this table to *show whether* each of the following equations is true or false.

> *Example:* $2! \times 5! = 10!$
> *Answer:* False because $2 \times 120 \neq 3{,}628{,}800$.

1. $3! + 3! = 6!$

2. $10! = 10 \times 9!$

3. $1! + 4! + 5! = 145$

4. $\dfrac{8!}{4!} = 2!$

5. $6! \times 7! = 10!$

$1! = 1$	
$2! = 2$	
$3! = 6$	
$4! = 24$	
$5! = 120$	
$6! = 720$	
$7! = 5{,}040$	
$8! = 40{,}320$	
$9! = 362{,}880$	
$10! = 3{,}628{,}800$	
$11! = 39{,}916{,}800$	
$12! = 479{,}001{,}600$	
$13! = 6{,}227{,}020{,}800$	
$14! = 87{,}178{,}291{,}200$	
$15! = 1{,}307{,}674{,}368{,}000$	

Use the fundamental counting principle to find the value of each of the following.

6. $_8P_3$.

7. $_{11}P_1$.

8. $_9P_5$.

Use the table of factorials and a calculator to find the value of each of the following.

9. $\dfrac{8!}{5!}$

10. $\dfrac{11!}{10!}$

11. $\dfrac{9!}{4!}$

Compare your answers to exercises 9–11 with your answers to exercises 6–8.

12. How do you think the value of $_{15}P_4$ could be expressed in terms of factorials?

The table of factorials will be useful in answering some of the following questions.

13. In how many different ways can six students sit in a row of six desks?

14. If the students sat in a different order each day, about how many years would it take them to sit in the row in every possible way?

15. In how many orders can the 13 cards in a bridge hand be arranged?

16. If a different arrangement of cards in the bridge hand were made each *second* without stopping, about how many *years* would it take to arrange the cards in every possible way?

The following sentences would make a reasonable paragraph if they were arranged in a different order.*

1. However, nobody had seen one for months.
2. He thought he saw a shape in the bushes.

*From the section on textual structure in *The Cambridge Encyclopedia of Language*, by David Crystal (Cambridge University Press, 1987.)

3. Mark had told him about the foxes.

4. John looked out of the window.

5. Could it be a fox?

17. In how many different orders can the sentences be arranged?

18. What do you think is the correct order?

A typical combination lock is opened by turning the dial through a sequence of three different numbers.

19. Find the number of possible lock combinations if the dial has the numbers 0 through 39.

20. Find the number of possible lock combinations if the dial has the numbers 0 through 49.

SET III

The order in which television programs are broadcast is carefully planned. The fall 1955 CBS television schedule for Monday evening between 7 and 11 o'clock was

7:00	"Studio One"
8:00	"Burns and Allen"
8:30	"Talent Scouts"
9:00	"I Love Lucy"
9:30	"December Bride"
10:00	"Mr. District Attorney"
10:30	News

1. In how many different orders could the programs be scheduled if the news had to be at 10:30?

2. In how many orders could they be scheduled if the news had to be at 10:30 and "Studio One" had to be at either 7:00, 8:00, or 9:00?

3

More on Permutations

The magazine published by the San Diego Zoo is called ZOONOOZ. This name is unusual because spelled backwards it is the same word. In other words, reversing the order of the letters ZOONOOZ leaves their order unchanged.

In how many *different* orders can the letters be arranged? This question is about the number of permutations of seven things, and it might seem that the answer would be 7!, or 5,040. This is clearly incorrect, however, because not all the letters of ZOONOOZ are different. We have just seen, for example, that reversing the order of the letters does not result in a new arrangement.

Even though not all the letters are different, we can find the number of orders by first labeling the letters with numbers to make them temporarily different. If the two Z's are labeled Z_1 and Z_2 and the four O's, O_1,

O_2, O_3, and O_4, the seven letters of

$$Z_1O_1O_2NO_3O_4Z_2$$

can be arranged in 7! different orders. Among these 7! orders are such arrangements as

$$Z_1Z_2NO_1O_2O_3O_4 \text{ and } Z_2Z_1NO_1O_2O_3O_4$$

Notice that, without the numbers, these arrangements are the same: ZZNOOOO. Some other orders that lead to the same pattern are

$$Z_1Z_2NO_1O_2O_4O_3,\ Z_1Z_2NO_1O_3O_2O_4,\text{ and } Z_1Z_2NO_1O_3O_4O_2$$

How many such orders leading to the same pattern are there altogether? Because the two Z's can be arranged in 2! ways and the four O's can be arranged in 4! ways, there are, by the fundamental counting principle, $2! \times 4!$ such orders.

By the same reasoning, it follows that not only are there $2! \times 4!$ numbered orders equivalent to the unnumbered order ZZNOOOO, but there are also $2! \times 4!$ numbered orders equivalent to *every* unnumbered order. This means that the number of different permutations of the seven letters of ZOONOOZ is not 7!, but

$$\frac{7!}{2! \times 4!}$$

This fraction can be simplified in the following way:

$$\frac{7!}{2! \times 4!} = \frac{7 \times 6 \times 5 \times \cancel{4!}}{2 \times 1 \times \cancel{4!}} = \frac{7 \times \overset{3}{\cancel{6}} \times 5}{\cancel{2}} = 105$$

There are only 105 different ways in which the letters of ZOONOOZ can be arranged. The arrangements are shown, in alphabetical order, on the next page. Merely counting them, as we have done, is much easier than making the list.

In general, the number of permutations of n things, of which a things are alike, another b things are alike, another c things are alike, and so forth, is

$$\frac{n!}{a! \times b! \times c! \ \dots}$$

1. NOOOOZZ
2. NOOOZOZ
3. NOOOZZO
4. NOOZOOZ
5. NOOZOZO
6. NOOZZOO
7. NOZOOOZ
8. NOZOOZO
9. NOZOZOO
10. NOZZOOO
11. NZOOOOZ
12. NZOOOZO
13. NZOOZOO
14. NZOZOOO
15. NZZOOOO
16. ONOOOZZ
17. ONOOZOZ
18. ONOOZZO
19. ONOZOOZ
20. ONOZOZO
21. ONOZZOO
22. ONZOOOZ
23. ONZOOZO
24. ONZOZOO
25. ONZZOOO
26. OONOOZZ
27. OONOZOZ
28. OONOZZO
29. OONZOOZ
30. OONZOZO
31. OONZZOO
32. OOONOZZ
33. OOONZOZ
34. OOONZZO
35. OOOONZZ

36. OOOOZNZ
37. OOOOZZN
38. OOOZNOZ
39. OOOZNZO
40. OOOZONZ
41. OOOZOZN
42. OOOZZNO
43. OOOZZON
44. OOZNOOZ
45. OOZNOZO
46. OOZNZOO
47. OOZONOZ
48. OOZONZO
49. OOZOONZ
50. OOZOOZN
51. OOZOZNO
52. OOZOZON
53. OOZZNOO
54. OOZZONO
55. OOZZOON
56. OZNOOOZ
57. OZNOOZO
58. OZNOZOO
59. OZNZOOO
60. OZONOOZ
61. OZONOZO
62. OZONZOO
63. OZOONOZ
64. OZOONZO
65. OZOOONZ
66. OZOOOZN
67. OZOOZNO
68. OZOOZON
69. OZOZNOO
70. OZOZONO

71. OZOZOON
72. OZZNOOD
73. OZZONOO
74. OZZOONO
75. OZZOOON
76. ZNOOOOZ
77. ZNOOOZO
78. ZNOOZOO
79. ZNOZOOO
80. ZNZOOOO
81. ZONOOOZ
82. ZONOOZO
83. ZONOZOO
84. ZONZOOO
85. ZOONOZO
86. ZOONOZO
87. ZOONZOO
88. ZOOONOZ
89. ZOOONZO
90. ZOOOONZ
91. ZOOOOZN
92. ZOOOZON
93. ZOOZNOO
94. ZOOZONO
95. ZOOZOON
96. ZOZNOOO
97. ZOZONOO
98. ZOZOONO
99. ZOZOOON
100. ZOZOOON
101. ZZNOOOO
102. ZZONOOO
103. ZZOONOO
104. ZZOOONO
105. ZZOOOON

EXERCISES

SET I

The word ANGERED is remarkable in that its letters can be rearranged to form another word with the same meaning, ENRAGED.

1. If every letter of ANGERED were different, its letters could be arranged in 7! different ways. How many is that?

2. Since ANGERED has two letters that are the same, the number of different arrangements of its letters is actually $\dfrac{7!}{2!}$. How many is that?

There was once a British race horse whose name was

POTOOOOOOOO.

This strange name was pronounced like the name of a common vegetable.

3. Why? (What follows the first three letters?)

4. Write an expression with factorials to show how many arrangements can be made of the letters of POTOOOOOOOO.

5. Find the value of the expression.

The letters of the word ANTEATER can be arranged in

$$\frac{8!}{2! \times 2! \times 2!}$$

different ways.

6. Explain where each of the four factorials in this expression comes from.

7. Find the value of the expression.

Compare your answers to exercises 1 and 7.

8. Why is $\dfrac{8!}{2! \times 2! \times 2!} = 7!$?

9. Write an expression for the number of ways in which the letters of the word ANTEATEREATER can be arranged.

10. Find the number of ways.

Riddle: A customer walks into a hardware store to buy something and asks the clerk how much 1 would cost. The clerk answers $1. The customer then asks how much 10 would cost. The clerk says $2. The customer says, "I'll buy 1515," and pays the clerk $4. What was the customer buying?

Answer: Digits for a house number.

11. Write an expression with factorials to show how many different four-digit house numbers can be made from the digits 1515.

12. Find the value of the expression.

13. Show that your answers are correct by making a list of all the different house numbers that can be made from two 1's and two 5's.

A bar code used by the United States Postal Service makes it possible for machines to sort mail. The bar code is printed along the lower right edge of envelopes and consists of bars arranged in groups of five. Each group of five bars contains two long bars and three short bars.

14. Write an expression with factorials for the number of different orders in which two long bars and three short bars can be arranged.

15. Find the value of this expression.

16. Check your answer by drawing all the possible orders. One of them is shown here.

Each group of five bars represents a digit of the ZIP code.

17. Are there enough orders so that every possible digit can be represented with a different order? Explain.

18. If the ZIP code used letters instead of digits, could every letter be represented with a different order of two long bars and three short bars? Explain.

Set II

It would be somewhat surprising to toss a coin 10 times and get five heads in a row followed by five tails in a row:

HHHHHTTTTT

This is due to the fact that there are so many ways in which the coin can turn up in 10 consecutive tosses. By the fundamental counting principle, the number of ways is

$$2 \times 2 \times 2 \times 2 \times 2 \times 2 \times 2 \times 2 \times 2 \times 2$$

1. How many ways is that?

The number of different orders in which you could get five heads and five tails is the number of possible permutations of the letters in

HHHHHTTTTT

2. Write an expression with factorials for the number of permutations of these letters.

3. Find the value of this expression to find the number of different orders in which five heads and five tails can turn up.

The number of different orders in which six heads and four tails can turn up is the number of possible permutations of the letters in

HHHHHHTTTT

4. Write an expression with factorials for the number of permutations of these letters.

5. Find the value of this expression to find the number of different orders in which six heads and four tails can turn up.

6. Express the number of different orders of getting seven heads and three tails in terms of factorials.

7. Find the number of orders by finding the value of the expression you wrote.

There are 10 orders of getting nine heads and one tail:

HHHHHHHHHT HHHHTHHHHH
HHHHHHHHTH HHHTHHHHHH
HHHHHHHTHH HHTHHHHHHH
HHHHHHTHHH HTHHHHHHHH
HHHHHTHHHH THHHHHHHHH

8. Express the number of orders of getting nine heads and one tail in terms of factorials.

9. Show why this expression is equal to 10.

There is only one order of getting ten heads and zero tails:

HHHHHHHHHH

10. Write an expression with factorials for the number of arrangements of these letters and show that it is equal to 1.

11. Use the same methods as in exercises 2–10 to copy and complete the following table.

Number of heads	10	9	8	7	6	5	4	3	2	1	0
Number of tails	0	1	2	3	4	5	6	7	8	9	10
Number of orders	1	10	▓	▓	▓	▓	▓	▓	▓	▓	▓

12. Add the numbers of orders in your table to see if you get the number you calculated in exercise 1.

A taste test is being planned to find out whether people can tell the difference between bottled water and tap water. Ten people are to be used for the test. Several small cups of water are prepared and given to each person one at a time for the person to taste and identify.

How many different orders can the cups of water be given if there are

13. four cups, with two cups of each type?

14. eight cups, with four cups of each type?

15. twelve cups, with six cups of each type?

Does it seem likely that some people in the group tested could identify all the cups correctly by merely guessing

16. in the test with four cups? Explain.

17. in the test with twelve cups? Explain.

SET III

Life magazine once contained a photograph of 16 children who were students at an elementary school in eastern Pennsylvania. What made the picture remarkable was that, although the school was small, the 16 children consisted of eight pairs of twins! Except for one pair of fraternal twins, the twins in the other seven pairs looked almost exactly alike.

On the assumption that the children were arranged in two rows of eight each as shown in this picture, in how many *different-looking* ways could they be seated for the photograph? (Assume that, except for the pair of fraternal twins, if any child changed places with his or her twin, the picture would not look different.)

Henry Groskensky, Life Magazine © Time Warner Inc.

By permission of Johnny Hart and Field Enterprises, Inc.

4

Combinations

Even the best bowler doesn't roll a strike every time. Suppose a bowler rolls the first ball and knocks over all the pins but three. In how many ways might this happen?

The locations of the ten pins in the game are numbered as shown in this diagram. If the three pins left standing are 3-5-6, it is fairly easy to knock them down with the second ball. If the pins are 4-7-10, knocking all three down is a challenge even for a pro. In how many ways might three pins be standing after the rest have been knocked over? There are ten choices for the first pin, nine choices for the second pin, and eight

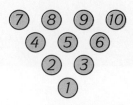

choices for the third. By the fundamental counting principle it would seem that there are

$$10 \times 9 \times 8 = 720$$

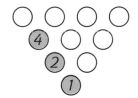

different sets of three pins.

Among the sets counted in this way, however, are the following. (The numbers indicate the locations of the pins.)

1-2-4	2-1-4	4-2-1
1-4-2	2-4-1	4-1-2

Notice that what we have counted as six sets of pins are actually only one set. They contain the same pins but in six different orders: six because the number of permutations of three different things is $3! = 6$.

The product $10 \times 9 \times 8$, then, counts the sets of three pins as if the order in which the pins are listed is important. Since the order is not important, the product $10 \times 9 \times 8$ is 6, or 3!, times as large as the actual number of different sets of pins. Dividing, we get

$$\frac{10 \times 9 \times 8}{3 \times 2 \times 1} = 120$$

The number of sets of three pins that are really different is 120.

The problem that we have just solved is called a *combination* problem.

| A **combination** is a selection of things in which the *order does not matter*.

Another way to describe the number of ways in which three bowling pins can be chosen from ten is to say that we are finding "the number of combinations of ten things taken three at a time." This can be written in symbols as

$$_{10}C_3$$

in which C stands for "combinations."

Counting the number of combinations of three pins that can be chosen from ten revealed that

$$_{10}C_3 = \frac{_{10}P_3}{3!}$$

| In general, the number of combinations of n things taken r at a time is

$$_nC_r = \frac{_nP_r}{r!}$$

EXERCISES

SET I

Five standbys hope to get seats on a flight but only three seats are available on the plane. If the standbys are represented by the letters A, B, C, D, and E, the different orders in which three names can be called are shown in the list below.

ABC	ABD	ABE	ACD	ACE
ACB	ADB	AEB	ADC	AEC
BAC	BAD	BAE	CAD	CAE
BCA	BDA	BEA	CDA	CEA
CAB	DAB	EAB	DAC	EAC
CBA	DBA	EBA	DCA	ECA
ADE	BCD	BCE	BDE	CDE
AED	BDC	BEC	BED	CED
DAE	CBD	CBE	DBE	DCE
DEA	CDB	CEB	DEB	DEC
EAD	DBC	EBC	EBD	ECD
EDA	DCB	ECB	EDB	EDC

1. How many orders are shown in this list?

2. What does the word *permutation* mean?

Recall that the symbol $_5P_3$ represents the number of permutations of five things taken three at a time and that $_5P_3 = 5 \times 4 \times 3$.

3. Find the value of $_5P_3$.

Notice that in the list of orders of standbys above, each set of three standbys appears six times.

4. How many different *sets* of three standbys can be chosen from the five people?

5. What does the word *combination* mean?

The symbol $_5C_3$ represents the number of combinations of five things taken three at a time; $_5C_3 = \dfrac{_5P_3}{3!}$.

6. Use your answer to exercise 3 to find the value of $_5C_3$.

Find the value of each of the following.

7. $_{10}P_2$.   8. $_{10}C_2$.   9. $_{12}P_3$.   10. $_{12}C_3$.

The Braille system of writing used by blind people consists of a code of characters that are made of raised dots. Each character is a combination of dots ranging from one to six.

Examples of letters written in Braille are shown below. The solid circles represent raised dots and the hollow circles show where the paper is flat.

The number of ways in which one of the six dots can be raised can be written as $_6C_1$.

11. To what number is $_6C_1$ equal?

The number of ways in which two of the six dots can be raised can be written as $_6C_2$.

12. To what number is $_6C_2$ equal?

Write and evaluate expressions for the number of ways in which

13. three of the six dots can be raised.

14. four of the six dots can be raised.

15. five of the six dots can be raised.

16. all six dots can be raised.

17. Use your answers to exercises 11 – 16 to copy and complete the following table.

Number of raised dots	1	2	3	4	5	6
Number of ways						

18. How many different symbols altogether are possible in Braille?

Astrologers believe that conjunctions of heavenly bodies influence events on the earth. A conjunction occurs when two or more heavenly bodies seem to come together in the same place in the sky.

Abraham ben Ezra, a mathematician of the twelfth century, figured out the number of possible conjunctions of the sun, moon, Mercury, Venus, Mars, Jupiter, and Saturn.

The number of possible conjunctions (combinations) of these seven heavenly bodies that contain two of them is $_7C_2$.

19. How many is that?

Find the number of possible conjunctions of the seven heavenly bodies that contain

20. three of them.

21. four of them.

22. five of them.

23. six of them.

24. all seven of them.

25. Use your answers to exercises 19–24 to find the total number of conjunctions of two or more of these seven heavenly bodies that are possible.

Set II

The members of the European Economic Community (EEC) use nine official languages in their work:

Danish	French	Italian
Dutch	German	Portuguese
English	Greek	Spanish

Because of this, the EEC has a massive translation problem.

Suppose that there is one translator for every possible pair of languages. For example, a translator between Danish and Dutch, a translator between Danish and English, a translator between Dutch and English, and so forth.

1. How many translators are needed?

2. If another language were added to the list, how many *more* translators would be needed?

A supermarket gave each of its customers a game card containing sixteen circular spots covered with silver film. Under the film, six of the spots were labeled as winners.

If you were given a card and the first four spots from which you scratched off the film were winners, you could turn in the card for $100.

3. In how many different ways could you choose four spots from which to scratch off the film?

If the first four spots from which you scratched off the film were winners, you could try for a bigger prize by scratching off two more spots in the hope that they were the other two winners. If they were, you could turn in the card for $500. If they weren't, you wouldn't get anything.

4. Having scratched four spots and found four winners, in how many ways could you choose two more spots to scratch?

Suppose that you have just bought a new car and that the radio has buttons that you can program to tune in three AM stations and five FM stations. There are twenty-four AM stations and fifteen FM stations from which you can choose.

5. How many different combinations of three AM stations can you choose from twenty-four?

6. How many different combinations of five FM stations can you choose from fifteen?

7. In how many different ways can you choose all eight stations?

The following description of combinations in a lottery appears in the book *A World of Luck.**

> Only by playing every possible combination of numbers can a bettor guarantee a win, and that technique appears to be a practical impossibility. To be sure of selecting the six winning numbers out of a thirty-six-number pool, a common pick-six bet, a player would have to buy 1,947,792 tickets.

Use a calculator to verify that this number of tickets is correct.

8. What calculations did you make?

9. If it took just 1 second to buy each ticket, how long would it take to buy 1,947,792 tickets? Express your answer in days.

This figure shows a set of matches that form five squares. It is possible to move four of the matches so that exactly four squares result.†

10. In how many different ways can four matches be chosen from this figure?

Exercise 10 can be solved in the same way as one of the previous exercises in this set.

11. Which one?

A World of Luck, by the editors of Time-Life Books (Time-Life Books, 1991.)

†*Mind Benders—Games of Shape,* by Ivan Moscovich (Vintage Books, 1986.)

12. To what do the matches of this exercise correspond in the other exercise?

13. Can you solve the puzzle?

A poker hand consists of 5 cards from a deck of 52 cards.

14. How many different poker hands are possible?

SET III

The menu of a Chinese restaurant lists 14 dishes. The number of dishes that you can choose depends on the number of people in the group you are eating with.

1. How many different combinations of 3 dishes can be chosen from the menu?

2. How many different combinations of 4 dishes can be chosen?

3. How many different combinations of 5 dishes can be chosen?

4. If you were to conclude from these examples that increasing the number of dishes chosen from the 14 always increases the number of choices that you have, what kind of reasoning would you be using?

5. How many different combinations of 13 dishes can be chosen?

6. What number of dishes chosen provides the most choices? What is the number of choices for that number of dishes?

Summary and Review

In this chapter we have become acquainted with:

The **fundamental counting principle** *(Lesson 1).* To find the number of ways in which a series of successive events can occur, *multiply* the numbers of ways in which each event can occur.

Permutations *(Lessons 2 and 3).* A permutation is an arrangement of things in a definite order.

The number of permutations of n different things taken r at a time is represented by the symbol $_nP_r$.

> *Example:* The number of permutations of 10 different things taken 3 at a time is
>
> $$_{10}P_3 = 10 \times 9 \times 8 = 720$$

The symbol $n!$ ("n factorial") means to multiply the consecutive numbers from n all the way down to 1.

The number of permutations of n different things is $n!$.

> *Example:* The number of permutations of five different things is
>
> $$5! = 5 \times 4 \times 3 \times 2 \times 1 = 120$$

The number of permutations of n things, of which a things are alike, another b things are alike, another c things are alike, and so forth, is

$$\frac{n!}{a! \times b! \times c! \, \ldots}$$

> *Example:* The number of permutations of seven things, of which two things are alike and another three things are alike is
>
> $$\frac{7!}{2! \times 3!} = \frac{5,040}{2 \times 6} = 420$$

Combinations *(Lesson 4).* A combination is a selection of things in which the order doesn't matter.

The number of combinations of n things taken r at a time is

$$_nC_r = \frac{_nP_r}{r!}$$

Example: The number of combinations of 10 things taken 3 at a time is

$$_{10}C_3 = \frac{10 \times 9 \times 8}{3!} = \frac{720}{6} = 120$$

EXERCISES

SET I

"THEN, AS YOU CAN SEE, WE GIVE THEM SOME MULTIPLE CHOICE TESTS."

A short multiple-choice test has six questions. In how many ways can the questions be answered if there are

1. four choices for each answer?

2. five choices for each answer?

The flags of 11 nations of the world have the pattern shown at the right with the three stripes in the colors listed below.

Nation	First stripe	Second stripe	Third stripe
Belgium	black	yellow	red
Chad	blue	yellow	red
France	blue	white	red
Guinea	red	yellow	green
Ireland	green	white	gold
Italy	green	white	red
Ivory Coast	gold	white	green
Mali	green	yellow	red
Nigeria	green	white	green
Peru	red	white	red
Romania	blue	gold	red

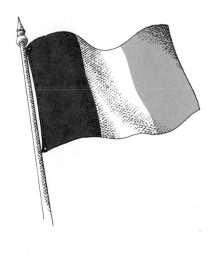

How many different flags having this pattern are possible if

3. the first stripe can be black, blue, gold, green, or red, the second stripe can be gold, white, or yellow, and the third stripe can be gold, green, or red?

4. each stripe may be any one of the seven colors but each stripe is a different color?

The call letters of radio and television stations in the United States begin with either K or W. Those west of the Mississippi River start with K and those east of it with W.

5. Some stations, such as KID in Idaho Falls, Idaho, and WOW in Omaha, Nebraska, have three call letters. How many sets of call letters having three letters are possible?

6. Other stations, such as KUZZ in Bakersfield, California, and WARF in Jasper, Alabama, have four call letters. How many sets of call letters having four letters are possible?

There are 11,160 radio and television stations in the United States.

7. Could every station be named with a different set of three letters beginning with K or W?

8. Could every station be named with a different set of four letters beginning with K or W?

A traveling salesman has to visit customers in the following cities: Boston, Chicago, Cleveland, Detroit, Milwaukee, New York, Philadelphia, and Washington.

9. In how many different orders could he travel to these cities if he visits each one exactly once?

10. How many different orders would there be if Baltimore and Pittsburgh are added to the list?

License plates in California used to consist of a digit, a letter, and four more digits.

11. How many license plates of this type were possible?

12. In 1956, the pattern was changed to three letters followed by three digits. How many license plates of this type were possible?

13. When this pattern was used up, the state began issuing license plates consisting of three digits followed by three letters. How many license plates of this type were possible?

14. In 1978, California began issuing plates consisting of a digit followed by three letters followed by three more digits. How many license plates of this type are possible?

During the baseball season, each of the 14 National League teams plays 81 home games.

15. How many games are played in all?

16. How many games would there be if each team played each of the other teams only once?

Set II

A toy train has an engine, six cars, and a caboose. Suppose that the engine is always put first and the caboose last. In how many different orders can the train be arranged using the engine, the caboose, and

1. one of the cars?

2. three of the cars?

3. five of the cars?

4. all six of the cars?

Anagrams, the rearrangements of the letters of words to form different words, are the basis for many word puzzles and games. In how many different orders can the letters of the names of the following states be arranged? (*Note:* The order of the letters does not have to result in another actual English word.)

5. ALABAMA

6. TENNESSEE

7. MISSISSIPPI

© 1978 Universal Press Syndicate

"Members of the jury, have you reached a verdict?"

Ordinarily, all twelve members of a jury must agree before a case can be decided. There are twelve different ways in which a jury can be deadlocked when one person disagrees with the rest, because any one of the jurors can be that person.

8. How many different combinations of two members of a jury can cause the jury to be deadlocked?

9. How many different combinations of three members of the jury can disagree with the rest?

In the game of Keno, the player is given a card containing 80 numbers. The player can mark and bet on as many as 15 numbers.

In how many ways can the player choose

10. 1 number?

11. 2 numbers?

12. 3 numbers?

13. 4 numbers?

The most popular bet is on 10 numbers. Ten numbers can be chosen in 1,646,492,110,120 different ways.

14. How was this number calculated?

SET III

The Pepsi-Cola Company had a "Matching Picture" contest in which the object was to match the pictures of four Miss Americas with their baby pictures. Contestants were allowed to enter as many times as they wished, and the first prize was $10,000.

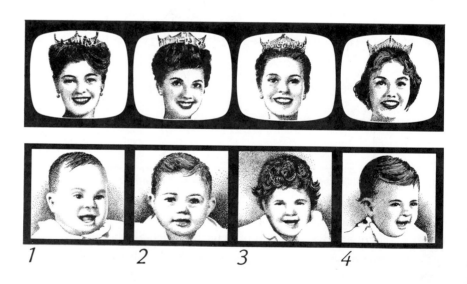

1. How many different entries would you have to send in to be sure of having all the pictures matched correctly?

2. Does it seem reasonable to think that you would win $10,000 if you sent in an entry with all the pictures correctly matched?

Further Exploration

LESSON 1

1. A company operating out of Las Vegas once ran a weekly crossword puzzle contest in which cash prizes were awarded.

The clues for the puzzles were deliberately made ambiguous so that many of the letters could be filled in in more than one way. For example, the clue "The mother quickly grew tired of her daughter asking her to _____ clothes" could be answered with MEND or SEND. Each puzzle had 10 clues of this type, each of which could be answered in two different ways.

You could send in as many different solutions to a given puzzle as you liked. The fee was $5 for one solution or $10 for a set of three solutions. For example, if you wanted to send in seven solutions, you would pay $10 + $10 + $5 = $25.

Puzzle No.2
Entry Fee $5.00

  a. How many solutions would you have to send in to be sure of getting the "correct" solution?

  b. How much money would you have to pay to send in every possible solution?

If more than one person sent in the correct solution, the prize was shared equally.

  c. If the prize was $3,500, would you enter the contest? Explain why or why not.

2. The Bell Telephone companies use a color code to keep track of the wires in their cables. The cables contain 50 wires, each of which is coded with two colors. The two colors appear as stripes on the wires: one color

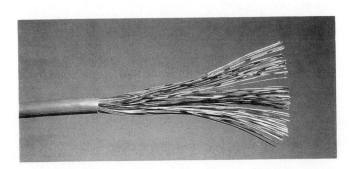

as a wide stripe and the other as a narrow one. The list below shows the colors of the wires with the color of the wide stripe listed first.

1. blue-white
2. gray-yellow
3. black-blue
4. orange-purple
5. green-white
6. yellow-orange
7. brown-black
8. orange-white
9. red-orange
10. blue-black
11. white-gray
12. purple-orange
13. brown-white
14. white-blue
15. black-orange
16. yellow-blue
17. green-yellow
18. blue-yellow
19. black-gray
20. white-orange
21. red-green
22. gray-white
23. orange-red
24. purple-green
25. brown-yellow
26. green-red
27. yellow-brown
28. red-brown
29. purple-blue
30. blue-red
31. black-green
32. brown-red
33. yellow-gray
34. red-blue
35. gray-black
36. black-brown
37. white-green
38. green-purple
39. purple-brown
40. yellow-green
41. brown-purple
42. gray-red
43. orange-yellow
44. green-black
45. blue-purple
46. red-gray
47. purple-gray
48. white-brown
49. gray-purple
50. orange-black

Can you figure out the system by which the colors are paired? If so, what is it and why are there exactly 50 possibilities according to the system?

Lesson 2

1. "Change ringing" is the practice of ringing a set of bells in every possible permutation. Each permutation is called a *change* and the maximum number of permutations possible for a given set of bells is called a *peal*.

A set of three bells—A, B, and C, for example—has a peal of six changes as shown in the tree diagram below. Bell ringers can ring a peal of three bells at the rate of about one change each second; a peal, then, takes about 6 seconds.

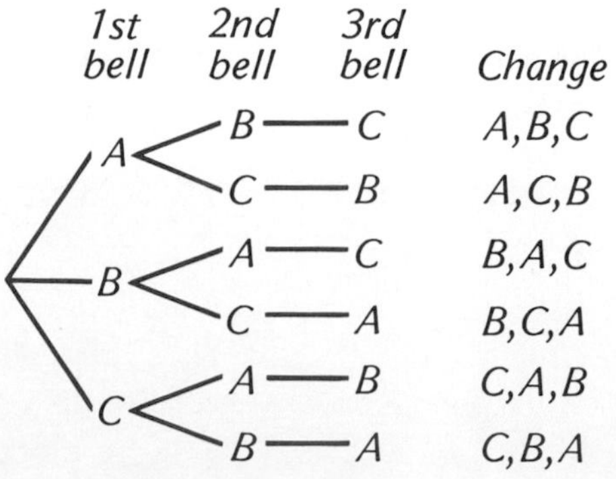

a. How many changes are there in a peal of four bells? About how long would such a peal take?

b. How many changes are there in a peal of seven bells? About how long would such a peal take?

c. How many changes are there in a peal of twelve bells? Could someone ring a peal of twelve bells? Explain.

2. This wreath of dolls was made in Guatemala. Every one of the 29 dolls is different.

If just 3 dolls were arranged in a circle, it might seem that there would be $3 \times 2 \times 1 = 6$ different arrangements possible.

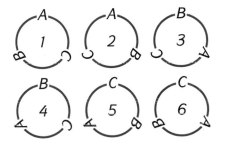

Notice, however, that if the wreath numbered 1 is rotated, it is actually identical to the wreaths numbered 3 and 5. If the wreath numbered 2 is rotated, it becomes wreaths 4 and 6. So there are actually only two different arrangements.

a. How many different arrangements would be possible for 4 dolls on a wreath?

b. How many different arrangements would be possible for 5 dolls?

The number of arrangements possible for the 29 dolls on the wreath in the picture is

304,888,344,611,713,860,501,504,000,000

c. How was this number found?

LESSON 3

1. The World Series is held early in October each year between the winners of the American and National League pennants. The two teams play until one team has won four games.

Suppose that the Yankees are playing the Dodgers. How many different orders of winners in the games played are possible?

First, consider the number of orders possible in which the Yankees win the series. The number of games that it takes to win the series — that is, four, five, six, or seven — depends on whether the Dodgers win zero, one, two, or three games.

If it takes only four games, the Yankees win all four, and there is only one possible order of winners: YYYY.

If the series lasts five games, and the Yankees win it, there are several possible orders of winners: DYYYY, YDYYY, YYDYY, and YYYDY. Because the winner of the series always wins the last game, the last letter in the row is Y. Then the problem is one of finding the number of orders of the other four letters in the row: three Y's and one D.

$$\frac{4!}{3!} = \frac{4 \times 3 \times 2 \times 1}{3 \times 2 \times 1} = 4$$

a. How many orders of winners are possible if the series lasts six games? (The Yankees win the last game, and so the problem is: In how many ways can five letters, three Y's and two D's, be arranged?)

b. How many orders of winners are possible if the Yankees win the series in the seventh game?

c. Copy and complete the following table:

Ways in which the Yankees win the series

Number of games in the series	4	5	6	7
Number of orders of winners	1	4	|||||||	|||||||

d. The sum of the numbers on the second line of the table is the number of orders of winners possible if the Yankees win. How many is that?

e. How many different orders of winners are possible if the Dodgers win the series?

f. How many different orders of winners in a World Series are possible altogether?

2. The map at the left shows part of downtown San Francisco. Someone who wants to walk from the corner of Taylor and O'Farrell to the corner of Powell and Sutter has a choice of 10 direct routes. They are shown in the figures on the facing page.

The fact that there are 10 routes can be discovered without making a diagram. Going from Taylor and O'Farrell to Powell and Sutter requires traveling three blocks north and two blocks east. If each block north is represented by an N and each block east by an E, each route can be represented by an arrangement of the letters NNNEE.

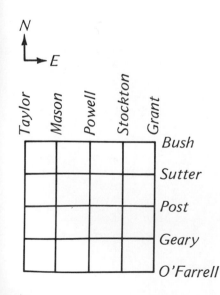

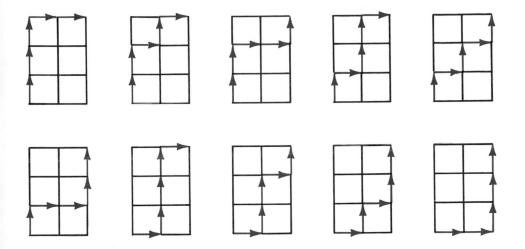

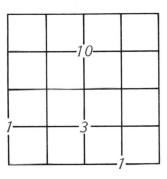

a. Show, using the fundamental counting principle, that there are 10 such arrangements.
b. Now copy the map and figure out the number of direct routes from Taylor and O'Farrell to each of the other corners. Write the numbers on the corners as illustrated in the map at the right.
c. What patterns do you see in the numbers on the finished map?

LESSON 4

1. Elaborate designs can be formed by winding string around a set of nails arranged in a circle. The figure below shows the pattern that would result from winding string around 21 nails in every possible way.

a. How many windings (lines) are in the figure? Explain how the answer can be found without counting them.

b. Suppose that the number of nails in the figure were doubled. How many windings would result from winding string around a set of 42 nails in every possible way?

2. A "double-six" set of dominoes contains the seven "doubles" dominoes shown below.

The rest of the dominoes, called "singles," have different numbers of spots on each square. Examples of singles are shown below.

a. How many singles dominoes are there if every possible combination of squares is included exactly once?

b. How many dominoes are in a complete "double-six" set of dominoes?

The squares of a "double-nine" set of dominoes either are blank or have from one to nine spots.

c. How many dominoes are in a complete "double-nine" set?

The squares of a "double-twelve" set of dominoes either are blank or have from one to twelve spots.

d. How many dominoes are in a complete "double-twelve" set?

8

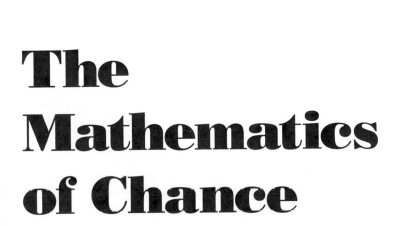

The
Mathematics
of Chance

Player's Advantage Against Various Up Cards.

Leigh Wiener, Life Magazine © Time Warner, Inc.

1

Probability: The Measure of Chance

While teaching mathematics at the Massachusetts Institute of Technology, Edward Thorp figured out how to consistently win at the game of blackjack, also called "twenty-one." He gave a talk on his method at a meeting of the American Mathematical Society and then proved that it worked by winning a lot of money in Las Vegas. Shown in the photograph above seated in front of a chart of the numbers he used, Thorp wrote a best-selling book titled *Beat the Dealer* about his method and his adventures in using it.*

The first book on the branch of mathematics called "probability theory" was written in the sixteenth century by an Italian physician and mathematician named Gerolamo Cardano. Cardano's book was titled

---

Beat The Dealer by Edward O. Thorp, Random House, 1966.

The Book on Games of Chance. Because games of chance have been played for thousands of years, it is not surprising that "the mathematics of chance" was at first used primarily to solve problems dealing with gambling. As the subject became better understood, scientists began to apply it to other fields of study. The applications have proved to be so significant that, according to one twentieth-century mathematician, probability theory has become "a cornerstone of all of the sciences."

Probabilities are expressed as numbers on a scale ranging from 0 to 1 or as percentages on a scale ranging from 0% to 100%. Both scales are shown in the figure below. The larger the probability of a given event, the more likely it is that the event will occur.

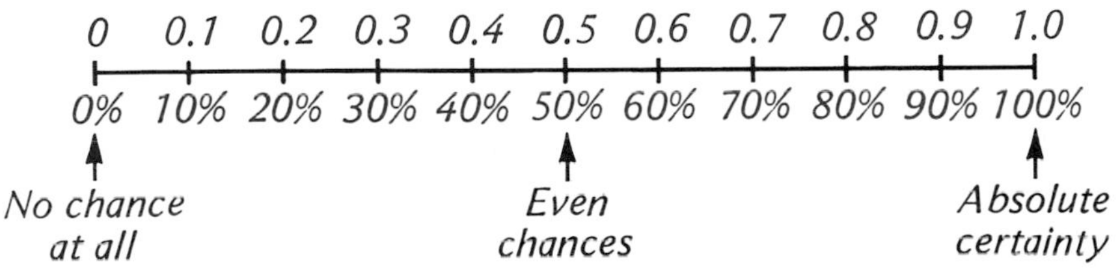

```
 0   0.1  0.2  0.3  0.4  0.5  0.6  0.7  0.8  0.9  1.0
 ├────┼────┼────┼────┼────┼────┼────┼────┼────┼────┤
 0%  10%  20%  30%  40%  50%  60%  70%  80%  90% 100%
 ▲                        ▲                        ▲
No chance                Even                  Absolute
  at all               chances                certainty
```

The probabilities of some events can be calculated as the ratio of two numbers. Consider, for example, the probabilities associated with dealing the first card from the deck in the game of blackjack. The deck has 52 cards and, if the game is played fairly, it is equally likely that any one of the 52 cards will be dealt. There are 4 cards in the deck numbered 7, so the probability of one of the 7's being dealt is $\frac{4}{52}$, or $\frac{1}{13}$. As a percentage, this probability is

$$\frac{1}{13} \times 100 = \frac{100}{13} \approx 8\%*$$

In general, this method of finding a probability requires finding two numbers and dividing one by the other:

> **Probability of an event** $= \dfrac{\text{number of ways in which the event can occur}}{\text{total number of equally likely outcomes}}$

This method for calculating probabilities works whenever there is a set of equally likely outcomes. There are many situations, however, in which another method is used.

---

*See pages 661–662 if you do not remember how to figure out percentages.

What, for example, is the probability that a firefighter will be injured on the job? During a given year, about 20 out of every 100 firefighters are injured at work. The probability of a firefighter being injured during a given year is about $\frac{20}{100}$ or $\frac{1}{5}$. As a percentage, this probability is

$$\frac{1}{5} \times 100 = 20\%$$

Notice that this method of finding a probability is also based on dividing one number by another.

Probability of an event $= \dfrac{\text{number of times the event occurred}}{\text{total number of possible occurrences}}$

We have considered two ways of finding the probability of an event, both of which require comparing numbers by division. A third way is to make an educated guess. You hear this type of probability all the time. For example, a weather report on the evening news that says there is a 60% chance of rain is giving a probability based on an educated guess that it is somewhat more likely to rain than not.

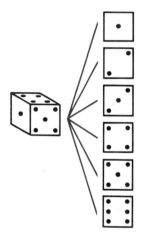

EXERCISES

SET I

When an ordinary die is thrown, it is equally likely that any one of its six faces will turn up.

If it is thrown once, what is the probability

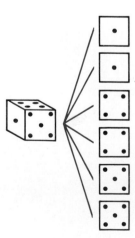

1. of getting a 3?

2. of getting a 4?

3. of an odd number turning up?

A die is misspotted with two faces marked with 1 spot, two faces marked with 4 spots, and two faces marked with 5 spots. If it is thrown once, what is the probability

4. of getting a 3?

5. of getting a 4?

6. of an odd number turning up?

A multiple-choice test has five choices for each answer.

If you have no idea of which answer to a question is correct, what is the probability

7. of guessing the correct answer?

8. of guessing a wrong answer?

If you realize that two of the five answers to a question are not correct, what is the probability

9. of guessing the correct answer for that question?

10. of guessing a wrong answer?

Some holidays always occur on the same day of the week and other holidays do not. If a year is chosen at random, what is the probability that

11. the Fourth of July is on a Saturday?

12. Thanksgiving is on a Thursday?

13. Easter is *not* on a Sunday?

14. Halloween is *not* on a Friday?

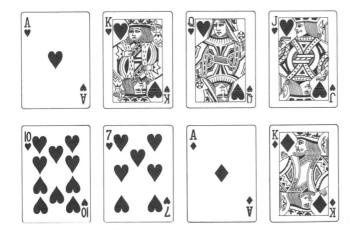

Suppose that the eight cards shown here are turned over and shuffled. If one card is then chosen at random, what is the probability that the card is

15. an ace?

16. a heart?

17. a face card?

18. the king of diamonds?

19. *not* the king of diamonds?

Out of every 80 telephone calls attempted, 8 result in busy signals and 10 result in no answer.

20. Which is more probable: getting a busy signal or getting no answer?

What is the probability of getting a busy signal

21. as a fraction?   22. as a percentage?

What is the probability of getting no answer

23. as a fraction?   24. as a percentage?

Probabilities for various events occurring in major league baseball games are listed below.

The losing team doesn't score	0.15
Both games of a double-header are won by the same team	0.47
The winning run is scored in the last inning	0.19
No bases are stolen during the game	0.36
A game goes into extra innings	0.09
The home team wins the game	0.53
A game has at least one home run	0.64

25. Which one of these events is the most likely to occur?

26. Which one is the least likely to occur?

27. Which two are the most unpredictable?

A nineteenth-century drawing of a shell game by A. B. Frost

In a shell game, the operator covers a pea with one of three shells and then moves the shells about. The player tries to keep track of the pea and then bets the operator that he knows where it is. Suppose that the operator moves the shells so rapidly that the player cannot follow them and must guess where the pea is.

28. If the game is an honest one, what is the probability that the player will guess the correct shell?

29. If the game is honest and you played it six times, how many times would you expect to win?

Shell games are usually not operated honestly.

30. What do you suppose is the probability that the player will win in a dishonest game?

SET II

Suppose that an ordinary coin is tossed. If it is assumed that the outcomes of heads and tails are equally likely, then the probability of it turning up heads is $\frac{1}{2}$, or 0.5.

1. By this reasoning, what is the probability of a coin turning up tails?

An English mathematician being held prisoner during World War II tossed a coin 10,000 times. The coin turned up heads 5,067 times.

2. Use these numbers to calculate the probability of the coin turning up heads.

3. By the same method, what is the probability of the coin turning up tails?

Suppose that the probability of a certain coin turning up heads is 1.

4. What would you conclude about the coin?

If someone in New York City is treated in an emergency room for a bite, the probability that the person was bitten by a dog is $\frac{9}{10}$.

5. Express this probability as a percentage.

The probability that the person was bitten by a cat is $\frac{1}{20}$.

6. Express this as a percentage.

The probability that the person was bitten by another person is $\frac{1}{25}$.

7. Express this as a percentage.

8. Do the three percentages that you have just determined add up to 100%?

9. Do you think they should? Explain.

An American roulette wheel has 38 compartments around its rim. Two of these are numbered 0 and 00 and are green; the others are numbered from 1 to 36, of which half are red and half are black.

Courtesy of *Gambling Times*

When the wheel is spun in one direction, a small ivory ball is rolled in the opposite direction along its rim. If the wheel is a fair one, the chances of the ball falling into any one of the 38 compartments as it slows down are equally likely.

Some typical bets in roulette are listed below. Express the probability of winning each bet both as a fraction and as a percentage.

10. The number 7.

11. A black number.

12. An odd number.

13. The numbers 3, 6, 9, 12, 15, 18, 21, 24, 27, 30, 33, and 36 (called a "column" bet).

14. Two adjoining numbers (called a "split" bet).

15. A red number, if all 26 numbers that had come up previously were black.*

16. Which one of these bets do you think has the biggest payoff? Why?

SET III

The board used in the game of darts is divided into 20 sectors, each of which has a point value between 1 and 20. If a dart lands within either the "doubles" ring or the "triples" ring, the number of points scored is doubled or tripled. As a result, the highest score possible with one dart is 60 points (a dart landing in the triples ring of the sector numbered 20).

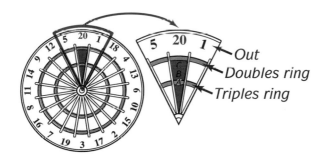

In an experiment, a player of average ability was asked to throw a dart 900 times, aiming it 300 times at A, 300 times at B, and 300 times at C. The results are shown in the table below.†

Spot aimed at	Resulting score									
	60	40	20	15	10	5	3	2	1	0
A	33	0	87	24	0	69	21	0	66	0
B	12	18	144	6	3	57	6	3	51	0
C	0	57	123	0	15	33	0	12	30	30

---

*This actually happened on August 18, 1913, at the casino in Monte Carlo.

†This problem is based on an experiment described in *Mathematics in Sport,* by M. Stewart Townend (Ellis Horwood Ltd., 1984).

Use the numbers in this table to calculate each of the following probabilities for this player. Express each probability as a fraction in decimal form.

If the dart is aimed at A, the probability of scoring

1. 60 points.

2. at least 20 points.

3. less than 20 points.

4. 0 points.

If the dart is aimed at B, the probability of scoring

5. 60 points.

6. at least 20 points.

7. less than 20 points.

8. 0 points.

If the dart is aimed at C, the probability of scoring

9. 60 points.

10. at least 20 points.

11. less than 20 points.

12. 0 points.

What do these results suggest about where an average player should aim the dart if the player wants

13. to score 60 points?

14. to score at least 20 points?

By permission of Johnny Hart and Field Enterprises, Inc.

Dice Games and Probability

Although games with dice have been played for thousands of years, the game of craps did not originate until about 1890. It has since become the most popular dice game in the world.

According to the rules of the game, the person throwing the dice wins if they come up either 7 or 11. If the dice come up 2 ("snake eyes"), 3, or 12, then the person immediately loses. How do the chances of getting each of these sums compare?

To determine the probabilities in this game, we will consider a pair of dice, one of which is white and the other red. Because there are six ways in which each die can turn up, there are

$$6 \times 6 = 36 \text{ ways}$$

in which the dice can turn up together. These 36 equally likely outcomes are shown in the tree diagram on the next page.

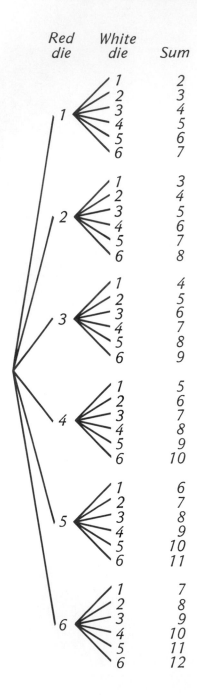

Red die   White die   Sum

A more convenient way of listing the 36 outcomes is shown in the table below.

			Red die			
White die	**1**	**2**	**3**	**4**	**5**	**6**
1	2	3	4	5	6	7
2	3	4	5	6	7	8
3	4	5	6	7	8	9
4	5	6	7	8	9	10
5	6	7	8	9	10	11
6	7	8	9	10	11	12

There are six ways in which a 7 can be thrown, and so the probability of rolling a 7 is

$$\frac{6}{36} = \frac{1}{6}.$$

There are only two ways in which an 11 can be thrown, and so the probability of rolling an 11 is

$$\frac{2}{36} = \frac{1}{18}.$$

Getting "snake eyes" is rather unlikely because there is only one way in which a 2 can be thrown; the probability is

$$\frac{1}{36}.$$

The percent probabilities of getting these sums are shown in the table below. (Each is rounded to the nearest percent.)

Sum	2	7	11
Probability	$\frac{1}{36}$	$\frac{1}{6}$	$\frac{1}{18}$
Percent probability	3%	17%	6%

These percentages tell us that, if a pair of dice is rolled 100 times, 2 should come up about 3 times, 7 about 17 times, and 11 about 6 times.

EXERCISES

SET I

When two dice are thrown, they can turn up in 36 equally likely ways, resulting in sums ranging from 2 to 12.

1. Use the information in the table on the facing page to copy and complete the following one.

Sum of two dice	2	3	4	5	6	7	8	9	10	11	12
Number of ways	1	▥	▥	▥	▥	6	▥	▥	▥	2	▥

In craps, the dice thrower wins on the first throw if a sum of 7 or 11 comes up.

2. How many ways are there altogether of getting these sums?

3. What is the probability, both as a fraction and as a percentage, that the dice thrower will win on the first throw?

The dice thrower loses if a sum of 2, 3, or 12 comes up.

4. How many ways are there altogether of getting these sums?

5. What is the probability, both as a fraction and as a percentage, that the dice thrower will lose on the first throw?

By rolling a 4, 5, 6, 8, 9, or 10, the dice thrower neither wins nor loses and must roll again. The number rolled is called the thrower's "point."

6. How many ways are there altogether of getting these sums?

7. What is the probability, both as a fraction and as a percentage, that the dice thrower will have to roll again?

Having rolled one of these sums, the dice thrower must "make the point" by rolling the same number again before rolling a 7 in order to win.

8. Which is more likely: that the same number will come up or that a 7 will?

9. Which of the sums 4, 5, 6, 8, 9, and 10 are least likely to turn up?

10. Which are most likely to turn up?

The probability that the dice thrower will win by rolling either a 7 or 11 on the first throw or by "making the point" is $\frac{244}{495}$.*

11. Express this probability as a percentage.

12. Which is more likely: that the dice thrower will *win* or *lose*?

It is unlikely that anyone would be fooled by the doctored dice shown in this illustration from *Mad* magazine. If all the faces of one die are numbered 4 and all the faces of the other die are numbered 3, the probability of rolling a 7 with them is 100%.

Professional cheats use misspotted dice that are much harder to detect. One type, called "tops and bottoms," consists of two dice numbered with only three different numbers on each die, each number being repeated on the die's opposite face.

---

*The numbers in this fraction were found by counting all the possibilities.

13. Copy and complete the following table of sums for two of these dice, one numbered 3, 4, 5 and the other numbered 1, 5, 6.

	First die					
Second die	3	4	5	3	4	5
1	4	?	?	?	?	?
5	?	?	?	?	?	?
6	?	?	?	?	?	?
1	?	?	?	?	?	?
5	?	?	?	?	?	?
6	?	?	?	?	?	?

What is the probability that someone throwing these dice will

14. win on the first throw by rolling a 7 or 11?

15. lose on the first throw by rolling a 2, 3, or 12?

16. neither win nor lose on the first throw by rolling a 4, 5, 6, 8, 9, or 10?

17. Having rolled one of these sums on the first throw, what is the probability that the dice thrower will win by getting that sum again before 7 turns up? Explain.

18. Is it possible for the dice thrower ever to lose as long as the other person doesn't realize that the dice are crooked?

Set II

Galileo became interested in probability when he was asked by some gamblers about the chances in a game played with three dice.

1. Show why three dice can turn up in 216 different ways.

2. What is the smallest sum that can turn up on three dice?

3. What is the largest sum?

All the ways and their sums are shown in the tables at the top of the next page.

Courtesy of the Burndy Library

First die: 1
Second die

Third die

	1	2	3	4	5	6
1	3	4	5	6	7	8
2	4	5	6	7	8	9
3	5	6	7	8	9	10
4	6	7	8	9	10	11
5	7	8	9	10	11	12
6	8	9	10	11	12	13

First die: 2
Second die

Third die

	1	2	3	4	5	6
1	4	5	6	7	8	9
2	5	6	7	8	9	10
3	6	7	8	9	10	11
4	7	8	9	10	11	12
5	8	9	10	11	12	13
6	9	10	11	12	13	14

First die: 3
Second die

Third die

	1	2	3	4	5	6
1	5	6	7	8	9	10
2	6	7	8	9	10	11
3	7	8	9	10	11	12
4	8	9	10	11	12	13
5	9	10	11	12	13	14
6	10	11	12	13	14	15

First die: 4
Second die

Third die

	1	2	3	4	5	6
1	6	7	8	9	10	11
2	7	8	9	10	11	12
3	8	9	10	11	12	13
4	9	10	11	12	13	14
5	10	11	12	13	14	15
6	11	12	13	14	15	16

First die: 5
Second die

Third die

	1	2	3	4	5	6
1	7	8	9	10	11	12
2	8	9	10	11	12	13
3	9	10	11	12	13	14
4	10	11	12	13	14	15
5	11	12	13	14	15	16
6	12	13	14	15	16	17

First die: 6
Second die

Third die

	1	2	3	4	5	6
1	8	9	10	11	12	13
2	9	10	11	12	13	14
3	10	11	12	13	14	15
4	11	12	13	14	15	16
5	12	13	14	15	16	17
6	13	14	15	16	17	18

Notice that the sum of 5 appears 3 times in the first table, 2 times in the second table, and 1 time in the third table: $3 + 2 + 1 = 6$ ways in all.

4. Refer to the tables above to copy and complete the following one.

Sum of three dice	3	4	5	6	7	8	9	10	11	12	13	14	15	16	17	18
Number of ways			6													

Check to see that the numbers of ways add up to 216.

5. What sums are least likely to turn up when three dice are thrown?

6. What sums are most likely?

What puzzled the gamblers about the game with three dice was that they had noticed that a sum of 10 turns up more frequently than a sum of 9. They reasoned that there were six ways of getting a 9:

$$1,2,6 \quad 1,3,5 \quad 1,4,4 \quad 2,2,5 \quad 2,3,4 \quad 3,3,3$$

and six ways of getting a 10:

$$1,3,6 \quad 1,4,5 \quad 2,2,6 \quad 2,3,5 \quad 2,4,4 \quad 3,3,4$$

Since there seemed to be the same number of ways of getting both numbers, they couldn't figure out why 10 turned up more frequently. Look at your table for exercise 4.

7. How many ways are there of actually getting each sum?

The gamblers' mistake was that they counted *combinations* of the three dice instead of *permutations.* For example, there is more than one way of getting a 9 if the dice turn up 1, 2, and 6:

$$1,2,6 \quad 1,6,2 \quad 2,1,6 \quad 2,6,1 \quad 6,1,2 \quad 6,2,1$$

8. How many ways does this list show?

In how many ways can the dice turn up

9. 1, 3, and 5?

10. 1, 4, and 4? (Make a list to check your answer.)

11. 2, 2, and 5?

12. 2, 3, and 4?

13. 3, 3, and 3?

14. Show that your answers to exercises 8–13 account for all the ways there actually are of getting a 9.

Now consider the different ways of getting a sum of 10. In how many ways can the dice turn up

15. 1, 3, and 6?

16. 1, 4, and 5?

17. 2, 2, and 6?

18. 2, 3, and 5?

19. 2, 4, and 4?

20. 3, 3, and 4?

21. Show that your answers to exercises 15–20 account for all the ways there actually are of getting a 10.

22. Graph the information in your table for exercise 4, representing the sums of the three dice on the x-axis and the numbers of ways of getting them on the y-axis. Convenient scales for the axes are shown here.

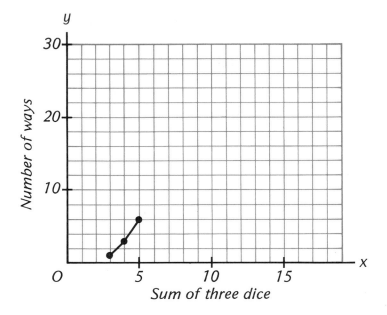

After you have plotted the points, use a ruler to connect them in order with straight-line segments. The first part of the graph is shown above.

A car dealer once made the following offer: a new car if you throw six dice and get six 6's. There was a catch, however: you had to pay $10 each time you threw the dice.*

23. How many equally likely outcomes are possible when six dice are thrown?

*"Making Guesses and Taking Risks" by Andrew Noble in *New Applications in Mathematics*, ed. by Christine Bondi (Penguin Books, 1991).

24. What is the probability, if you throw the dice once, that you would win the car?

Suppose that you really want to win the car and decide to throw the six dice as many as 5,000 times.

25. How much money would you have to pay the dealer if you throw the dice 5,000 times?

26. If you were willing to throw the dice 5,000 times, do you think you would be absolutely certain of winning the car?

Set III

A dice game invented by Bradley Efron, a mathematician at Stanford University, has such a strange set of outcomes that even an experienced gambler would find them hard to believe.*

The faces of four dice are numbered as shown in the list at the right.

Die A   1, 2, 3, 9, 10, 11
Die B   0, 1, 7, 8, 8, 9
Die C   5, 5, 6, 6, 7, 7
Die D   3, 4, 4, 5, 11, 12

Each of two players chooses a die, and the two dice are thrown. The player who gets the larger number wins. If both players get the same number, the tie is broken by rolling the dice again.

Suppose that one player chooses die A and the other player chooses die B. The player who has the greater probability of winning can be determined from the table of possible outcomes shown below. The A's indicate the outcomes in which die A wins, the B's the outcomes in which

		Die A					
		1	*2*	*3*	*9*	*10*	*11*
Die B	*0*	A	A	A	A	A	A
	1	T	A	A	A	A	A
	7	B	B	B	A	A	A
	8	B	B	B	A	A	A
	8	B	B	B	A	A	A
	9	B	B	B	T	A	A

---

*Described in *Wheels, Life, and Other Mathematical Amusements,* by Martin Gardner (W. H. Freeman, 1983).

die B wins, and the T's the outcomes in which there is a tie. Because there are more outcomes in which A wins than B, the player with die A has the greater probability of winning.

1. What is the probability, both as a fraction and as a percentage, that A will beat B on the first throw?

2. Make a table showing the possible outcomes if one player chooses die B and the other player chooses die C.

3. Which player is more likely to win?

4. Make a table showing the possible outcomes if one player chooses die C and the other player chooses die D.

5. Which player is more likely to win?

6. Make a table showing the possible outcomes if one player chooses die D and the other player chooses die A.

7. Which player is more likely to win?

8. What is strange about the outcomes of this game?

Georges de la Tour, *Le Tricheur à l'As de Carreau* [The cheater with the ace of diamonds]; courtesy Musée du Louvre

Probabilities of Successive Events

LESSON

Playing cards are thought to have originated in China, perhaps as early as the seventh century. The designs on the cards now used throughout the English-speaking world were introduced in France in the sixteenth century, about the time that the picture of the card players above was painted. Titled *The Cheater with the Ace of Diamonds,* the picture shows three card cheats surrounding an unsuspecting player.

An experienced cheat can take a deck that has three aces at the top, shuffle it, and then deal the cards so that he gets all three aces. If the game were played honestly, what would be the probability of being dealt three cards and all three of them turning out to be aces?

One way to calculate the probability of being dealt three cards all of which are aces is to count the number of ways of getting three of the four

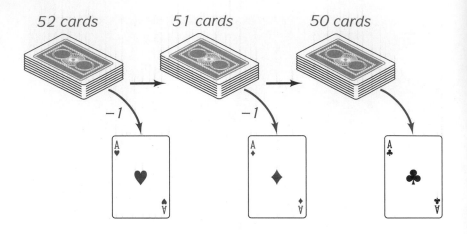

aces in the deck,

$$_4C_3 = \frac{4 \times 3 \times 2}{3!} = 4$$

and the number of ways of getting 3 cards out of the 52 cards in the deck,

$$_{52}C_3 = \frac{52 \times 51 \times 50}{3!} = 22{,}100*$$

Dividing the numbers of ways, we find that the probability is

$$\frac{4}{22{,}100} = \frac{1}{5{,}525}$$

Another way to figure out this probability is to reason as follows. The probability that the first card is an ace is $\frac{4}{52}$ because 4 of the 52 cards in the original deck are aces. After the first ace has been dealt, the probability that the second card is an ace is $\frac{3}{51}$ because there are 3 aces left out of 51 cards. After the second ace has been dealt, the probability that the third card is an ace is $\frac{2}{50}$ because there are 2 aces left out of 50 cards. If these three probabilities are multiplied, we get

$$\frac{4}{52} \times \frac{3}{51} \times \frac{2}{50} = \frac{24}{132{,}600} = \frac{1}{5{,}525}$$

---

*Notice that the order in which the cards are dealt is not important, so we count *combinations,* not permutations.

The fact that this result agrees with the previous one suggests that probabilities can be multiplied in the same way that numbers of choices are multiplied.

| To find the probability of several events occurring in succession, multiply the probabilities of the individual events.

In figuring out the probability of being dealt three aces, the probability of each successive event is affected by the occurrence of the previous one.

| Events that are influenced by other events are said to be **dependent**.

Suppose, instead, that you are playing a card game in which the deck is shuffled and dealt several times during the game. What is the probability that, on three successive deals, the first card you are dealt is an ace? In this case, the probability of each successive event is *not* affected by the occurrence of the previous one.

| Events that have no influence on each other are said to be **independent**.

On each deal, the probability of getting an ace is the same because each deal starts with all 4 aces in a complete deck of 52 cards. The probability is

$$\frac{4}{52} \times \frac{4}{52} \times \frac{4}{52} = \frac{64}{140,608} = \frac{1}{2,197}$$

The fact that this result is larger than the previous one shows that being dealt an ace at the beginning of three successive deals is more likely that getting three aces all in one deal.

EXERCISES

SET I

Suppose that a coin is tossed 2 times in succession.

1. What is the probability that it turns up tails on the first toss?

2. What is the probability that it turns up tails on the second toss?

3. What is the probability that it turns up tails on both tosses?

4. Are the events of the coin turning up tails on the first toss and on the second toss *dependent* or *independent*?

First toss	Second toss	Outcome
T	T	T, T
T	H	T, H
H	T	H, T
H	H	H, H

Suppose that a bag contains three coins: a penny, a nickel, and a dime.

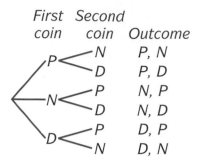

5. What is the probability that if a coin is chosen at random, it will be the penny?

Suppose a coin is chosen and it turns out to be the penny.

6. What is the probability that if a second coin is chosen, it will turn out to be the nickel?

7. What is the probability that the first coin will be the penny and the second coin will be the nickel?

8. Are the events of getting a penny and then a nickel *dependent* or *independent*?

A girl is told by her boyfriend that she is "one in a billion." She has a dimple in her chin (probability, 1%), eyes of different colors (probability, 0.1%), and is absolutely crazy about mathematics (probability, 0.01%).

Suppose that these probabilities are true for *everyone*. Because $1\% = 0.01 = \dfrac{1}{100}$, having a dimple in your chin would make you "one in a hundred."

9. What would having eyes of different colors make you?

10. What would being "absolutely crazy about mathematics" make you?

11. Why is the girl "one in a billion?"

The probability that a high school athlete in baseball, basketball, or football goes on to play at the college level is $\dfrac{1}{20}$. The probability that one of these athletes goes from college to professional sports is $\dfrac{17}{1,000}$.

12. What is the probability that a high school athlete goes on to play in both college and professional sports?

13. Out of 20,000 high school athletes, about how many go on to both college and professional sports?

In the United States, the probability that a man will live to the age of 75 is about $\frac{3}{10}$ and the probability that a woman will live to the age of 75 is about $\frac{1}{2}$. On the assumption that these probabilities are true for a person's grandparents, find each of the following probabilities.

14. The probability that both grandparents on a person's father's side will live to the age of 75.

15. The probability that both grandparents on a person's mother's side will live to the age of 75.

16. The probability that all four of a person's grandparents will live to the age of 75.

17. Do you think it is correct to assume that all four events are independent?

The actor Sean Connery once bet on the number 17 three times in succession in a roulette game in the St. Vincent casino. All three times the number 17 came up, and Mr. Connery won $20,000.

The wheel had 37 compartments, numbered 0 and 1 through 36.* What is the probability that the number 17 would come up

18. on a single spin?

19. on two successive spins?

20. on three successive spins?

A restaurant has four cooks, each of whom has a 7% probability of being absent from work.

21. Express the probability of 7% as a decimal fraction.

22. What is the probability that two of the cooks will be absent on the same day?

23. What is the probability that all four cooks will be absent at the same time?

24. Do you think it is reasonable to assume that the event of each cook being absent is always independent of the others? Explain.

---

*European roulette wheels have 37 compartments, unlike those in the United States, which have 38.

Set II

The following bet is from a book titled *Never Give a Sucker an Even Break.**

Take a small opaque bottle and seven olives, two of which are green, five black. The green ones are considered the "unlucky" ones. Place all seven olives in the bottle, the neck of which should be of such a size that it will allow only one olive to pass through at a time. Ask the sucker to shake them and then wager that he will not be able to roll out three olives without getting an unlucky green one amongst them. If a green olive shows, he loses.

1. Are the events that each successive olive to come out of the bottle is black *dependent* or *independent?*

2. What is the probability that the first olive that comes out of the bottle is black?

3. Would it be better to bet in favor of its being black or against it?

4. What is the probability that the second olive also is black?

5. Would it be better to bet in favor of its being black or against it?

6. What is the probability that the third olive also is black?

7. Would it be better to bet in favor of its being black or against it?

8. What is the probability that all three olives are black?

9. Would it be better to bet in favor of all three being black or against it?

*By John Fisher (Pantheon Books, 1976).

In 1976, a bowler in Toledo, Ohio made 33 strikes in a row!

Suppose that, each time the ball is rolled, the probability that a bowler will get a strike is 0.5.

10. What is the probability that this bowler will get two strikes in a row?

11. What is the probability that this bowler will get three strikes in a row?

To bowl a perfect game, a bowler needs to make 12 consecutive strikes.

12. Use a calculator to find the probability that this bowler will bowl a perfect game.

What would be the probability of a perfect game if the probability of each strike is

13. 0.7?

14. 0.9?

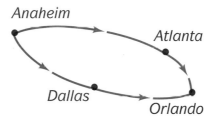

You want to fly from Anaheim to Orlando and find that you must change planes in either Dallas or Atlanta. You also find out that the probabilities of the four possible flights being on time are:

From Anaheim to Dallas:	70%
From Dallas to Orlando:	75%
From Anaheim to Atlanta:	60%
From Atlanta to Orlando:	80%

Suppose that both the flights you take must be on time for you to arrive in Orlando on time. What is the probability of arriving in Orlando on time if you go by way of

15. Dallas?

16. Atlanta?

17. Do you think it is reasonable to assume that the events of both flights arriving on time are always independent?

Suppose a poker player has been dealt three spades and two other cards from a deck of fifty-two cards. The player plans to throw away the two other cards and draw two more cards from the deck.

18. Why is the probability $\frac{10}{47}$ that the first card drawn will be a spade?

19. If the first card drawn is a spade, what is the probability that the second card drawn will also be a spade?

If both cards turn out to be spades, the player will have a hand called a "flush."

20. What is the probability that both of the cards drawn will be spades? Express your answer as a percent.

Set III

In a court case on a charge of overtime parking, a policeman observed the positions of the valves of the tires on one side of a parked car as shown in this figure. He recorded them as being at "one o'clock" and "six o'clock."

Later, when the allowed time had expired, he observed that the car was still in the same parking space with the valves of the two tires in the same positions as before. The owner of the car claimed that he had left the parking space before the time had expired and returned to it later. He said that the valves being in the same position as before must have been a coincidence.

1. If the position of a tire valve is recorded to the nearest "hour," how many different positions are possible?

2. What is the probability that one of the valves would be in its original position if the owner of the car was telling the truth?

3. Do you think that the event of each valve returning to its original position is independent or dependent of the others?

4. If the events are independent, what would the probability be of both valves returning to their original positions?

5. If the events are dependent, what effect might that have on the probability of both valves returning to their original positions?

The judge acquitted the defendant but said that he would have been convicted if all four wheels had been checked and found to have been in their original positions.

6. If the events are independent, what would the probability be of all four valves returning to their original positions?

7. Do you think this result, which the judge accepted, applies to the situation? Explain.

By permission of Johnny Hart and Field Enterprises, Inc.

4 Binomial Probability

Which of the following is the better bet? If a coin is tossed four times, would you bet that it would turn up heads twice and tails twice or would you bet that it would not?

If a coin is tossed once, we will assume that the outcomes of turning up heads or tails are equally likely. Each outcome, then, has a probability of $\frac{1}{2}$.

If the coin is tossed twice, there are four equally likely outcomes. They are illustrated in the tree diagram at the left. The probability that the coin will come up heads once and tails once is 50% because there are two ways out of the four of this happening and

First toss	Second toss	Outcome
H	H	H,H
	T	H,T
T	H	T,H
	T	T,T

$$\frac{2}{4} = \frac{1}{2} = 50\%$$

Continuing the tree diagram to illustrate four tosses, we find that there are 16 equally likely outcomes.

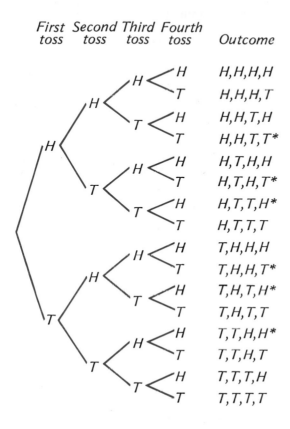

First toss	Second toss	Third toss	Fourth toss	Outcome
			H	H,H,H,H
			T	H,H,H,T
			H	H,H,T,H
			T	H,H,T,T*
			H	H,T,H,H
			T	H,T,H,T*
			H	H,T,T,H*
			T	H,T,T,T
			H	T,H,H,H
			T	T,H,H,T*
			H	T,H,T,H*
			T	T,H,T,T
			H	T,T,H,H*
			T	T,T,H,T
			H	T,T,T,H
			T	T,T,T,T

Of these 16 possible outcomes, there are exactly 6 (marked with asterisks in the diagram) in which the coin turns up heads twice and tails twice. This means that the probability of this occurring is

$$\frac{6}{16} = \frac{3}{8} = 37.5\%$$

The probability that the coin will *not* turn up heads twice and tails twice is

$$\frac{10}{16} = \frac{5}{8} = 62.5\%$$

So betting that a coin will not turn up heads twice and tails twice when it is tossed four times is the better bet.

The chances of getting heads or tails when a coin is tossed is one example of *binomial* probability.

The probabilities in a situation in which there are *two possible outcomes* for each part are called **binomial.**

The two possible outcomes in binomial probability do not have to be equally likely. For example, suppose that you are playing a game in which the probability of winning is $\frac{4}{5}$; that is, you can expect to win four out of every five games you play. It follows that your probability of losing is $\frac{1}{5}$.

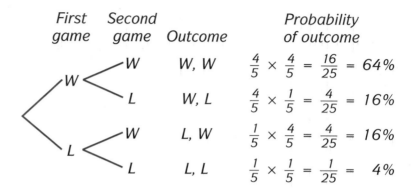

First game	Second game	Outcome	Probability of outcome
W	W	W, W	$\frac{4}{5} \times \frac{4}{5} = \frac{16}{25} = 64\%$
	L	W, L	$\frac{4}{5} \times \frac{1}{5} = \frac{4}{25} = 16\%$
L	W	L, W	$\frac{1}{5} \times \frac{4}{5} = \frac{4}{25} = 16\%$
	L	L, L	$\frac{1}{5} \times \frac{1}{5} = \frac{1}{25} = 4\%$

Now suppose that you play the game twice. The four possible outcomes, shown in this tree diagram, are not equally likely. The probability that you will win both games is 64%. The probability that you will lose both games is 4%. Because there are two outcomes in which you win exactly one game, each having a probability of 16%, the probability that you will win exactly one game is

$$16\% + 16\% = 32\%$$

These results are summarized in the table below.

Number of games won	0	1	2
Number of possibilities	1	2	1
Percent probability	4%	32%	64%

EXERCISES

SET I

The tree diagram at the left shows the possibilities of boys and girls in a family with two children. We will assume that the chances of a child being a boy or girl are equally likely.*

First child	Second child	Children in family
B	B	B,B
	G	B,G
G	B	G,B
	G	G,G

---

*According to birth records, in the United States the probability that a child is a boy is 0.512 and the probability that a child is a girl is 0.488.

1. Show why, if the probability of a child being a boy is 0.50, or 50%, the probability of two boys in a row is 0.25, or 25%.

2. Show why the probability of a boy followed by a girl is also 25%.

The diagram shows that there are two ways in which the family can consist of one boy and one girl: BG and GB, so the probability of a boy and a girl is 25% + 25% = 50%.

3. Refer to the diagram to copy and complete the following table.

Number of boys	0	1	2
Number of possibilities	1	▨	▨
Percent probability	25%	▨	▨

What is the percent probability that, in a family with two children,

4. both children are of the same sex?

5. at least one child is a girl?

This tree diagram shows the possibilities for three successive times that a baseball player comes up to bat. (H stands for hit and O for no hit.)

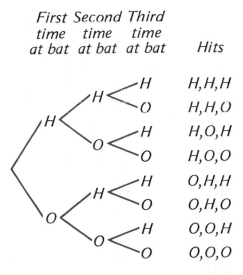

First time at bat	*Second time at bat*	*Third time at bat*	*Hits*
		H	H,H,H
	H	O	H,H,O
H		H	H,O,H
	O	O	H,O,O
		H	O,H,H
	H	O	O,H,O
O		H	O,O,H
	O	O	O,O,O

We will assume that the player's batting average is 0.400, so that the probability of a hit each time at bat is 0.4 and the probability of no hit is 0.6.

6. Show why the probability of the player getting three hits in three times at bat, shown in the diagram as HHH, is 0.064, or 6.4%.

7. Show why the probability of the player getting two hits followed by a miss, shown in the diagram as HHO, is 0.096, or 9.6%.

The diagram shows that there are three ways in which the player can get exactly two hits: HHO, HOH, and OHH. Thus the probability of getting exactly two hits is 9.6% + 9.6% + 9.6% = 28.8%.

8. What is the probability of the player getting a hit followed by two misses?

9. How many ways does the diagram show in which the player gets exactly one hit?

10. What is the probability of the player getting exactly one hit?

11. What is the probability of the player getting no hits?

12. Use the diagram and your answers to exercises 6–11 to copy and complete the following table.

Number of hits		0	1	2	3
Number of possibilities	1	▨	▨	▨	
Percent probability	▨	▨	▨	▨	

13. Why should the four numbers in the second line of your table add up to 8?

14. What should the four numbers in the third line of your table add up to?

Use your table to answer the following questions.

15. What is the most probable number of hits the player will get in three times at bat?

16. What is the least probable number of hits?

The probability that the player will get more than one hit is

$$28.8\% + 6.4\% = 35.2\%$$

17. What is the probability that the player will get *fewer* than two hits?

The tree diagram at the top of the next page shows the possibilities of its snowing during four consecutive winter days at a ski resort. (S stands for snow and O for no snow.)

Suppose that the probability of snow on any given day is 70%, or 0.70, so that the probability of no snow is 30%, or 0.30.

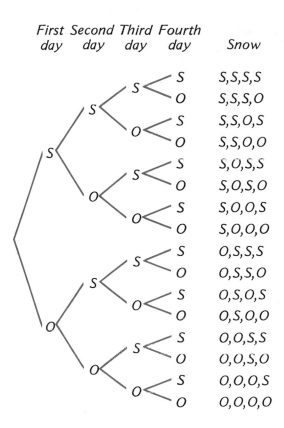

First Second Third Fourth
day   day   day   day   Snow

			S	S,S,S,S
		S	O	S,S,S,O
		O	S	S,S,O,S
			O	S,S,O,O
	S	S	S	S,O,S,S
			O	S,O,S,O
	O	S	S	S,O,O,S
			O	S,O,O,O
	S	S	S	O,S,S,S
			O	O,S,S,O
	O	S	S	O,S,O,S
			O	O,S,O,O
	S	S	S	O,O,S,S
			O	O,O,S,O
	O	S	S	O,O,O,S
			O	O,O,O,O

18. Show why the probability of its snowing all 4 days, shown in the diagram as SSSS, is 0.2401, or 24.01%.

19. Show why the probability of snow on the first three days followed by no snow on the fourth day, shown in the diagram as SSSO, is 0.1029, or 10.29%.

20. In how many ways does the diagram show it can snow on exactly three days?

21. Show why the probability that it will snow on exactly three days is 41.16%.

22. What is the probability that it will snow on the first two days, followed by no snow on the third and fourth days, shown in the figure as SSOO?

23. In how many ways does the diagram show it can snow on exactly two days?

24. What is the probability that it will snow on exactly two days?

25. What is the probability that it will snow on the first day, followed by no snow on the other three days?

26. In how many ways does the diagram show it can snow on exactly one day?

27. What is the probability that it will snow on exactly one day?

28. What is the probability that it will not snow on any of the four days?

29. Use your answers to exercises 18–28 to copy and complete the following table. Round each probability to the nearest whole number.

Number of days it snows	0	1	2	3	4
Number of possibilities	1	▦	▦	▦	▦
Percent probability	▦	▦	▦	▦	▦

30. What should the five numbers in the second line of your table add up to?

31. What should the five numbers in the third line of your table add up to?

32. What is the most likely number of days that it will snow?

33. What is the probability that it will snow on more than two days?

SET II

The Austrian monk Gregor Mendel founded the science of genetics in the nineteenth century when he experimented with the crossbreeding of plants. His discoveries, described in a report titled "Experiments with Plant Hybrids," were perhaps the first in which the theory of probability was applied to science.

In an experiment with garden peas, Mendel crossbred plants having round peas with plants having wrinkled peas. All the offspring had round peas. When these plants were bred with each other, 75% of the offspring had round peas and the remaining 25% had wrinkled peas.

To explain the appearance of both types of peas among these offspring, Mendel assumed that a pair of genes in each plant determined the type of pea that it produced. The pairs and types are shown below.

Czechoslovakian Academy of Sciences

Genes	RR	RW	WR	WW
Type of pea	◉	◉	◉	◉

The genes are labeled R for round and W for wrinkled in the diagram for simplicity. Note that a plant produces wrinkled peas only if *both* of its genes are W.

Each parent plant contributes one of its genes to each of its offspring. The crossbreeding of the first generation to produce the second is shown in this tree diagram.

1. According to this diagram, what is the probability of getting a plant with round peas in the second generation?

2. What is the probability of getting a plant with wrinkled peas?

The answers to these questions explain why all the offspring in the first stage of Mendel's experiment had round peas.

The crossbreeding of the second generation to produce the third is shown in the second tree diagram.

3. According to this diagram, what is the probability of getting a plant with round peas in the third generation?

4. What is the probability of getting a plant with wrinkled peas in the third generation?

Several decades after Mendel's work, the German botanist Karl Correns did an experiment with Japanese four-o'clocks. He crossbred plants having red flowers with plants having white flowers. All the offspring plants had pink flowers. When the plants with pink flowers were bred with each other, 25% of the plants in the third generation had red flowers, 50% had pink flowers, and 25% had white flowers.

Suppose that a pair of genes in each plant determines the color of its flowers and that R and W represent red- and white-color genes, respectively. The pairs and types are shown below.

Genes	RR	RW	WR	WW
Color of flowers	Red	Pink	Pink	White

Look again at the first tree diagram above.

5. According to the diagram, what is the probability of getting a plant with pink flowers in the second generation?

6. What is the probability of getting a plant with red or white flowers in the second generation?

Look again at the second tree diagram above.

7. According to this diagram, what is the probability of getting a plant with pink flowers in the third generation?

8. What is the probability of getting a plant with red flowers in the third generation? A plant with white flowers?

First generation / Second generation

Gene from first plant — Gene from second plant — Offspring

		Offspring
R	W	RW
	W	RW
R	W	RW
	W	RW

Second generation / Third generation

Gene from first plant — Gene from second plant — Offspring

		Offspring
R	R	RR
	W	RW
W	R	WR
	W	WW

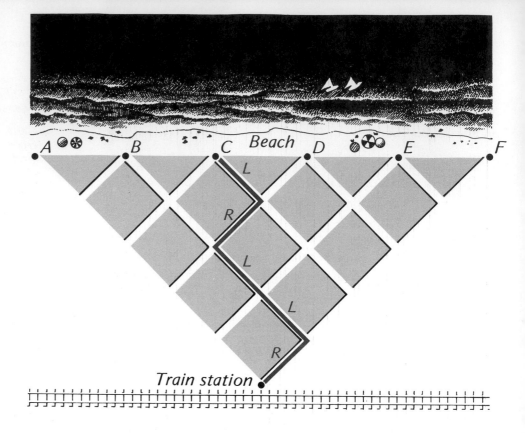

Set III

This map represents the streets of a beach town that lead from a train station to the beach.* An excursion train arrives at the station and the passengers head for the beach. No one knows whether to take the street at the left or right at each intersection, so we will assume that the probability of each person going in either direction is the same: 0.50, or 50%. One possible path is shown on the map. This path corresponds to going "right, left, left, right, left."

The tree diagram on the facing page shows the possibilities of someone going left or right at each of the five intersections encountered between the station and the beach.

1. Refer to the diagram to copy and complete the following table. Round each probability to the nearest whole number.

Number of intersections at which path goes to the left	0	1	2	3	4	5
Number of possibilities	1	‖‖‖	‖‖‖	‖‖‖	‖‖‖	‖‖‖
Percent probability	3%	‖‖‖	‖‖‖	‖‖‖	‖‖‖	‖‖‖

---

*Adapted from a story in *Mathematics in Your World,* by K. W. Menninger (Viking, 1962).

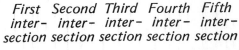

First Second Third Fourth Fifth
inter- inter- inter- inter- intersection section section section section

Path

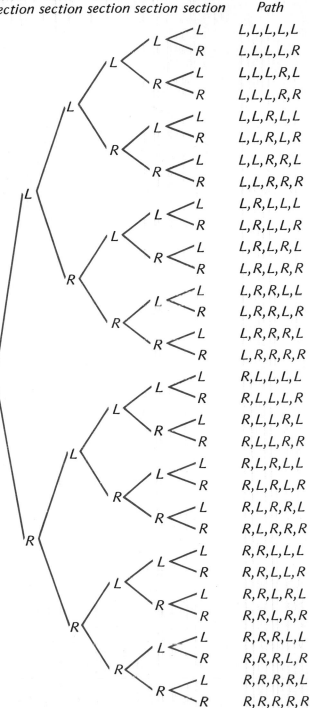

Path
L,L,L,L,L
L,L,L,L,R
L,L,L,R,L
L,L,L,R,R
L,L,R,L,L
L,L,R,L,R
L,L,R,R,L
L,L,R,R,R
L,R,L,L,L
L,R,L,L,R
L,R,L,R,L
L,R,L,R,R
L,R,R,L,L
L,R,R,L,R
L,R,R,R,L
L,R,R,R,R
R,L,L,L,L
R,L,L,L,R
R,L,L,R,L
R,L,L,R,R
R,L,R,L,L
R,L,R,L,R
R,L,R,R,L
R,L,R,R,R
R,R,L,L,L
R,R,L,L,R
R,R,L,R,L
R,R,L,R,R
R,R,R,L,L
R,R,R,L,R
R,R,R,R,L
R,R,R,R,R

Look at the map of the town. At what point on the beach will someone end up who takes the street

2. to the left at all five intersections?

3. to the right at all five intersections?

4. How do the distances walked by the people from the train station to each of the six marked points on the beach compare with one another?

5. If each person from the excursion train remains at the point at which he or she arrives at the beach, what do you think the arrangement of the people along the beach will be like?

LESSON

5

Pascal's Triangle

All eight children in this remarkable family are girls. Because the probabilities of a child being a boy or girl are about the same, it would seem that in a family of this size, four girls and four boys would be much more likely. Exactly how do the probabilities of a family having eight girls and a family having four girls and four boys compare?

Although this could be figured out from a tree diagram, there is an easier way. It is based on a pattern of numbers so old that no one knows who first discovered it. The pattern, called "Pascal's triangle," is named after the great seventeenth-century mathematician Blaise Pascal. One of the originators of probability theory, Pascal wrote a book about the triangle and its properties. Pascal's triangle looks like the figure at the top of the next page.

Each number within the triangle is found by adding the pair of numbers in the row above it at the left and right. The triangle can be

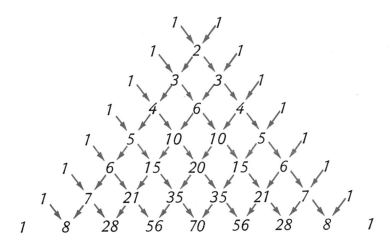

continued indefinitely by writing a 1 at both ends of each new row and then adding each pair of numbers in the preceding row.

Each row of the triangle contains the numbers of ways of getting each possible outcome for a situation in which the probabilities are binomial. The *fourth* row, for example, contains the numbers of ways in which a coin can turn up when it is tossed *four* times.*

Number of heads	0	1	2	3	4
Number of ways	1	4	6	4	1

The question about the probabilities for a family with *eight* children can be answered by looking at the *eighth* row of the triangle.

Number of girls	0	1	2	3	4	5	6	7	8
Number of ways	1	8	28	56	70	56	28	8	1

If we assume that the probabilities of a child being a boy or girl are each $\frac{1}{2}$, the probability of boys and girls in any particular sequence of eight children is

$$\frac{1}{2} \times \frac{1}{2} \times \frac{1}{2} \times \frac{1}{2} \times \frac{1}{2} \times \frac{1}{2} \times \frac{1}{2} \times \frac{1}{2} = \frac{1}{256}$$

There is only one sequence of eight girls, so the probability of a family having eight girls is $1 \times \frac{1}{256}$, or approximately 0.4%. There are 70 sequences of four girls and four boys, each having probability $\frac{1}{256}$, so the

---

*The ways are illustrated in the tree diagram on page 477.

probability of a family having four girls and four boys is

$$70 \times \frac{1}{256} = \frac{70}{256}$$

or about 27.3%.

EXERCISES

SET I

The first four rows of Pascal's triangle are shown below. Also shown is the number of each row and the sum of the numbers in each row.

Number of row							Sum of numbers in row
1			1	1			2
2		1	2	1			4
3	1	3	3	1			8
4	1	4	6	4	1		16

1. Copy this figure and then add six more rows. Also continue the numbers in the two columns.

2. What kind of sequence do the numbers in the column of sums form?

The probability that a teenager will have an accident during his or her first year of driving is 40%, or 0.4. The probability of not having an accident, then, is 60%, or 0.6.

Suppose that a group of five teenagers get their driver's licenses at the same time. There is one way in which none of them will have an accident during the first year, five ways in which exactly one of them will have an accident (it could be any one of the five teenagers), and so on. The numbers of ways in which the group might have various numbers of accidents is given by the fifth row of Pascal's triangle.

3. Use Pascal's triangle to copy and complete the following table.

Number of teenagers having an accident	0	1	2	3	4	5
Number of ways	‖‖‖	‖‖‖	‖‖‖	‖‖‖	‖‖‖	‖‖‖

Since the probability of not having an accident is 0.6, the probability that none of the five teenagers will have an accident is

$$0.6 \times 0.6 \times 0.6 \times 0.6 \times 0.6$$

4. Find this probability as a percent to the nearest tenth.

Since there are five ways in which exactly one of the teenagers can have an accident, the probability that exactly one of the five teenagers has an accident is

$$5 \times (0.4 \times 0.6 \times 0.6 \times 0.6 \times 0.6)$$

5. Find this probability as a percent to the nearest tenth.

6. According to your table for exercise 3, how many ways are there in which exactly two of the teenagers can have accidents?

7. What is the probability, to the nearest tenth of a percent, that exactly two of the five teenagers will have accidents?

8. How many ways are there in which exactly three of the teenagers can have accidents?

9. What is the probability, to the nearest tenth of a percent, that exactly three of the five teenagers will have accidents?

10. How many ways are there in which exactly four of the teenagers can have accidents?

11. What is the probability, to the nearest tenth of a percent, that exactly four of the five teenagers will have accidents?

12. What is the probability, to the nearest tenth of a percent, that all five teenagers will have accidents?

13. What is the most likely number of the five teenagers to have an accident during the first year of driving? Why?

In professional basketball, Rick Barry holds the record for the highest free throw percentage: 90%. This means that his probability of getting a basket on a free throw was 0.9 and his probability of not getting a basket was 0.1.

14. Use Pascal's triangle to copy and complete the following table for the numbers of ways of getting baskets for six free throws.

Number of baskets	0	1	2	3	4	5	6
Number of ways							

15. Show that the probability of Barry making baskets on six consecutive free throws was about 53.1%.

16. Show that the probability of his making five consecutive baskets and then missing the last basket was about 5.9%.

17. In how many ways could he make exactly five baskets on six consecutive free throws?

18. What was the probability that he made exactly five baskets on six consecutive free throws?

19. What was the probability that he made *at least* five baskets on six consecutive free throws?

A copy store has three color copiers, but they frequently break down. The probability that a given copier will work on a given day is 80%.

20. What is the probability that a given copier will *not* work?

21. Use Pascal's triangle to copy and complete the following table for the numbers of copiers that work on a given day.

Number of copiers that work	0	1	2	3
Number of ways				

What is the probability that, on a given day,

22. all three copiers are working?

23. none of the copiers are working?

24. two of the copiers are working?

25. only one copier is working?

26. *at least* one copier is working?

Suppose that a true-or-false quiz has eight questions and that, since you haven't studied for it, you have to guess all eight answers.

27. Use Pascal's triangle to copy and complete the following table.

Number of correct answers	0	1	2	3	4	5	6	7	8
Number of ways									

The probability of guessing each answer correctly is $\frac{1}{2}$.

28. What is the probability of guessing all eight answers correctly?

29. In how many ways can you answer exactly half of the questions correctly?

30. What is the probability of answering exactly half of the questions correctly?

Suppose that to get a passing grade on the quiz, you have to answer at least six questions correctly.

31. In how many ways can you answer *six or more* questions correctly?

32. What is the probability that you will get a passing grade by guessing?

Set II

This painting shows an unlikely situation: a set of pennies is being tossed onto a table in which every coin has turned up heads.

Painting by Stanley Meltzoff. Copyright © 1965 by Scientific American, Inc. All rights reserved.

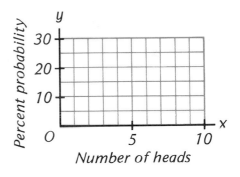

1. Use Pascal's triangle to complete the following table of probabilities of getting various numbers of heads when 10 coins are tossed. Round each probability to the nearest percent.

Number of heads	0	1	2	3	4	5	6	7	8	9	10
Number of ways	1	10	▓	▓	▓	▓	▓	▓	▓	▓	▓
Percent probability	0%	1%	▓	▓	▓	▓	▓	▓	▓	▓	▓

2. Graph the information in your table, representing the numbers of heads on the *x*-axis and the percent probabilities on the *y*-axis. Convenient scales for the axes are shown above.

3. If 10 coins are tossed, what number of heads is most likely to turn up?

4. Would it be better to bet *in favor of* or *against* that number turning up? Explain.

Here is a graph of the probabilities of getting various numbers of heads when 20 coins are tossed.

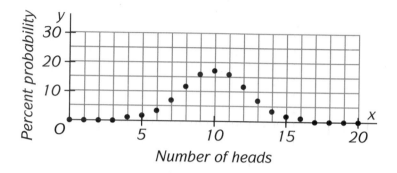

5. In what way is this graph similar to the one that you drew?

6. In what ways is it different?

Refer to the graph above to answer the following questions.

7. If 20 coins are tossed, what number of heads is most likely to turn up?

8. What is the probability of that number turning up? (Make an estimate from the graph.)

9. Which would be more surprising: equal numbers of heads and tails turning up when *10* coins are tossed or when *20* coins are tossed?

10. What happens to the probability of equal numbers of heads and tails turning up as the number of coins being tossed increases?

SET III

Pascal's triangle has other interesting properties in addition to those related to probability. In fact, it has so many that Pascal said that more properties had been omitted from his book on the triangle than had been included.

1. Find the values of 11^2, 11^3, and 11^4.

2. What do these numbers have to do with Pascal's triangle?

3. Now find the value of 11^5. Does it fit the pattern?

One sloping row of numbers in Pascal's triangle is shown in color in the figure below.

```
                1       1
            1       2       1
        1       3       3       1
    1       4       6       4       1
1       5       10      10      5       1
    1       6       15      20      15      6       1
1       7       21      35      35      21      7       1
```

4. Starting at the top of this row, what is the sum of:

> its first two numbers?
> its first three numbers?
> its first four numbers?
> its first five numbers?
> its first six numbers?

5. What do you notice about these answers?

Another sloping row of numbers is shown in color in the figure below.

```
                1       1
            1       2       1
        1       3       3       1
    1       4       6       4       1
1       5       10      10      5       1
    1       6       15      20      15      6       1
1       7       21      35      35      21      7       1
```

6. Starting at the top of this row, what is the sum of:

   its first and second numbers?
   its second and third numbers?
   its third and fourth numbers?
   its fourth and fifth numbers?
   its fifth and sixth numbers?

7. These sums form a certain number sequence. What sequence is it?

The numbers in the figure below have been separated into a set of sloping rows.

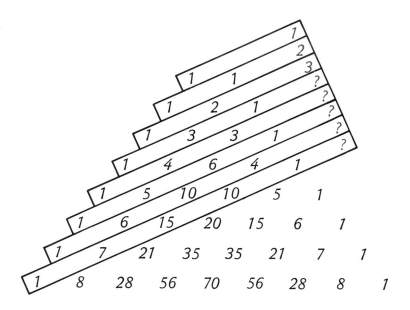

The sums of the numbers in the first three rows are 1, 2, and 3.

8. What are the sums of the numbers in the next five rows?

9. These eight sums form a certain number sequence. What sequence is it?

By permission of Johnny Hart and Field Enterprises, Inc.

LESSON
6

The Birthday Problem: Complementary Events

An amazing problem in probability is that of the "coinciding birthdays." The problem concerns the number of people needed in a group in order for the odds to favor two of them sharing the same birthday. The answer, just 23 people, is quite surprising considering the number of days in a year.

The solution of the "coinciding birthdays" problem is based on the probabilities of *complementary events*. An example of such probabilities is shown in the tree diagram below. If two coins are tossed, the probability

First coin	Second coin	Outcome	Probability
H	H	HH	$\frac{1}{4}$
	T	HT	
T	H	TH	$\frac{3}{4}$
	T	TT	

of both coins turning up heads is $\frac{1}{4}$. The probability that both coins do *not* turn up heads is $\frac{3}{4}$. These two probabilities add up to 1:

$$\frac{1}{4} + \frac{3}{4} = \frac{4}{4} = 1$$

The event that something happens and the event that it does not happen are called **complementary events.**

Probabilities of complementary events always add up to 1.

This means that the probability of one of the complementary events can be found by subtracting the probability of the other one from 1. In the "coinciding birthdays" problem, either there are two people in a group who share the same birthday or no two people do. Because these events are complementary, the probability that two people share the same birthday can be found by subtracting the probability that no two people share the same birthday from 1.

What is the probability that no two people in the group share the same birthday? For simplicity, we will ignore leap year and assume that the probability of someone being born on any one of the 365 days of an ordinary year is the same.

The first person's birthday can be any day. The probability that the second person's birthday is different is $\frac{364}{365}$, the probability that the third person's birthday is different is $\frac{363}{365}$, the probability that the fourth person's birthday is different is $\frac{362}{365}$, and so on.

Because the probability of several things happening in succession is found by multiplying the probabilities of all of them, the probability that the second person's birthday is different *and* the third person's birthday is different *and* the fourth person's birthday is different, and so on, is

$$\frac{364}{365} \times \frac{363}{365} \times \frac{362}{365} \times \dots$$

This product, then, is the probability that *no* two people in the group share the same birthday. The probability that two people in the group *do* share the same birthday is

$$1 - \frac{364}{365} \times \frac{363}{365} \times \frac{362}{365} \times \dots$$

How this probability varies with the number of people in the group will be considered in some of the exercises in this lesson.

EXERCISES

SET I

Racing statistics show that the probability that the favored horse will win in a given race is $\frac{1}{3}$.

1. What is the probability that the favored horse will *not* win?

2. What are the events that the favored horse will win and that the favored horse will *not* win called?

What is the probability that, in two consecutive races,

3. the favored horses will win both races?

4. the favored horses will *not* win both races?

What is the probability that, in three consecutive races,

5. the favored horses will win all three races?

6. the favored horses will *not* win all three races?

7. the favored horses will win *none* of the races?

8. the favored horses will win *at least one* of the races?

The American artist Rube Goldberg (1883 – 1970) was so well known for his cleverly ridiculous inventions that, according to one dictionary, his name has come to mean "having a fantastically complicated, improvised appearance" and "deviously complex and impractical."

The invention for sharpening pencils shown here makes use of an opossum and a woodpecker and includes an emergency knife in case either animal "gets sick and can't work."

From *Rube Goldberg vs. the Machine Age* by Reuben L. Goldberg. © 1968 King Features Syndicate, Inc. Reprinted by permission of Hastings House, Publishers, Inc.

Suppose that the probability that the opossum will get sick is 30% and the probability that the woodpecker will get sick is 10%. What is the probability that

9. the opossum will *not* get sick?

10. the woodpecker will *not* get sick?

11. *both* the opossum and woodpecker will get sick?

12. *neither* the opossum *nor* the woodpecker will get sick?

13. Are the events that both will get sick and neither one will get sick complementary? Explain.

The makers of Cracker Jack began putting a prize in each box in 1912. Some of the early prizes are now worth as much as $2,000!*

Suppose that the company is using 100 different prizes at present and that the probability of any one of the prizes being included in a box is the same as any other.

You open a box of Cracker Jack and find a particular prize. What is the probability that, when you open a second box,

14. you will find the same prize?

15. you will find a different prize?

The probability that, if you open *three* boxes, there will be a different prize in each box, is

$$\frac{99}{100} \times \frac{98}{100}$$

16. Find this probability to the nearest percent.

What is the probability that there will be a different prize in each box if you open

17. four boxes?   18. five boxes?

The probabilities of getting the same prize at least twice from different numbers of boxes of Cracker Jack are shown in the graph at the top of the next page.

Would you expect to find a different prize in each box if you opened

19. 10 boxes?   20. 15 boxes?

Sketched from a photograph by John Snedeker and C. L. Haines, appearing in Cracker Jack Prizes by Alex Jaramillo, Abbeville Press, Inc. 1989.

Cracker Jack Prizes, by Alex Jaramillo (Abbeville Press, 1989).

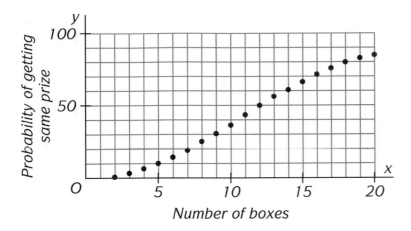

21. What is the smallest number of boxes you would have to open to have a 50% probability of getting the same prize in two boxes?

Set II

The Arthur Murray Dance Studios once offered $25 worth of dancing lessons to anyone who had a "lucky dollar." A "lucky dollar" was a bill whose serial number included a 2, 5, or 7. If you have any dollar bills with you, check to see whether or not you would have been a winner.

Courtesy of the U.S. Department of the Treasury

1. The probability that any given digit is a lucky one is $\frac{3}{10}$. Explain.

2. What is the probability that any given digit is *not* a lucky one?

There are eight digits in the serial number of a dollar bill.

3. Show why the probability that all eight digits will not be lucky ones is about 6%.

4. What is the probability that at least one of the digits in the serial number is a lucky one?

5. Out of 100 people with dollar bills, how many would you expect to be winners?

6. Why do you suppose the Arthur Murray Dance Studios made this offer?

The probabilities that there are two people in a group who share the same birthday and that no two people do are shown below. *Each is rounded to the nearest percent.*

Number of people in group	Probability that all birthdays are different	Probability that there are two people who share the same birthday
2	100%	0%
4	98%	2%
6	96%	4%
8	93%	7%
10	88%	12%
12	83%	17%
14	78%	22%
16	72%	28%
18	65%	35%
20	59%	41%
22	52%	48%
24	46%	54%
26	40%	60%
28	35%	65%
30	29%	71%
32	25%	75%
34	20%	80%
36	17%	83%
38	14%	86%
40	11%	89%
42	9%	91%
44	7%	93%
46	5%	95%
48	4%	96%
50	3%	97%

7. Graph the information in the table, representing the number of people in the group on the x-axis and the probabilities on the y-axis. Convenient scales for the axes are shown at the top of the next page.

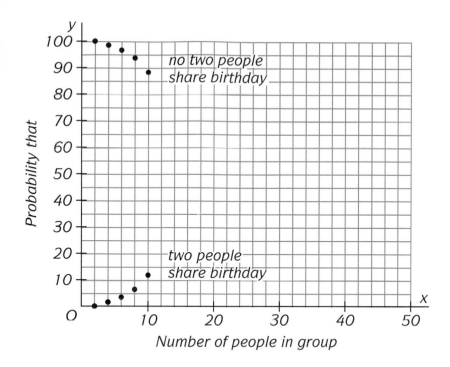

The first part of the graph is shown in the figure.

8. How does your graph support the answer "23 people" that was given for the "coinciding birthdays" problem?

The birthdays of the first 30 presidents of the United States are included in the list below and on the next page.

9. Before looking through the list, would you bet *in favor of* or *against* two of them having the same birthday? Explain.

10. Now go through the list carefully. Do two share the same birthday? If so, which two?

President	Birthday
1. Washington	Feb. 22
2. J. Adams	Oct. 30
3. Jefferson	Apr. 13
4. Madison	Mar. 16
5. Monroe	Apr. 28
6. J. Q. Adams	July 11
7. Jackson	Mar. 15
8. Van Buren	Dec. 5
9. W. H. Harrison	Feb. 9
10. Tyler	Mar. 29

President	Birthday	President	Birthday
11. Polk	Nov. 2	21. Arthur	Oct. 5
12. Taylor	Nov. 24	22. Cleveland	Mar. 18
13. Fillmore	Jan. 7	23. B. Harrison	Aug. 20
14. Pierce	Nov. 23	24. McKinley	Jan. 29
15. Buchanan	Apr. 23	25. T. Roosevelt	Oct. 27
16. Lincoln	Feb. 12	26. Taft	Sept. 15
17. A. Johnson	Dec. 29	27. Wilson	Dec. 28
18. Grant	Apr. 27	28. Harding	Nov. 2
19. Hayes	Oct. 4	29. Coolidge	July 4
20. Garfield	Nov. 19	30. Hoover	Aug. 10

352 Varaşoğlu - Varol

Varaşoğlu Osman Suat	
Merter Sitesi	23 73 61
Varavir Şevki Kyalı Bağdat C 181	52 12 64
Varaylı Zeki Ev Slmy Eczane S 50	33 24 32
Varbarbut Avram	
Ev Şiş Sıracevizler C 51	48 78 22
Varbarut Yako	
Ev Şiş Sıracevizler C 35	40 83 52
Varboz Muammer	
Ev Fth Sarıgüzel C 111/1	23 71 21
Varcın Kutsiye	
Ev Fth Sofular C 163	23 97 74
Ev Aks Atatürk Bulv 146	28 26 98
Vardal Dursun Ali	
Ayvans Demir C 49/2	21 70 77
Vardal Enver Ev Atıkalı Lodos S 2	25 30 35
Vardal İhsan	
Ev Gayrett Yıldızposta C	66 01 32
Vardal Kenan	
Mcköy Kervangeçmez S	47 61 23
Vardal Muammer	
Ev Kuşt Bestekârşakırağa S 22 ..	46 15 07
Vardal Nebile Ev Yköy Naima S	73 91 16
Vardal Nebile	
Ev Gayrett Yıldızposta C	66 14 64
Vardal Necati Ev Ayvans Çember S	23 84 31
Vardal Refik Ev Üsk Kuşatçı S 1/3	33 02 52
Vardal Sabri	
Ev Bahçelie Basın Sit Bl B 2	71 42 94
Vardal Veysel Bkapı Liman H 101	27 95 67
Vardar Abdülriza	
Krköy Nohut Çk 3/2	45 84 25
Vardar Abdürrezak Tarlab Çukur S 6	43 05 73

Suppose that a telephone book is opened to a page at random and any set of consecutive telephone numbers (such as that from a Turkish telephone book shown here) is chosen from that page. What is the probability that the last two digits of two of the telephone numbers in the set are the same? This problem is identical with the one about getting the same prize twice from a series of Cracker Jack boxes, described in Set I beginning with exercise 14.

11. What aspect of this problem makes it identical with the Cracker Jack problem?

Because the probabilities in the two problems are the same, the graph for the Cracker Jack problem can be used for the telephone digits problem.

12. How should the x-axis be relabeled to represent the telephone problem?

13. How should the y-axis be relabeled?

14. If you looked at 20 consecutive telephone numbers, would you expect to find the last two digits of two of them to be the same?

Choose a set of consecutive telephone numbers from the telephone book for your city.

15. How many did you look through before finding two numbers that end in the same two digits? If possible, show the set of numbers.

SET III

In the seventeenth century, a wealthy Frenchman known as the Chevalier de Méré was fond of betting that, if a die is rolled four times, the number 6 will turn up at least once.

1. If a die is rolled once, what is the probability that it will *not* turn up 6?

2. What is the probability that, if a die is rolled four times, it will not turn up 6 on any of the rolls? Express it to the nearest percent.

3. What is the probability to the nearest percent that, if a die is rolled four times, the number 6 will turn up at least once?

4. Would it be better to bet *in favor of* or *against* this happening?

After winning a lot of money with this bet, de Méré switched to a different bet. He bet that if two dice are rolled 24 times, a sum of 12 will turn up at least once.

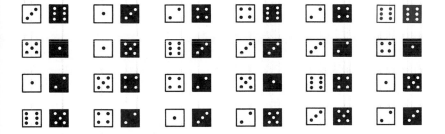

5. If a pair of dice is rolled once, what is the probability that it will *not* turn up 12?

6. Write an expression to represent the probability that if two dice are rolled 24 times, they will not turn up 12 on any of the rolls.

Expressed as a percentage, the value of this expression is 51%.

7. What is the probability that if two dice are rolled 24 times, a sum of 12 will turn up at least once?

8. Would it be better to bet *in favor of* or *against* this happening?

9. Do you think de Méré won a lot of money with this bet?

CHAPTER

8

Summary and Review

Drawing by Chas. Addams; © 1979 The New Yorker Magazine, Inc.

In this chapter we have become acquainted with:

Probability: the measure of chance *(Lessons 1 and 2).* The probability of an event in which there is a set of equally likely outcomes can be calculated as the ratio:

$$\frac{\text{number of ways in which the event can occur}}{\text{total number of equally likely outcomes}}$$

The probability of an event can also be calculated as the ratio:

$$\frac{\text{number of times the event occurred}}{\text{total number of possible occurrences}}$$

Probabilities can be expressed as numbers on a scale ranging from 0 to 1 or as percentages on a scale ranging from 0% to 100%.

Probabilities of successive events *(Lesson 3).* To find the probability of several events occurring in succession, multiply the probabilities of the individual events.

Events that are influenced by other events are said to be *dependent*. Events that have no influence on each other are said to be *independent*.

Binomial probability and Pascal's triangle *(Lessons 4 and 5)*. The probabilities in a situation in which there are two possible outcomes for each part are called "binomial."

Pascal's triangle is a pattern of numbers from which binomial probabilities can be easily determined. Each number within the triangle is found by adding the pair of numbers directly above it at the left and right. The triangle begins like this:

$$
\begin{array}{ccccccc}
 & & & 1 & & 1 & & \\
 & & 1 & & 2 & & 1 & \\
 & 1 & & 3 & & 3 & & 1 \\
1 & & 4 & & 6 & & 4 & & 1
\end{array}
$$

Complementary events *(Lesson 6)*. The event that something happens and the event that it does not happen are called *complementary events*.

Probabilities of complementary events always add up to 1.

EXERCISES

SET I

If a frog catches a fly, the probability that the fly was alive when the frog ate it is 100%.

1. What does that mean?

If you touch a toad, the probability that you will get warts as a result is 0%.

2. What does that mean?

Probabilities about smoking are listed below:

A smoker wants to quit	66%
A smoker has tried to quit	84%
A smoker will succeed in quitting	21%

3. Do most people who smoke want to quit?

4. What is the probability that someone who smokes does not want to quit?

5. What is the probability that someone who smokes has never tried to quit?

6. Do most people who try to quit smoking succeed?

7. What is the probability that someone who tries to quit smoking will not succeed?

Somebody once said that if enough monkeys were given enough typewriters and they hit the keys at random long enough, they might eventually write every book that had ever been written!

Suppose the typewriter has 44 keys and that a monkey is equally likely to strike each key. What is the probability that if a monkey hits

8. one key, it would type T?

9. two keys, it would type TH?

10. three keys, it would type THE?

This figure appeared at the front of *Ssu Yuan Yü Chien,* a book written in China by Chu Shih-Chieh in 1303.

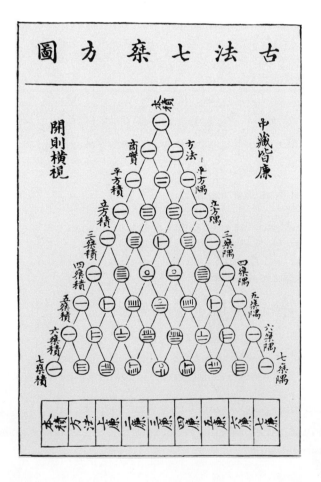

11. What do you think it looks like?

12. Translate the symbols in the circles into symbols more familiar to you.

13. What type of probabilities can be determined from the figure?

A popular song of the 1920s began like this:

> Keep your sunny side up, up!
> Hide the side that gets blue.
> If you have nine sons in a row
> Baseball teams make money, you know!*

14. If you assume that the probability of a child being a boy is $\frac{1}{2}$, then the probability of a family having two sons in a row is $\frac{1}{4}$. Explain.

15. What is the probability of a family having nine sons in a row?

When people are given lie detector tests, the probability that the lie detector gives an incorrect result is 0.2.

16. What is the probability that the lie detector gives a correct result?

Suppose four people applying for a job are given a lie detector test and that all four tell the truth.

17. What is the probability that the lie detector gives correct results for all four people?

18. What is the probability that the lie detector falsely indicates that at least one person is lying?

Set II

In horse racing, the "daily double" is a bet on the winners of the first and second races.

Suppose there are ten horses in the first race and eight horses in the second race and that your bets are based on nothing more than guessing.

*"Sunny Side Up," copyright © 1929 by DeSylva, Brown, & Henderson, Inc. Copyright renewed, assigned to Chappell & Co., Inc. International copyright secured. All rights reserved. Used by permission.

What is the probability

1. of correctly guessing the winner of the first race?
2. of correctly guessing the winner of the second race?
3. of winning the daily double by guessing both winners?

A "triple bet" at the race track is a bet on the first three horses to come in in a given race (the "win," "place," and "show.")

Suppose there are nine horses in a race. What is the probability

4. of correctly guessing the horse that comes in first?
5. of, having guessed the first horse correctly, correctly guessing the horse that comes in second?
6. of, having guessed the first two horses correctly, correctly guessing the horse that comes in third?
7. of winning the triple bet by guessing all three winners?

To test for extrasensory perception (ESP), a deck of 25 cards is usually used in which there are 5 cards marked with each of the symbols shown below.

Suppose the deck is shuffled, a card is drawn at random, and the card has a star on it.

8. What is the probability that someone could correctly guess this simply by luck?

Suppose each time a card is drawn, it is put back and the deck reshuffled before the next card is drawn.

9. What is the probability that someone could correctly guess three cards in succession?

Suppose this is done once with 500 people, each being asked to name three cards in succession.

10. If several people out of the 500 name all three cards correctly, does this show that they have ESP? Explain.

A soft-drink company held a promotion in which some of the cans contained prizes instead of cola. Of 25,000,000 cans distributed to dealers during the promotion, 750,000 contained prizes.

If you bought one can of cola, what was the probability that

11. it would contain a prize?

12. it would not contain a prize?

If you bought a six-pack of the cola, what was the probability that

13. none of the six cans contained a prize?

14. at least one of the cans contained a prize?

Suppose someone offers to bet you even money that, if three dice are thrown, all three numbers that turn up will be different.

15. In how many different ways can three different numbers turn up? (*Hint:* Use the fundamental counting principle.)

16. What is the probability that if three dice are thrown, all three numbers are different?

17. If you accepted the bet, would you be more likely to win or to lose?

Suppose that the probability that a jet engine will fail during a flight is 0.0001.

18. What is the probability that the engine will *not* fail?

If an airplane has two jet engines, what is the probability that

19. *neither* engine will fail?

20. *both* engines will fail?

Suppose the airplane can safely complete the flight as long as both engines do not fail.

21. What is the probability that a twin-engine jet will safely complete a flight?

Set III

The Counterfeiter's Club is having a raffle, and each of the 25 members buys a ticket. The 25 stubs are put into a box from which 1 will be drawn to determine which member will win the prize.

1. What is the probability that Fraudulent Fred, one of the members, will win the prize?

2. Fred decides to improve his chances by forging seven copies of his stub and sneaking them into the box. Now what is the probability that he will win?

3. Now what is the probability that any one of the other members will win?

Just before the drawing, Fred finds out that, after the first stub has been drawn to determine the winner, a second stub will be drawn to decide who will get the booby prize.

4. What is the probability that Fred will win both prizes, so that his forgery is discovered?

5. The first stub has been drawn, and it is not one of Fred's. What is the probability that he will still get the booby prize?

Further Exploration

LESSON 1

1. The young son of a doctor in Springfield, Massachusetts, once discovered a simple test for left-handedness. He got the idea from observing the V-shaped sign that Mr. Spock gave on the "Star Trek" television series.

The test requires spreading apart the fingers of each hand. The fingers are stretched to form a large V with two fingers on each branch of the V. The possible outcomes of this test and the corresponding predictions are shown below.

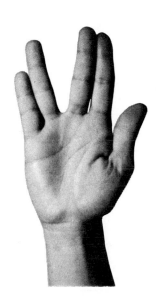

Hand whose fingers stretch farther	Handedness of person
Right hand	Left-handed
Left hand	Right-handed
No difference	Right-handed

When an article about the test appeared in a national magazine, 3,225 readers tried it and reported the results summarized in the following table.

Handedness	Number of people	Number for whom test predicted correctly
Right-handed	2,880	2,097
Left-handed	345	289

From this information, find the probability to the nearest percent that a person in this group is

  a. right-handed.
  b. left-handed.

What is the probability that a person the test predicts to be

  c. right-handed actually is right-handed?
  d. left-handed actually is left-handed?

e. For which group of people does the test seem to be more accurate?

f. What is the probability that the test predicts the handedness of a person correctly?

2. Suppose that nine tennis players are ranked in ability by the numbers 1 through 9, with the best player given the number 1 and the worst player the number 9.*

The players are organized into three teams as shown below:

Team A: Players 1, 6, 8
Team B: Players 3, 5, 7
Team C: Players 2, 4, 9

Suppose each pair of teams has a round-robin tournament in which each member of one team plays each member of the other team exactly once. For example, the tournament between team A and team B is illustrated by the tree diagram shown here.

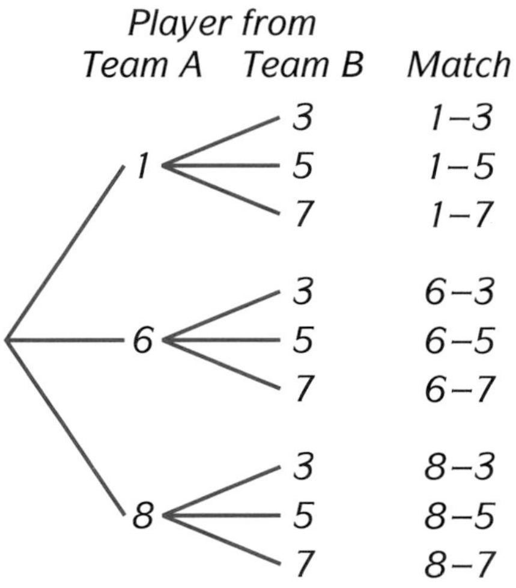

Player from

Team A	Team B	Match
1	3	1–3
	5	1–5
	7	1–7
6	3	6–3
	5	6–5
	7	6–7
8	3	8–3
	5	8–5
	7	8–7

a. How many of these matches would you expect the player from team A to be more likely to win?

The winner of this tournament is the team that wins the most matches. What is the probability that

b. team A will win the tournament?

c. team B will win the tournament?

d. On the basis of these probabilities, which team seems like the better team: team A or team B?

e. Make a tree diagram for a round-robin tournament between team B and team C.

*Based on a problem by Martin Gardner in his book *Time Travel and Other Mathematical Bewilderments* (W. H. Freeman, 1988).

What is the probability that

   f. team B will win this tournament?

   g. team C will win it?

   h. On the basis of these probabilities, which team seems like the better team: team B or team C?

   i. On the basis of your answers to parts d and h, which team of the three seems like the *best* team?

   j. Make a tree diagram for a round-robin tournament between team A and team C.

What is the probability that

   k. team A will win this tournament?

   l. team C will win it?

   m. What is strange about your answers to parts k and l in comparison to your answer to part i?

LESSON 2

1. When a pair of normal dice are thrown, they can come up any sum from 2 through 12, with 7 the most likely. The 36 possible ways of getting these sums are shown in the table on page 458.

    Is it possible to make a special pair of dice for which each sum from 1 through 12 is *equally likely?** The answer is "yes" if at least one of the faces on the dice is left blank.

*Two dice with
a sum of 1*

   a. If each of the sums from 1 through 12 is equally likely, how many ways would there have to be to get each sum?

One die could be numbered in the normal way with the numbers from 1 through 6.

   b. Copy and complete this table of outcomes for the two dice so that each sum from 1 through 12 appears the same number of times.

   c. How would the faces of the special die have to be numbered?

		Normal die				
	1	2	3	4	5	6
0	1	2	3	4	5	6
?	?	?	?	?	?	?
?	?	?	?	?	?	?
?	?	?	?	?	?	?
?	?	?	?	?	?	?
?	?	?	?	?	?	?

Special die (row labels at left)

---

*This puzzle is from *100 Brain-Twisters,* by D. St. P. Barnard (Van Nostrand, 1966).

2.   Suppose that in a crap game, a pair of dice is thrown 1,980 times and that the sums on the dice turn up *exactly as often* as in the table on page 458.*

Since there is 1 way out of 36 ways to get a sum of 2, we would expect a sum of 2 to come up

$$\frac{1}{36} \times 1,980 = \frac{1,980}{36} = 55$$

times.

Since there are 2 ways out of 36 ways to get a sum of 3, we would expect a sum of 3 to come up

$$\frac{2}{36} \times 1,980 = \frac{3,960}{36} = 110$$

times.

a. Use the same method to figure out how many times each of the sums from 4 through 12 would be expected to come up. Copy the table shown here, filling these numbers into the second column.

Sum	Times thrown	Winning rolls
2	55	0
3	110	0
4	▓▓▓	55
5	▓▓▓	88
6	▓▓▓	▓▓▓
7	▓▓▓	▓▓▓
8	▓▓▓	▓▓▓
9	▓▓▓	▓▓▓
10	▓▓▓	▓▓▓
11	▓▓▓	▓▓▓
12	▓▓▓	▓▓▓

If a sum of 2 comes up, the dice thrower loses, so of the 55 times a sum of 2 comes up, none are winning rolls. The same is true for the 110 times a sum of 3 comes up.

---

*The number 1,980 is chosen for convenience to avoid fractions in the calculations that follow.

If the sum of 4 comes up, the thrower must throw it again before a sum of 7 comes up. There are 3 ways to get a sum of 4 and 6 ways to get a sum of 7, so the thrower will win $\dfrac{3}{3+6} = \dfrac{3}{9} = \dfrac{1}{3}$ of the time when trying to repeat the 4:

$$\frac{1}{3} \times 165 = \frac{165}{3} = 55 \text{ times}$$

b. Use the same method to show that the dice thrower can expect to win 88 times when a sum of 5 comes up.

c. Figure out how many times the dice thrower can expect to win if a sum of 6 comes up.

d. Figure out the number of ways the dice thrower can expect to win if each of the sums from 7 through 12 comes up. (Remember that 7 and 11 win and 12 loses.) Fill these numbers into the third column of the table you made for part a.

e. Add the numbers in the third column to find out how many times the dice thrower can expect to win out of the 1,980 times the dice are thrown.

f. Use your answer to part e to calculate the probability that the person throwing the dice in a crap game will win. Express your answer to the nearest tenth of a percent.

Lesson 3

1. Most of the world's major countries are linked by submarine telephone cables. Some of them are shown in this map.

Amplifiers, called "repeaters," are spaced evenly along each cable to keep the sound loud enough to hear. A transatlantic cable connecting North America to Great Britain has 51 of these repeaters. According to Arthur C. Clarke.

> Each repeater is a link in a chain; if a single one fails, the whole chain is useless. . . . To make reasonably sure that the system will operate for . . . twenty years without failure, the degree of reliability required for each item is fantastically high.*

The "reliability" of a repeater is the probability that it will work without failure. Suppose that the probability that the individual repeaters along the transatlantic cable work over a period of 20 years is 99%.

   a. Use a calculator to find the probability that all 51 repeaters work over this period. A shortcut for doing this if your calculator does not have a "power" button is shown here:

$$\underbrace{0.99 \times 0.99 \times 0.99 \times 0.99 \times 0.99 \times 0.99 \times \ldots}_{51\ times} =$$

$$\underbrace{(0.99 \times 0.99 \times 0.99) \times (0.99 \times 0.99 \times 0.99) \times \ldots}_{17\ times} =$$

$$\underbrace{0.970299 \times 0.970299 \times \ldots}_{17\ times} = ?$$

Store 0.970299 in the memory of the calculator and then recall it as needed to do the multiplication.

   b. Find the probability that all 51 repeaters work over a period of 20 years if the probability that the individual repeaters work is 98%.

Would you bet that the cable works for 20 years if the reliability of the individual repeaters is

   c. 99%?

   d. 98%?

Find the probability that all 51 repeaters work over a period of 20 years if the probability that the individual repeaters work is

   e. 97%.

   f. 90%.

Voice Across the Sea, by Arthur C. Clarke (Harper and Row, 1974).

2.   A special die has two white and four black faces. Each one of the six faces is equally likely to turn up. The die will be rolled several times, and you are asked to choose one of the following sequences of colors.

Sequence A: white, black, white, white, white
Sequence B: white, black, white, white, white, black
Sequence C: black, white, white, white, white, white

If the first rolls of the die come up in the sequence you have chosen, you will win $25.*

   a. Which sequence would you bet on?
   b. What is the probability that a white face will turn up when the die is rolled once?
   c. What is the probability that a black face will turn up?
   d. Find the probability of the first sequence of colors turning up by multiplying the appropriate probabilities.
   e. Find the probability of the second sequence turning up.
   f. Find the probability of the third sequence turning up.
   g. Which sequence would actually be the best choice to win the $25? Explain.

LESSON 4

1.   Which pattern of heads and tails would you bet on occurring first if a coin is tossed repeatedly:

TTH or HTT?

Try tossing a coin repeatedly to find out. For example, if the coin is tossed and turns up THHTT, then HTT has turned up first.

   a. Try this 10 times, keeping a record of which pattern occurs first in a table like this.

	What happened	Which pattern occurred first
(Example)	THHTT	HTT

You *probably* found that the pattern HTT occurred first more than half the time!

---

*This problem is based on a psychological experiment performed by A. Tversky and D. Kahneman and is adapted from an exercise in *Introduction to the Practice of Statistics,* by David S. Moore and George P. McCabe (W. H. Freeman, 1989).

This may seem strange to you if you realize that, if a coin is tossed just three times, the patterns TTH and HTT are *equally likely.* They are just two of eight equally probable outcomes shown in the tree diagram below.

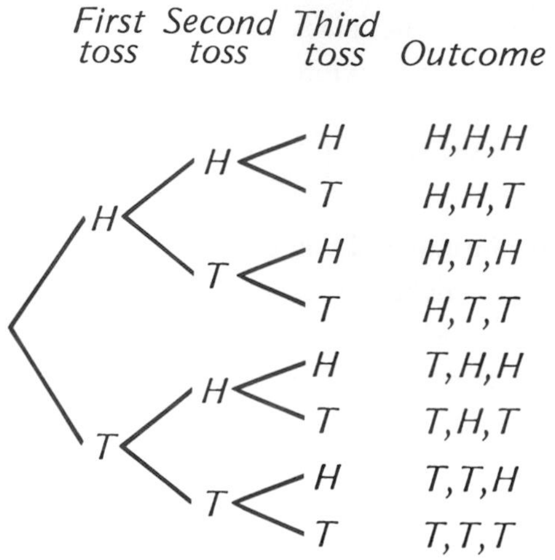

| *First toss* | *Second toss* | *Third toss* | *Outcome* |

A rather surprising game has been invented in which each of two players bets on which one of these eight outcomes will occur first.* They then toss a coin repeatedly until one of the two outcomes occurs. The player who chose that outcome wins.

The surprising thing about the game is that, *regardless of what outcome the first player chooses, the second player can choose one that is more likely to occur first!*

The table below shows what the second player should choose for each choice that the first player might make.

If the first player chooses	The second player should choose
HHH	THH
HHT	THH
HTH	HHT
HTT	HHT
THH	TTH
THT	TTH
TTH	HTT
TTT	HTT

---

*The game, invented by a mathematician named Walter Penney, is described in *Time Travel and Other Mathematical Bewilderments,* by Martin Gardner (W. H. Freeman, 1988), pp. 60–64.

b. Try playing the game with someone, following the strategy in this table. Play the game at least 10 times and keep a record as shown below of what happens.

Other player	Me	What happened	Winner
H,T,H	H,H,T	T,T,H,H,T	Me

(If you do not have someone to play the game with, play it by yourself, making up choices for an imaginary opponent.) Regardless of what the other person does, the odds are in your favor so that you should win most of the time.

2.   Here is another heredity experiment that can be explained with probability. Several blue parakeets are mated with some yellow ones, and every one of the offspring is green. When the all-green generation is mated, the colors of its offspring are green, blue, yellow, and white.

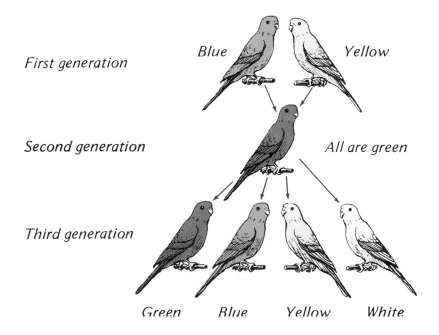

First generation — Blue — Yellow

Second generation — All are green

Third generation — Green   Blue   Yellow   White

To explain the appearance of parakeets with four different colors, it is necessary to suppose that pairs of two genes determine the color. Again we will represent each gene with a letter. Suppose that the genes of the first generation are those given in the table at the right. Then the breeding of this first generation will result in a second generation of all-green

Color of parakeet	Genes
Blue	BB-WW
Yellow	YY-WW

parakeets, as shown in the tree diagram below. (We will assume that, if the genes of a parakeet include both B and Y, it is green, and that W does not affect the color unless it appears without the other letters.)

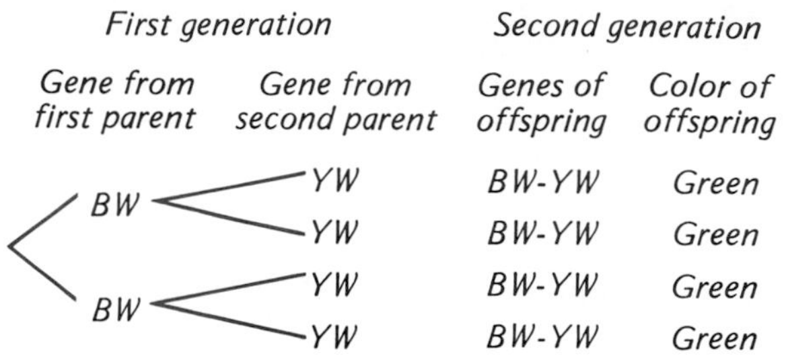

	First generation		Second generation	
	Gene from first parent	Gene from second parent	Genes of offspring	Color of offspring
	BW	YW	BW-YW	Green
		YW	BW-YW	Green
	BW	YW	BW-YW	Green
		YW	BW-YW	Green

The breeding of the all-green generation of parakeets to produce the green, blue, yellow, and white parakeets in the third generation is shown in this tree diagram.

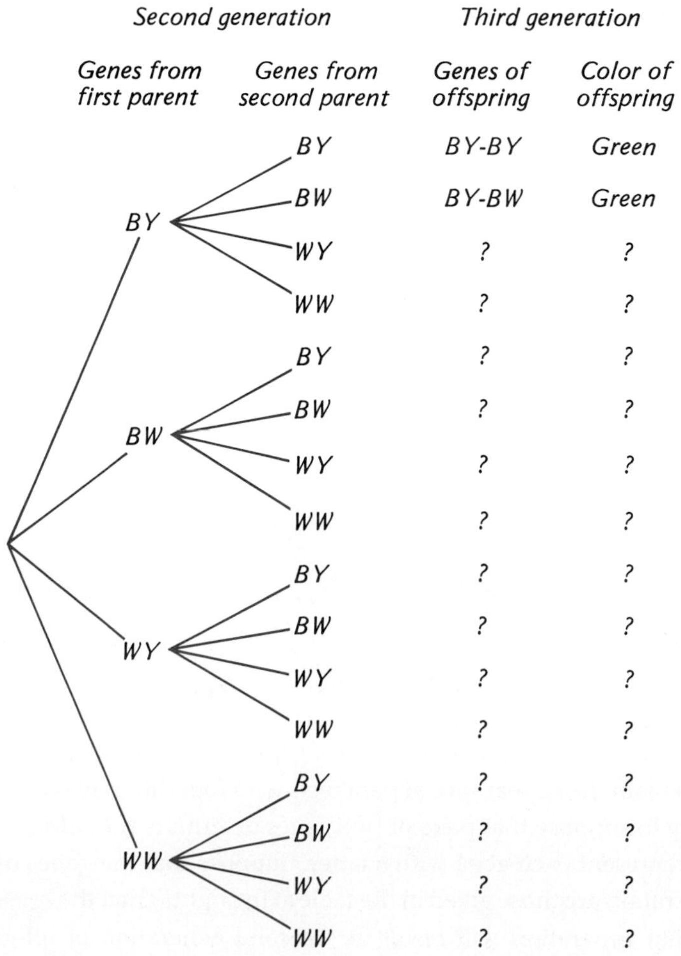

	Second generation		Third generation	
	Genes from first parent	Genes from second parent	Genes of offspring	Color of offspring
	BY	BY	BY-BY	Green
		BW	BY-BW	Green
		WY	?	?
		WW	?	?
	BW	BY	?	?
		BW	?	?
		WY	?	?
		WW	?	?
	WY	BY	?	?
		BW	?	?
		WY	?	?
		WW	?	?
	WW	BY	?	?
		BW	?	?
		WY	?	?
		WW	?	?

a. Copy and complete the columns of the diagram representing the genes and color of the parakeets in the third generation.
b. What color of parakeet is most common in the third generation? What color is least common?
c. Copy and complete the following table of probabilities for the third generation.

Color of parakeet	Green	Blue	Yellow	White
Probability	▊▊▊	▊▊▊	▊▊▊	▊▊▊

LESSON 5

1. The numbers of possible combinations of things appear in Pascal's triangle in an interesting way.

Recall, for example, that the various symbols in Braille are formed by choosing combinations from six dots of any number of dots from one to six. For example, from six dots one dot can be chosen in six different ways:

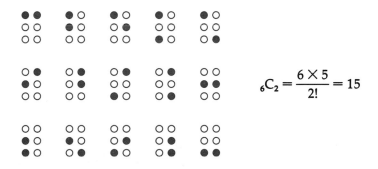

$$_6C_1 = \frac{6}{1!} = 6$$

From six dots, two dots can be chosen in 15 different ways:

$$_6C_2 = \frac{6 \times 5}{2!} = 15$$

Both these numbers, 6 and 15, appear on the same row of Pascal's triangle.

a. On what row do they both appear?
b. If the 1 at the beginning of the row is not counted, where do 6 and 15 first appear in the row?
c. Recalling that $_6C_3 = \dfrac{6 \times 5 \times 4}{3!}$, evaluate $_6C_3$.
d. Where does this number first appear in the triangle? (Again, do not count the 1 at the beginning of the row.)

Find the number to which each of the following is equal and tell where each one appears in Pascal's triangle.

   e. $_4C_2$.
   f. $_7C_3$.
   g. $_8C_4$.
   h. Use Pascal's triangle to guess a combination that is equal to 21. (That is, guess numbers for n and r so that $_nC_r = 21$.)
   i. Check your guess by calculating the value of the combination.
   j. Find the value of $_9C_5$.
   k. Where would you expect it to appear in Pascal's triangle?

2. There is a remarkable connection between Pascal's triangle and a problem that you encountered earlier.* The problem concerns choosing two or more points on a circle and connecting every pair of points with a straight line segment. For a given number of points, what is the greatest number of regions that can be formed in this way?

The figures below illustrate solutions to the problem for circles on which from two to eight points have been chosen.

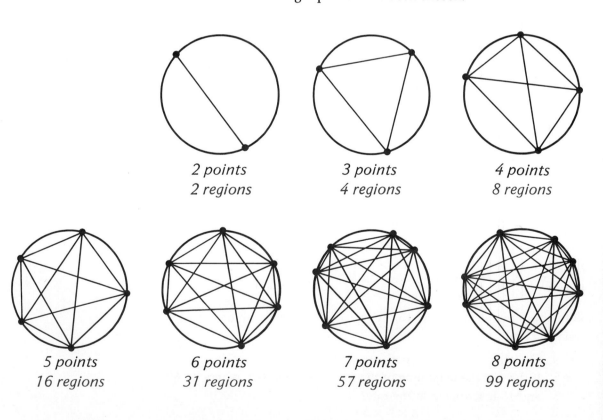

| 2 points | 3 points | 4 points |
| 2 regions | 4 regions | 8 regions |

| 5 points | 6 points | 7 points | 8 points |
| 16 regions | 31 regions | 57 regions | 99 regions |

*See page 28.

Compare the numbers of regions with the numbers in Pascal's triangle below.

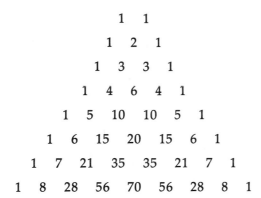

```
            1   1
          1   2   1
        1   3   3   1
      1   4   6   4   1
    1   5   10   10   5   1
  1   6   15   20   15   6   1
1   7   21   35   35   21   7   1
1   8   28   56   70   56   28   8   1
```

a. What is the connection between them?
b. How many regions do you think can be formed by choosing nine points on a circle? Explain.

LESSON 6

1. If a survey contains some potentially embarrassing questions, many people will not answer those questions. To get around this, a method called "randomized response" is sometimes used.*

For each embarrassing question, the person answering the question is asked to toss a coin in private. If the answer to the question is "yes" *and* the coin lands heads, the person is told to answer "yes." Otherwise, they are to answer "no." Even though only the person giving the answers knows whether or not they are true, the answers are useful.

Suppose, for example, that a question in the survey is: "Did you vote in the last election?" The following tree diagram shows the possible responses.

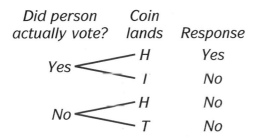

Did person   Coin
actually vote? lands   Response

Yes ⟨ —H _Yes_
       —I _No_

_No_ ⟨ —H _No_
       —T _No_

---

*Adapted from an exercise in *Introduction to the Practice of Statistics,* by David S. Moore and George P. McCabe (W. H. Freeman, 1989).

Suppose that the actual probability that a person voted is 42%.

  a. What is the probability that someone answering the question will answer "yes"?

  b. What is the probability that they will answer "no"?

Suppose instead that the actual probability that a person voted is 34%.

  c. Now what is the probability that someone answering the question will answer "yes"?

  d. What is the probability that they will answer "no"?

Now suppose that the survey is taken and 26% of the answers to the question on voting are "yes."

  e. What would you conclude the percentage of the people who actually voted to be?

2. In the novel *The Bridge on the River Kwai* by Pierre Boule, an air force officer talking about the chances of three commandos successfully parachuting into the jungle says:

> If they do only one jump, you know, there's a fifty percent chance of injury. Two jumps, it's eighty percent. The third time, it's dead certain they won't get off scot free.*

If the chance of injury on one jump is 50%, the other probabilities stated by the officer are incorrect. Find the actual probability that at least one person will be injured if

  a. two commandos make the jump.

  b. three commandos make the jump.

  c. Would an injury be "dead certain" if many commandos made the jump? Explain.

---

*This exercise is based on a discussion of the problem by Darrell Huff in his book *How to Take a Chance* (Norton, 1959).

An Introduction to Statistics

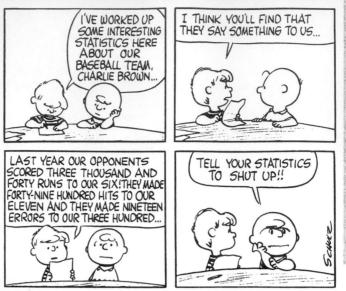

© 1960 United Feature Syndicate, Inc.

1

Organizing Data: Frequency Distributions

More than any other sport, baseball is associated with statistics. Many people, in fact, think of the word *statistics* in the sense that baseball fans use it—as meaning numbers or tables of numbers. Mathematicians, however, use the word in a more general sense. The subject of statistics is a branch of mathematics that deals with the collection, organization, and interpretation of numerical data. We will begin our survey of statistics by considering some of the ways in which numerical data can be organized.

In the past 50 years or so of baseball, no major league player has matched Ted Williams's batting average of .406.* Have batting averages changed through the years? The averages of the top National League batter and top American League batter for each year from 1901 to 1910 are listed on the next page.

---

*A baseball player's batting average is found by dividing the number of base hits by the number of times at bat and is always expressed to three decimal places.

.382   .377   .339   .355   .350
.357   .339   .331   .381   .324
.355   .350   .422   .306   .377
.349   .354   .376   .358   .385

To see what pattern these numbers may have, we can organize them in a **frequency distribution.** The distribution is made by first making a list of equal intervals into which the numbers fall. The intervals chosen in the table below are convenient. We then go through the 20 batting averages in the order in which they appear in the list above and make tally marks indicating the intervals in which the averages fall. The numbers of tally marks, called the *frequencies,* are then listed in the third column of the table.

Batting average	Tally marks	Frequency
.300 – .309	I	1
.310 – .319		0
.320 – .329	I	1
.330 – .339	III	3
.340 – .349	I	1
.350 – .359	NHL II	7
.360 – .369		0
.370 – .379	III	3
.380 – .389	III	3
.390 – .399		0
.400 – .409		0
.410 – .419		0
.420 – .429	I	1

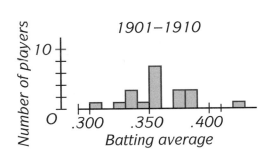

The information in this table can be presented in a type of graph called a **histogram.** Each interval in the frequency distribution is represented by a bar in the histogram. All of the bars have the same width, determined by the size of the intervals in the distribution. The heights of the bars are determined by the frequencies of the intervals.

It is easy to see from the histogram that only one of the players had a batting average over .400. Seven players had averages between .350 and .359.

This example illustrates the method by which a set of numbers can be organized by means of a frequency distribution and how the frequency distribution can then be pictured by means of a histogram. If the same method is applied to the averages of the top National League batter and top American League batter for each year from 1901 to 1930, from 1931 to 1960, and from 1961 to 1990, the following histograms result.

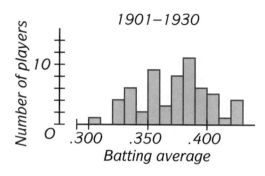

1901–1930

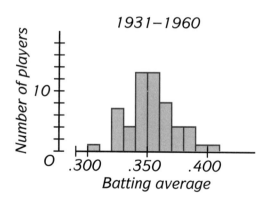

1931–1960

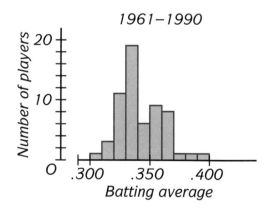

1961–1990

These histograms suggest that the number of top batters with averages over .400 has decreased through the years and that batting averages in general have decreased. The degree of reasonableness of these conclusions can be determined by more advanced statistical methods.

The subject of statistics has become an important tool with which to learn more about our physical, biological, social, political, and economic worlds. H. G. Wells once said:

> Statistical thinking will one day be as necessary for efficient citizenship as the ability to read and write.

EXERCISES

SET I

The numbers of games played in the World Series has varied from four to eight. (The rule limiting the series to seven games was established in 1922.)

The distribution of the numbers of games played in the first 25 World Series is shown in the graph below.

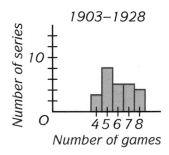

1. What is this type of graph called?

2. According to this graph, what was the most frequent number of games played?

3. How many World Series lasted this number of games?

The graphs below show the distribution of games in the second 25 years and the third 25 years of the World Series.

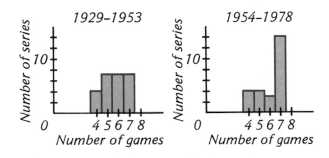

Compare these two graphs and the previous one to answer the following questions.

4. Has the likelihood that the World Series will last only four games changed much over the years?

5. What has happened over the years to the likelihood that the World Series will last the maximum number of games?

This photograph shows Bob Beamon making his record-breaking long jump in the 1968 Olympics.*

The Olympic records in the long jump are listed in the table below.

Photograph by E. D. Lacey

Year	Winner	Meters
1896	Ellery Clark	6.35
1900	Alvin Kraenzlein	7.18
1904	Myer Prinstein	7.34
1908	Francis Irons	7.48
1912	Albert Gutterson	7.60
1920	William Pettersson	7.15
1924	De Hart Hubbard	7.44
1928	Edward Hamm	7.73
1932	Edward Gordon	7.63
1936	Jesse Owens	8.06
1948	William Steele	7.82
1952	Jerome Biffle	7.57
1956	Gregory Bell	7.83
1960	Ralph Boston	8.12
1964	Lynn Davies	8.07
1968	Robert Beamon	8.90
1972	Randy Williams	8.24
1976	Arnie Robinson	8.35
1980	Lutz Dombrowski	8.54
1984	Carl Lewis	8.54
1988	Carl Lewis	8.72
1992	Carl Lewis	8.67

**Distance
in meters**

6.01 – 6.50

6.51 – 7.00

7.01 – 7.50

7.51 – 8.00

8.01 – 8.50

8.51 – 9.00

6. Make a frequency distribution of the first 11 distances (the distances from 1896 to 1948). Use the intervals shown in the list on the left.

7. Make a frequency distribution of the second 11 distances (the

---

*At the time, one sportswriter predicted that this record would last into the twenty-first century, but it was broken by Mike Powell in Tokyo, 1991.

distances from 1952 to 1992). Use the same intervals as in exercise 6.

8. Make a histogram to illustrate your frequency distribution for exercise 6. Label the axes as shown here.

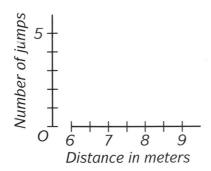

9. Make a histogram to illustrate your frequency distribution for exercise 7.

The first histogram shows that most of the record distances from 1896 to 1948 were between 7 and 8 meters.

10. What does the second histogram reveal?

This histogram shows the number of hours that adults sleep each night. The bars are centered over the numbers of hours they represent. The first bar, for example, represents the percentage of adults who sleep 5 hours each night.

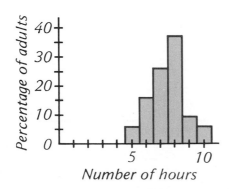

11. Approximately what percentage of adults sleep 8 hours each night?

12. Which is more common: for an adult to sleep *more* than 8 hours each night or *less* than 8 hours each night?

13. How does the number of adults who sleep 5 hours a night seem to compare with the number who sleep 10 hours a night?

The methods of statistics have been used to try to determine authorship. For example, some scholars think that Shakespeare's plays may actually have been written by Sir Francis Bacon. In an attempt to determine whether this might be true, someone counted the lengths of large numbers of words from Shakespeare's plays and from the writings of Bacon. The results are shown in the histograms below.

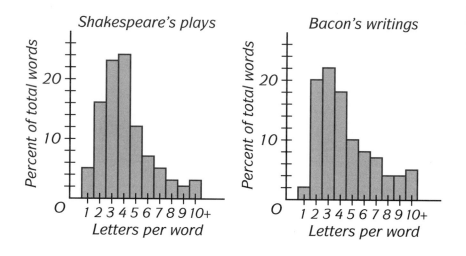

14. Which contains the greater percentage of two-letter words: Shakespeare's plays or the writings of Bacon?

15. Which contains the greater percentage of four-letter words?

16. What do the histograms reveal about each author's use of words that are more than five letters long?

Observations such as these suggest that Bacon did not write Shakespeare's plays.

Set II

The men and women who have won the Oscars for best actor and best actress have varied in age from 21 to 80.*

The ages of the winners for the first 30 years and the second 30 years of the Academy Awards are listed on the next page.

*Based on an idea by Richard Brown and Gretchen Davis.

	1928–1957						1958–1987				
Year	Actor	Actress	Year	Actor	Actress	Year	Actor	Actress	Year	Actor	Actress
1928	42	22	1943	49	24	1958	51	41	1973	48	37
1929	40	36	1944	41	29	1959	35	38	1974	56	42
1930	62	28	1945	40	37	1960	47	28	1975	38	39
1931	53	62	1946	49	30	1961	31	27	1976	60	35
1932	35	32	1947	56	34	1962	46	31	1977	30	31
1933	34	24	1948	41	34	1963	36	37	1978	40	39
1934	33	29	1949	38	33	1964	56	29	1979	40	33
1935	52	27	1950	38	28	1965	41	25	1980	37	31
1936	41	27	1951	52	38	1966	44	34	1981	76	72
1937	37	28	1952	51	45	1967	42	58	1982	39	33
1938	38	30	1953	35	24	1968	43	59	1983	52	49
1939	34	26	1954	30	26	1969	62	35	1984	45	38
1940	32	29	1955	38	47	1970	43	34	1985	35	61
1941	40	24	1956	41	41	1971	41	34	1986	61	21
1942	43	34	1957	43	27	1972	48	26	1987	43	41

1. Make a frequency distribution of the ages of the winning actors for the first 30 years of the awards. Use the intervals shown at the right.

2. Make a frequency distribution of the ages of the winning actresses for the first 30 years of the awards. Use the same intervals as in exercise 1.

3. Make histograms illustrating your frequency distributions for exercises 1 and 2. Label the axes as shown at the right below.

4. Follow the directions for exercise 1 but this time for the second 30 years of the awards.

5. Follow the directions for exercise 2 but for the second 30 years of the awards.

6. Make histograms illustrating your frequency distributions for exercises 4 and 5.

7. Compare the four histograms you have drawn. What conclusions about the winners of the Oscars for best actor and best actress do you draw from them?

Age in years

20–29
30–39
40–49
50–59
60–69
70–79
80–89

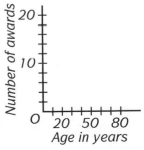

This histogram represents the distribution of the earth's surface area by elevation above (+) and below (−) sea level.

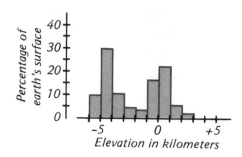

8. Within what interval of elevation is most of the earth's dry land?

9. Within what two intervals of elevation is most of the ocean floor?

Approximately what percentage of the earth's surface is

10. above sea level?   11. below sea level?

Hawaiian, a Polynesian language, is considered one of the most musical in the world. Although this language has been spoken for more than a thousand years, its written form was devised only as recently as 1822. Here is a sample of the language, a story about a menehune and a shark:

AKA AME KA MANO*

O Aka he menehune uuku momona.
I kekahi la ua luu o Aka,
Ua nahu ka mano i kona manamana wawae nui.
Alaila ua huhu loa ka mano,
No ka mea ua pololi loa oia.

Ua huhu pu no ka menehune,
No ka mea he hoaloha o Aka no lakou.
Nolaila haha lakou i hinai kaoli,
A ua hoopiha ia me ka maunu.

Me keia ua paa ia lakou ka mano.
Ua huki ia iluna o ke one e make.
Ua hoike mai na iwi keokeo
"Mai hoopa i ka menehune."

---

*From *Where the Red Lehua Grows,* by Jane Comstock Clarke (Honolulu Star-Bulletin, Ltd., 1943).

12. The first paragraph contains 117 letters. Write the alphabet in a column and make a frequency distribution of the letters in this paragraph.

13. State some conclusions about the Hawaiian language that seem reasonable on the basis of your frequency distribution.

14. The second paragraph contains 28 words. Make a frequency distribution of the *last* letters of these words.

15. What does your frequency distribution suggest?

16. Is this conclusion supported by the other two paragraphs? Explain.

17. The third paragraph also contains 28 words. Make a frequency distribution of the *first* letters of these words.

18. What does your frequency distribution suggest?

19. Is this conclusion supported by the other two paragraphs? Explain.

SET III

The frequency distribution on the next page shows the ages of the population of Ghana as reported in the 1960 census of that country. The ages of people older than 80 are not shown.*

Look at the frequencies of the people whose ages were reported as 30, 40, 50, 60, 70, and 80.

1. What is strange about these frequencies?

2. Do you see any other patterns of this sort in the table? If so, what are they?

3. Why do you suppose these patterns exist?

---

*Adapted from a problem in *Beginning Statistics with Data Analysis*, by Frederick Mosteller, Stephen E. Fienberg, and Robert E. K. Rourke (Addison-Wesley, 1983).

Age in years	Population (in 1000s)	Age in years	Population (in 1000s)	Age in years	Population (in 1000s)
Under 1	277	27	88	54	22
1	211	28	137	55	26
2	258	29	84	56	29
3	289	30	232	57	15
4	257	31	52	58	23
5	232	32	95	59	15
6	244	33	49	60	72
7	193	34	58	61	9
8	192	35	120	62	14
9	156	36	85	63	9
10	173	37	42	64	13
11	101	38	77	65	23
12	156	39	48	66	8
13	120	40	155	67	7
14	130	41	31	68	13
15	120	42	70	69	7
16	106	43	30	70	33
17	88	44	24	71	3
18	131	45	78	72	9
19	94	46	39	73	4
20	176	47	23	74	5
21	106	48	48	75	12
22	110	49	28	76	5
23	82	50	97	77	2
24	111	51	15	78	6
25	166	52	29	79	3
26	111	53	14	80	19

The Breaking of Ciphers and Codes: An Application of Statistics

In one of the Sherlock Holmes stories, "The Adventure of the Dancing Men," the methods of statistics are applied to deciphering a series of mysterious messages. The messages were written as rows of little stick figures in which each figure stands for a letter of the alphabet. The first message looked like this:

As you can see, some of the figures are holding flags, and Sherlock Holmes concluded that the flags were used to indicate the ends of words.

This message, then, consists of four words. Of the fifteen figures that it contains, one appears four times:

Holmes, knowing that the letter E is used more frequently than any other letter in English, guessed that this figure represented E. This seems especially likely because two of these figures are holding flags and E is the last letter of many words.

Later in the story more messages appeared and Holmes, by putting a number of clues together, was able to decipher what they said. For example, the first word of the longest message

is five letters long and begins and ends with the letter E. The name of the woman to whom it was sent was Elsie, and so it seemed probable that this word was ELSIE. Therefore, the figures

stand for L, S, and I, respectively.

Reasoning in a similar fashion with some other messages, Holmes decided that the first one said

AM HERE ABE SLANEY

and that the other message shown here said

ELSIE, PREPARE TO MEET THY GOD

Holmes trapped the villain, Abe Slaney, by writing a message to him with the stick figures. Slaney, assuming that only Elsie could read the cipher, thought that the message was from her and so he went to her home, where he was arrested.*

---

*If you are interested in reading the entire story, "The Adventure of the Dancing Men" is included in the series of adventures titled *The Return of Sherlock Holmes* by Sir Arthur Conan Doyle.

The use of coded messages today has advanced far beyond detective stories and puzzles. In our age of electronic communications, coding has become very important not only in national security but also in protecting private information stored in computers. The U.S. Department of State receives and sends several million coded words every week. Banks protect the movement of trillions of dollars of funds and securities over the telephone each day by means of codes. Businesses prevent their competitors from eavesdropping on their electronic mail by sending it in coded form. The National Security Agency, which develops and breaks codes, has more than 10,000 employees and, because modern codes are so complex, is thought to have more computers than any other organization in the world.

As one mathematician has written:

> At the level where cryptography affects you and me, it is important that data relevant to us, whether medical, financial, or whatever, can be transmitted securely.*

Cryptography, the science of code writing that helps protect the transmission of such data, makes use of many mathematical ideas, especially in algebra and statistics.

EXERCISES

SET I

The messages in "The Adventure of the Dancing Men" are written in a *simple substitution cipher*, a cipher in which each letter is always represented by the same symbol. Here is a list of the letters and symbols used in the two messages included in this lesson.

---

*Nelson Stephens in the chapter on cryptography in *New Applications of Mathematics*, edited by Christine Bondi (Penguin Books, 1991).

1. What does the following message say?

Here is the message that Sherlock Holmes sent to Abe Slaney. Although it begins with a symbol that does not appear in the list, you should be able to figure out what it says.

2. What does the message say?

The frequencies of the letters in any written language are always about the same in every large sample of it. Here is a graph of the frequencies (in percent) of the letters in ordinary English.

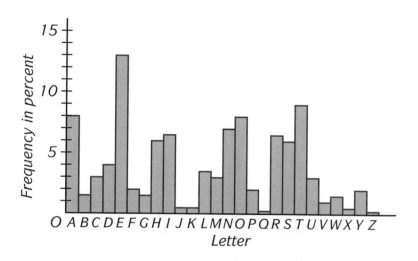

3. What two letters are used most frequently in English?

4. What two letters are used least frequently?

The graph shows that the letter E appears about 13 times in a typical sample of 100 letters. In the first 100 letters of *A Christmas Carol* by Charles Dickens, the letter E appears 13 times.

5. According to the graph, about how many times does the letter A appear in a typical sample of 100 letters?

6. About how many times would you expect the letter A to appear in the first 300 letters of the twenty-third psalm?

It actually appears 22 times (in the King James Version).

7. According to the graph, about how many times does the letter C appear in a typical sample of 100 letters?

8. About how many times would you expect the letter C to appear in the first 1,000 letters of Lincoln's Gettysburg Address?

It actually appears 31 times.

A company that produces commercial lettering sheets says that the "letters are provided in the frequency they are most commonly used." The adjoining figure shows the alphabet as it appears on one of the company's sheets. The sheet contains 100 letters in all.

Look at this figure and the graph of the frequencies of the letters in ordinary English on the previous page.

AAAAABBCCCCC
DDDDEEEEEEFFFGG
GHHHHIIIIIIIJJKKLLLLL
LMMMMNNNNNNO
OOOOOPPPQQRRR
RRSSSSSSTTTTTTUU
UVVWWXXYYZZ

9. Do the numbers of the most frequently used letters in English,

   E, T, A, and O

   that appear on this sheet seem correct?

10. Explain.

11. Do the numbers of the most rarely used letters in English,

    J, K, X, Q, and Z

    that appear on this sheet seem correct?

12. Explain.

13. How do you think the sheet might be improved?

A sentence sometimes used as a typing exercise is:

   Pack my box with five dozen liquor jugs.

14. Write the alphabet in a column and make a frequency distribution of the letters in this sentence.

15. Why is the sentence used as a typing exercise?

16. Does the letter E appear *more* or *less* frequently in the "liquor jugs" sentence than in ordinary English?

17. Does the letter Q appear *more* or *less* frequently in the "liquor jugs" sentence than in ordinary English?

18. If the "liquor jugs" sentence were written in code, would it be easy or difficult to decipher? Why?

The digits 0 and 1 appear by themselves on the telephone dial. The other eight digits appear with three letters each; the digit 2, for example, appears with ABC. This means that

$$8 \times 3 = 24$$

letters of the 26 letters of the alphabet appear on the telephone dial.

19. Which two letters do you think are missing from the telephone dial?

20. Why do you suppose the telephone company chose not to use these two letters?

Set II

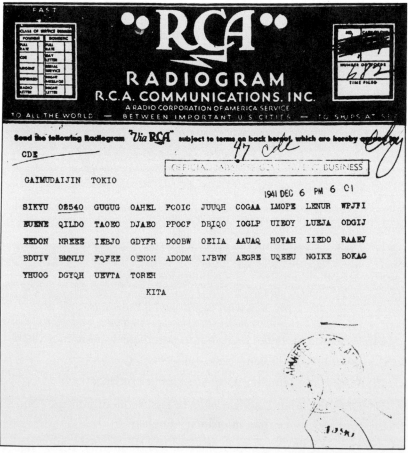

U. S. Department of the Navy

Ciphers and codes are particularly important when countries are at war. The telegram on the facing page was sent to Tokyo from the Japanese consulate in Honolulu on the evening before the attack on Pearl Harbor. Written in a cipher in which the words were run together and the letters grouped in blocks of five, the telegram reported on the ships of the United States fleet that were in port.

1. Here is another cipher in which the words are not separated so that there are no clues about their lengths. For convenience, the message has been written in groups of five letters each.

```
B F C O N    A N Y K F    I X K U S    I X H U C
O N G F B    C N I A C    H C N A A    K S T N I
C O N P H    B F Y K I    N K M C O    N R H W H
F N A N F    H T L C O    B A W S H    L N I H F
B P W K U    C H F C U    K S N B F    C O N H P
N U B Y H    F T B Y C    K U L H C    C O N I N
Y B A B T    N D H C C    S N K M P    B I X H L
```

a. Copy the cipher on your paper, leaving two blank lines between each line of letters. Also write the letters of the alphabet in a column so that you can record what each letter stands for as you figure it out.

A graph of the frequencies of the letters in the cipher is shown below.

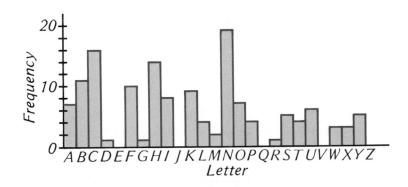

b. The most frequent letter is probably E, so write an E underneath each place this letter appears in the cipher.
c. The second most frequent letter is probably T. Write a T underneath each place this letter appears.
d. A very common word in English is THE. On the assumption that THE appears more than once in the cipher, guess what letter represents H.

e. A strong clue in solving a cipher is knowledge of certain words that are likely to appear in it. For example, this cipher mentions the Second World War. Find a place where the words THE SECOND WORLD WAR can appear in order to figure out what several more letters represent.

f. Solving for the remaining letters in the cipher is left to you. What does the cipher say?

Hagelin cryptograph; courtesy of Crypto AG

The machine in this photograph can be used both to translate information into coded form and to translate it back again. Manufactured in Switzerland, it has been used by many governments and commercial organizations.

The printed tapes shown below are similar to those produced by the machine.

A

```
THISW ISWAW SECRE TWMES SAGEW WRITT ENWON WAWHA GELIN WCR

HUTIH QLIYZ OQYRR SVOED TRUDK XRWAF UVHQG AAOZT QRUYO YRR
```

B

```
HUTIHQLIYZOQYRRSVOEDTRUDKXRWAFUVHQGAAOZTQRUYOYRRLJZSUTDHM

THIS IS A SECRET MESSAGE  RITTEN ON A HAGELIN CRYPTOGRAPH
```

The first line of tape A shows what is typed on the machine, and the second line shows it in coded form. (The machine automatically breaks the message into five-letter groups.)

2. What seems to be the purpose of the W's in the first line?

3. Is each letter on the first line always replaced by the same letter in the coded form?

The first line of tape B shows what is typed on an identical machine when the code is received, and the second line shows the deciphered message produced by the machine.

4. Why do you suppose the word *written* on the second line turns out misspelled?

5. Do you think a cipher produced by this machine could be solved by making a frequency distribution of the letters contained in it?

SET III

The following mystery story is from *Games* magazine.* The idea is not to guess "who did it," because it is the author "who's done it." The problem is to figure out exactly what the author has done.

A SIN OF OMISSION

Around midnight, a sly-looking man slips into a luxury city building. A woman occupant, watching his actions from a fourth-floor window, grows suspicious and dials 911 for a patrol car. This lady complains, "A man in a brown suit, with shaggy hair, a slight build, and a criminal air is prowling through my lobby."

Fairly soon two young cops, Smith and Jarvis, pull up. Looking for an unknown vagrant, Smith spots Jim Oats walking out a front door. Oats, a minor burglar, is bold as brass, arrogant, and calm. Smith grabs him by his collar.

"O.K., Oats," snarls Smith, "what brings you to this location?"

Fixing his captor with a chilly look and frosty indignation, Oats quips, "I can go on a short stroll. Lift your filthy hands off my shirt. I'm not guilty of anything."

Smith drops his hands limply. This haughty air is too much for him to swallow. Angrily Smith says, "What a story. I'm nobody's fool, you punk. I just wish I could put you back in jail, but I can't obtain any proof against you. You know all about why I'm at this building — a station log full of burglary, arson, and muggings."

"Now, now," Oats laughs, "think of my rights. How can you talk this way?" Smith's probing hands start to frisk Oats for guns, narcotics, anything unlawful or contraband. Nothing shows up — only a small bound book. "What's this?" Smith asks.

Oats, tidying up his clothing, pluckishly says, "That's my political study of voting habits in this district. Why don't you look at my lists? I work for important politicians now — guys with lots of clout." An ominous implication lurks in this last thrust.

"Don't talk down to us," Smith snaps. But studying Oats's book, Jarvis finds nothing unusual. Smith finally hands him back his lists. Our cops can't hold him. Jarvis admits Oats can go. Just as a formality, Jarvis asks him, "Did you commit any criminal act in this building? Anything at all of which a courtroom jury could find you guilty?"

"No," Oats says flatly. "No way," and jauntily skips off. Halting six blocks away, Oats digs a tiny picklock from his sock and a diamond ring from his shaggy hair.

Games magazine, November/December 1977.

$180,000

$80,000

$35,000 EACH

$24,000 EACH

$18,000

$15,000 EACH

3

Measures of Central Tendency

The book *Working,* by Studs Terkel, contains interviews with many different people about their jobs. In one of the interviews, a steelworker remarks:

> I don't know where they got the idea that we make so much. The lowest class payin' job there, he's makin' two dollars an hour if he's makin' that much. It starts with jobs class-1 and then they go up to class-35. But no one knows who that one is. Probably the superintendent. So they put all these class jobs together, divide it by the number of people workin' there, and you come up with a fabulous amount. But it's the big bosses who are makin' all the big money and the little guys are makin' the little money.*

To illustrate the situation with the salaries described, consider a company with just 15 people. Suppose that their annual salaries are those shown in the figure above.

*Steve Dubi, quoted in *Working,* by Studs Terkel (Pantheon Books, 1974).

The sum of the 15 salaries is $525,000. If this number is divided by 15, the number of people, the result is

$$\frac{\$525,000}{15} = \$35,000$$

This number, called the *mean,* or *arithmetic average,* is the amount of money that each person in the company would make if the total amount of money spent on salaries was shared equally among all employees.

> The **mean** of a set of numbers is found by adding them and dividing the result by the number of numbers added.

Although it would be correct to say that the "average" salary in the company is $35,000, in this case this number gives a distorted picture of the situation. If $35,000 is average, then most of the salaries in the company are below average.

Perhaps a more realistic way of illustrating the situation in this company would be to use the salary in the middle. This number, $18,000, is the *median.* Just as many people have salaries above this number as have salaries below it.

> The **median** of a set of numbers is the number in the middle when the numbers are arranged in order of size.*

Another look at the salaries reveals that more people in the company make $15,000 than any other number. This number is called the *mode.*

> The **mode** of a set of numbers is the number that occurs most frequently, if there is such a number.

The "average" of a set of numbers may refer to any one of the three numbers: *mean, median,* or *mode.* These numbers are called *measures of central tendency* because they are numbers about which the numbers in a given set tend to center. Which measure of central tendency is the most appropriate for a given set of numbers depends on the numbers in the set and the purpose that the measure is to serve.

---

*If the set of numbers contains an even number of numbers, the median is found by taking the mean of the two middle numbers. For example, the median of the set of numbers 3, 6, 10, 15 is $\dfrac{6+10}{2} = \dfrac{16}{2} = 8.$

EXERCISES

SET I

By permission of Johnny Hart and Field Enterprises, Inc.

B. C. and Thor are playing golf. The scores on the first nine holes are shown in the table below.

	Hole									
	1	2	3	4	5	6	7	8	9	Total
B. C.	4	7	5	2	4	7	3	6	7	45
Thor	3	5	4	2	3	7	3	5	16	48

On the ninth hole, Thor got stuck in a sand trap and lost the game.

1. What was B. C.'s mean, or average, score on the nine holes?

2. What was Thor's mean, or average, score?

3. Which player did better on most of the holes?

4. Is this obvious from their mean scores?

B. C.'s scores on the nine holes arranged in order from smallest to largest are:

$$2 \quad 3 \quad 4 \quad 4 \quad 5 \quad 6 \quad 7 \quad 7 \quad 7$$

5. What was his median score?

6. Arrange Thor's scores on the nine holes in order from smallest to largest.

7. What was his median score?

8. What was the mode of B. C.'s scores?

9. What was the mode of Thor's scores?

10. Which measure of central tendency — the mean, median, or mode — do you think gives the best comparison of the abilities of the two players?

The Funk and Wagnalls *New Encyclopedia* is sold in supermarkets at the rate of one volume per week. The first volume costs 9¢, the second volume costs 99¢, and the remaining 27 volumes are $5.99 each.

11. How much does the entire set cost? (Ignore tax.)

12. What is the mean price per volume?

13. What is the median price per volume?

14. What is the mode price per volume?

A typical sign on a gasoline pump reads:

MINIMUM OCTANE RATING
(R + M)/2 METHOD
87

The sign indicates that the octane rating is found by adding two numbers, R and M, and dividing the result by 2.

15. Which measure of central tendency is found by doing this?

16. The sign says that the octane rating of this gasoline is 87. If R = 85 for this gasoline, what would M equal?

The greatest recorded snowfall in a 24-hour period occurred at Silver Lake, Colorado, in April 1921. It was 192 centimeters.

17. What was the mean hourly snowfall during that period?

18. Does this mean that that number of centimeters of snow fell during each hour of that period?

19. What was the least amount of snow that could have fallen there in a single hour?

20. What was the greatest amount of snow that could have fallen there in a single hour?

A useful fact that can sometimes be used to simplify the calculation of a mean, or average, can be discovered from the following exercises.

Find the mean of each of these sets of numbers.

21. 4, 6, 7, 15.

22. 104, 106, 107, 115.

23. Exactly how do the numbers in the second set compare in size with those in the first set?

24. How does the mean of the numbers in the second set compare with the mean of the numbers in the first set?

25. What shortcut does this suggest for finding the mean of the second set of numbers?

In 1861, the British scientist Sir William Crookes discovered the element thallium. He made ten measurements of its atomic weight, getting the numbers shown here.

203.63	203.63	203.64	203.64	203.64
203.64	203.64	203.65	203.65	203.67

26. Use the shortcut suggested by exercises 21 through 25 to find the mean of these measurements. Round it to the nearest hundredth.

Set II

In the 1990 Super Bowl, the San Francisco 49ers beat the Denver Broncos with a score of 55–10: the point spread was $55 - 10 = 45$. In the 1991 Super Bowl, the New York Giants beat the Buffalo Bills with a score of 20–19: the point spread was $20 - 19 = 1$.

The point spreads for the first 25 Super Bowls are listed below.

14	21	18	5	19
19	7	17	10	32
9	17	4	29	4
14	10	12	22	45
3	4	17	36	1

1. Find the mean point spread for the first 25 Super Bowls.

2. Find the median point spread.

3. Why is there no modal point spread?

In his *History of the Peloponnesian War,* the Greek historian Thucydides described the efforts of the Plateans, who wanted to climb over their enemy's walls. He wrote:

> They made ladders equal in height to the enemy's wall, getting the measure by counting the layers of bricks at a point where the . . . wall . . . happened not to have been plastered over. Many counted the layers at the same time, and while some were sure to make a mistake, the majority were likely to hit the true count. . . . The measurement of the ladders, then, they got at in this way, reckoning the measure from the thickness of the bricks.*

4. Which measure of central tendency — mean, median, or mode — did the Plateans use in deciding how long to build their ladders?

A motor-vehicle bureau survey has revealed that, 30 years ago, each car on the road contained an average of 3.2 persons.† Twenty years ago, occupancy had declined to 2.1 persons per car. Ten years ago, the average was down to 1.4 persons. If this trend continues, 10 years from now every third car going by will have nobody in it.

5. What sort of measure of central tendency are the numbers in this survey?

Thucydides, translated by Charles Forster Smith (Harvard University Press, 1965.)

†Adapted from an anecdote in *Mathematical Circles Revisited,* by Howard W. Eves (Prindle, Weber and Schmidt, 1971.)

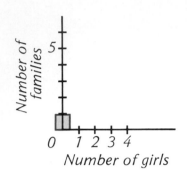

According to probability theory, the expected numbers of girls in 16 families having four children each are

0   1   1   1   1   2   2   2   2   2   2   3   3   3   3   4

6. Make a histogram of this information, labeling the axes as shown at the left. The bar shown represents the only family having no girls.

7. What kind of symmetry do the five bars in the histogram have?

8. Find the mean, median, and mode numbers of girls per family.

The *actual* numbers of girls in a sample* of 16 families having four children each are

2   2   2   2   0   2   3   2   3   2   3   1   1   4   3   2

9. Make a histogram of this information, labeling the axes as in exercise 6.

10. Find the mean, median, and mode numbers of girls per family.

11. How does the shape of the histogram suggest that not all these measures of central tendency are the same number?

Ten of his friends decided to throw a surprise birthday party for Uncle Fletcher. The mean age of the first eight people to arrive is 21 years. The ninth person to arrive, Cora Bucksaddle, is 30 years old.

12. What does the mean age become after she joins the party?

The tenth person to arrive, Ole Chinbunny, is also 30 years old.

13. What does the mean age become after he joins the party?

14. Given that Ole and Cora are the same age, why doesn't Ole's joining the party raise the average age by the same amount that Cora's did?

When the guest of honor, Uncle Fletcher, joins the party, the mean age becomes 30 years.

15. What birthday is Uncle Fletcher celebrating?

---

*The first 16 families having four children listed in *Who's Who 1980* (London, Black.)

SET III
REDISTRIBUTION OF THE WEALTH IN OILARIA*

The Sheik of Oilaria has proposed the following share-the-wealth program for his sheikdom. The population is divided into five economic classes. Class 1 is the poorest, class 2 is the next-poorest, and so on, to class 5, which is the richest. The plan is to average the wealth by pairs, first averaging the wealth of classes 1 and 2, then averaging the new wealth of class 2 with class 3, and so forth. Averaging the wealth of a pair of classes means that the total wealth of the two classes is redistributed evenly to everyone in them.

The Sheik's Grand Vizier approves the plan but suggests that averaging begin with the two richest classes, then proceed down the scale instead of up.

Suppose that there are the same number of people in each economic class and that the table at the right shows the amount of money possessed by each class before the wealth is redistributed.

Figure out how much money each of the five classes would have if the wealth were redistributed according to

1. the Sheik's plan.

2. the Grand Vizier's plan.

3. Which plan would the poorest class prefer?

4. Explain why this choice would be preferred by the poorest class even if the amount of money possessed by each class was unknown.

5. Which plan would the richest class prefer?

6. Explain why this choice would be preferred by the richest class even if the amount of money possessed by each class were unknown.

Class	Amount of money in billions
1	1
2	3
3	4
4	7
5	13

---

*This problem by Walter Penney originally appeared in Martin Gardner's "Mathematical Games" column in *Scientific American*, December 1979.

Drawing by John Gallagher

"Oh! Oh!"

4

Measures of Variability

The basketball team from Dribble High has arrived for its game with Wembly, and its prospects of winning look pretty good. Although Wembly's coach had been told that the average height of the 11 players on the Dribble team was 180 centimeters, he did not know about Joe Dunkshot, its most promising player. This is the first game of the season, and Dribble's coach had been keeping Joe a secret.

A measure of central tendency, by itself, does not tell everything that someone might want to know about a set of numbers. In addition to a measure of central tendency, it is often helpful to have information about how the numbers vary. One *measure of variability* is the *range*.

Here is a list of the heights, in centimeters, of the 11 members of the Dribble team, in order from shortest to tallest:

168   170   171   176   178   178   180   180   181   183   215

This list shows that the players vary in height from 168 centimeters to 215 centimeters. The *range* in heights is

$$215 \text{ cm} - 168 \text{ cm} = 47 \text{ cm}$$

The **range** of a set of numbers is the difference between the largest and the smallest numbers in the set.

Although the range of a set of numbers is easy to figure out, it is determined by only two of the numbers and does not give us any information about how the rest of the numbers in the set vary.

Another very commonly used (and very helpful) *measure of variability* is the *standard deviation.*

The **standard deviation** of a set of numbers is determined by finding:

1. the mean, or average, of the numbers,
2. the difference between each number in the set and the mean,
3. the squares of these differences,
4. the mean of the squares, and
5. the square root of this mean.

The calculation of the standard deviation in heights of the members of the Dribble team is shown below.

Height	Difference from mean	Square of difference
168	−12	144
170	10	100
171	9	81
176	4	16
178	2	4
178	2	4
180	0	0
180	0	0
181	1	1
183	3	9
+ 215	35	+1,225
1,980		1,584

$$\text{Mean} = \frac{1,980}{11} = 180 \qquad \begin{array}{c}\text{Mean of}\\\text{squares}\end{array} = \frac{1,584}{11} = 144 \qquad \begin{array}{c}\text{Square root}\\\text{of mean}\end{array} = \sqrt{144} = 12$$

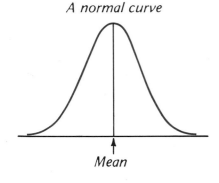

Height in centimeters

The mean height of the 11 members of the Dribble team is 180 centimeters, and the standard deviation for these heights is 12 centimeters.

For comparison, consider the variation in the heights of the approximately 2 million 17-year-old boys in the United States. The mean height is 176 centimeters, and the standard deviation is 6 centimeters. If a histogram were made of the heights and a smooth curve drawn through the midpoints of the tops of the bars, it would look like the adjoining figure.

The curve is bell-shaped with its peak above the mean, 176 centimeters. The standard deviation, 6 centimeters, has been used as the unit for numbering the horizontal axis in each direction from the mean. The heights of

$$34\% + 34\% = 68\%$$

of the 17-year-old boys in the United States are within one standard deviation of the mean; in other words, 68% of the boys are between 170 centimeters and 182 centimeters tall. The heights of

$$14\% + 34\% + 34\% + 14\% = 96\%$$

are within two standard deviations of the mean (between 164 centimeters and 188 centimeters), and nearly 100% are within three standard deviations of the mean (between 158 centimeters and 194 centimeters). This graph is an example of a *normal curve* and the percentages are significant because many large sets of numbers are distributed in the same way.

> A **normal curve** is a bell-shaped curve that closely matches the distribution of many large sets of numbers. For such sets of numbers, 68% of the numbers are within one standard deviation of the mean, 96% are within two standard deviations of the mean, and nearly 100% are within three standard deviations of the mean.

A normal curve

Mean

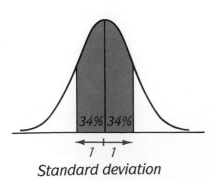

Standard deviation
from mean

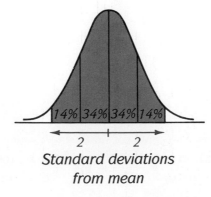

Standard deviations
from mean

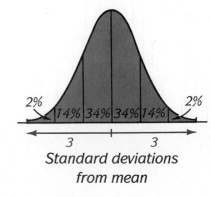

Standard deviations
from mean

EXERCISES

SET I

The times, in seconds, that it takes 10 runners to run an 800-meter race are

114   116   116   119   120   120   121   121   125   128.

1. What is the range of the times in seconds?
2. Find the mean time.
3. Copy and complete the table at the right.
4. Find the mean of the numbers in the last column of the table.
5. Find the standard deviation of the times by taking the square root of this average.
6. What percentage of the 10 times are one standard deviation or less from the mean?
7. What percentage are two standard deviations or less from the mean?

Time	Difference from mean	Square of difference
114	6	36
116		
116		
119		
120		
120		
121		
121		
125		
128		

Suppose that the time of the fastest runner was decreased by 10 seconds and that the time of the slowest runner was increased by 10 seconds, so that the 10 times are

104   116   116   119   120   120   121   121   125   138

8. Now what is the range of the times?
9. What is the mean time?
10. Copy and complete the table at the right.
11. Find the mean of the numbers in the last column of the table.
12. Find the standard deviation in the times by taking the square root of this average.

Time	Difference from mean	Square of difference
104	16	256
116		
116		
119		
120		
120		
121		
121		
125		
138		

On the basis of your answers to exercises 1 through 12, if the smallest and largest numbers in a set of numbers are increased and decreased by equal amounts, what do you think happens to

13. the range of the numbers?
14. the mean of the numbers?
15. the standard deviation of the numbers?

The scores of the winning teams in the 50 Rose Bowl games played from 1931 to 1980 are arranged from smallest to largest in the list below.

$$
\begin{array}{cccccccccccc}
7 & 7 & 7 & 7 & 9 & 10 & 10 & 13 & 13 & 14 & 14 & 14 & 14 \\
14 & 14 & 17 & 17 & 17 & 17 & 17 & 17 & 18 & 20 & 20 & 20 & 21 \\
21 & 21 & 21 & 23 & 24 & 25 & 27 & 27 & 27 & 28 & 29 & 29 & 34 \\
34 & 35 & 35 & 38 & 40 & 42 & 42 & 42 & 44 & 45 & 49 \\
\end{array}
$$

16. What is the range of the scores?

The sum of the numbers in the list is 1,150.

17. Find the mean score of the winning team in these games.

The standard deviation of these numbers is 11.

18. How many of the scores are one standard deviation or less from the mean?

19. What percentage of the scores are one standard deviation or less from the mean?

20. How many of the scores are two standard deviations or less from the mean?

21. What percentage of the scores are two standard deviations or less from the mean?

SET II

Michael Crichton's novel *Jurassic Park* includes a figure showing the height distribution of some dinosaurs called procompsognathids (or "compys").

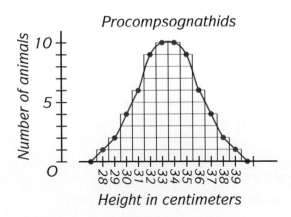

Procompsognathids

Number of animals / *Height in centimeters*

Commenting on the appearance of this graph, one of the characters in the story, a mathematician named Ian Malcolm, says:

"You see here we have a normal . . . distribution for the animal population. It shows that most of the animals cluster around an average central value, and a few are either larger or smaller than the average, at the tails of the curve. . . . Any healthy biological population shows this kind of distribution."*

1. According to this graph, what does the "average central value" of the heights seem to be?

2. What is the range of heights of the animals?

The first bar of the graph shows that 1 "compy" is 28 centimeters tall.

3. How many dinosaurs are represented in the graph altogether?

The normal curve first appeared in a paper by the French mathematician Abraham de Moivre in 1733. Since then, it has been found to have many useful applications.

It is related, for example, to binomial probabilities. Compare the histograms below showing the relative frequencies of various numbers of heads when a set of coins is repeatedly tossed with the normal curve shown at the top of the next page.

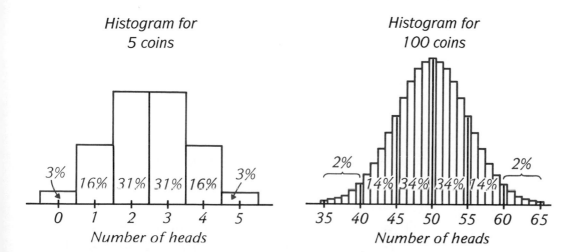

Histogram for
5 coins

3%   16% 31% 31% 16%   3%

0   1   2   3   4   5
Number of heads

Histogram for
100 coins

2%   2%
        14% 34% 34% 14%

35   40   45   50   55   60   65
Number of heads

Jurassic Park by Michael Crichton (Alfred A. Knopf, 1990).

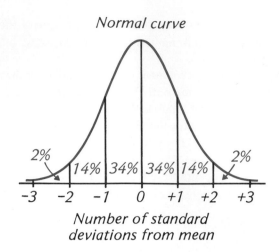

Normal curve

2% 14% 34% 34% 14% 2%

−3 −2 −1 0 +1 +2 +3

Number of standard deviations from mean

The first histogram at the bottom of the preceding page shows that, when five coins are repeatedly tossed, two or three heads turn up

$$31\% + 31\% = 62\%$$

of the time.

4. When 100 coins are repeatedly tossed, what percentage of the time do between 45 and 55 heads turn up?

5. When a large set of numbers fits a normal curve, what percentage of the numbers are within one standard deviation of the mean?

6. When five coins are repeatedly tossed, what percentage of the time do between one and four heads turn up?

7. When 100 coins are repeatedly tossed, what percentage of the time do between 40 and 60 heads turn up?

8. When a large set of numbers fits a normal curve, what percentage of the numbers are within two standard deviations of the mean?

9. When five coins are repeatedly tossed, what percentage of the time do between zero and five heads turn up?

10. When 100 coins are repeatedly tossed, what percentage of the time do between 35 and 65 heads turn up?

11. When a large set of numbers fits a normal curve, what percentage of the numbers are within three standard deviations of the mean?

12. What happens to the shape of a "coin-tossing histogram" as the number of coins increases?

The Scholastic Aptitude Test (SAT) is the test most commonly required for college entrance. It was originally designed so that the distribution of scores looked like this.

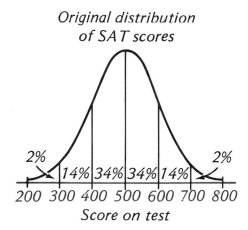

13. What does the curve in this graph look like?
14. What was the mean score?

The scale of scores extends three standard deviations below the mean and three standard deviations above the mean.

15. What is the range of the scores?
16. What is the standard deviation of the scores?

What percentage of the scores were between

17. 500 and 600?
18. 500 and 700?
19. 500 and 800?

Some colleges do not admit applicants whose scores are less than 600.

20. According to this distribution, what percentage of students would be expected to have scores of 600 or more?

In recent years, the mean score on the SAT has declined, although the standard deviation has remained about the same.

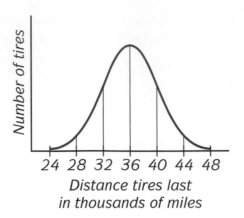

A company produces car tires that last an average (mean) distance of 36,000 miles. The distances are normally distributed with a standard deviation of 4,000 miles.*

21. What percentage of the tires last more than 40,000 miles?

The company wants to guarantee the tires so that only 2% will fail before the guaranteed number of miles.

22. For how many miles should the company guarantee the tires?

The numbers of children of the first 25 presidents of the United States are shown in this histogram.

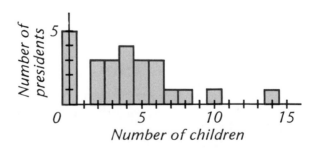

The mean of the numbers is 4.1 and the standard deviation is 3.3.†

23. How many of the first 25 presidents had a number of children within one standard deviation of the mean?

---

*Based on a problem in *Elementary Statistics,* by Mario Triola (Addison-Wesley, 1992).

†Each of these numbers has been rounded to the nearest tenth.

24. What percentage of these presidents had a number of children within one standard deviation of the mean?

25. How many of the first 25 presidents had a number of children within two standard deviations of the mean?

26. What percentage of these presidents had a number of children within two standard deviations of the mean?

27. How many of the first 25 presidents had a number of children within three standard deviations of the mean?

28. What percentage of these presidents had a number of children within three standard deviations of the mean?

Look again at the histogram of the numbers of children and at your answers to exercises 24, 26, and 28.

29. What is interesting about the answers?

SET III

The curves in this figure show the variations in the weights of quarters that were minted and put into circulation at the same time.*

One curve shows the weight distribution when the coins were new and the other two show the distributions when they had been in circulation for five years and for ten years.

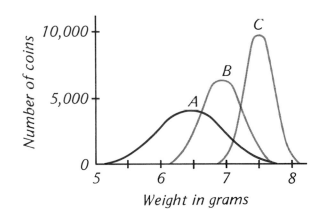

1. Which curve do you think shows the weights of the newly minted quarters, which curve the coins after five years, and which curve the coins after ten years?

2. What happens to the average weight of the coins as time passes?

3. What happens to the standard deviation of the weight of the coins as time passes?

*Adapted from a graph in the article "Scientific Numismatics" by D. D. Kosambi, *Scientific American*, February 1966.

Reprinted by the Saturday Evening Post. Bursino/Cartoonists & Writer's Syndicate

"They said you wanted to see me the minute I got back from my vacation."

LESSON

5

Displaying Data: Statistical Graphs

Statistical information is often presented in graphical form. Graphs reveal patterns in tables of numbers in a very simple way — so simple, in fact, that the patterns can often be seen in a single glance.

The pattern in the "Weekly Business Chart" in this cartoon, for example, suggests that the fellow returning from his vacation may be in for trouble. It looks as if the amount of business done by the company stayed the same from week to week until he went on his vacation. While he was away, business seems to have improved.

Even though there are no scales on the graph, the title above it suggests that the units from left to right represent weeks. Although we might guess from the graph that the fellow's vacation lasted two weeks, it is impossible to tell how much the weekly business increased. Compare, for example, the two graphs below.

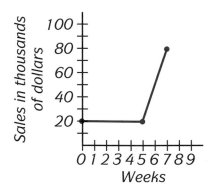

If the scales on the first graph are correct, weekly sales went from $20,000 to $80,000 in two weeks. Because

$$\frac{80,000}{20,000} = 4,$$

the weekly sales became 4 times what they were.

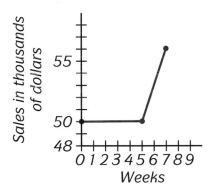

If the scales on the second graph are correct, the weekly sales went from $50,000 to $56,000, a considerably smaller change.

One thing, then, that determines the information given by a graph is the numbering of its axes. The scales are important and, if they are removed, such a graph becomes almost meaningless.

Although graphs are used to present statistical information because of the simple picture that they can give of the patterns it contains, they can also be deliberately drawn to give a false impression. And even a graph that is meant to give an honest representation can be misinterpreted.

EXERCISES

SET I

Temperatures during the year vary considerably from city to city. The following tables give the normal low and high Fahrenheit temperatures in Baghdad, Acapulco, and Melbourne for each month of the year.

Baghdad			Acapulco			Melbourne		
Month	Low	High	Month	Low	High	Month	Low	High
Jan.	39	60	Jan.	70	85	Jan.	57	78
Feb.	42	64	Feb.	70	87	Feb.	57	78
Mar.	48	71	Mar.	70	87	Mar.	55	75
Apr.	57	85	Apr.	71	87	Apr.	51	68
May	67	97	May	74	89	May	47	62
June	73	105	June	76	89	June	44	57
July	76	110	July	75	89	July	42	56
Aug.	76	110	Aug.	75	89	Aug.	43	59
Sept.	70	104	Sept.	75	88	Sept.	46	63
Oct.	61	92	Oct.	74	88	Oct.	48	67
Nov.	51	77	Nov.	72	88	Nov.	51	71
Dec.	42	64	Dec.	70	87	Dec.	54	75

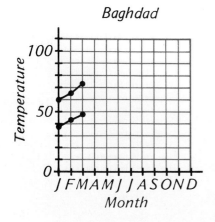

Baghdad

1. Make a line graph for Baghdad to show how the low and high temperatures vary during the year, labeling the axes as shown at the left. The temperatures for the months of January, February, and March are shown as an example.

2. Make a line graph showing the variations of low and high temperatures in Acapulco during the year, labeling the axes in the same way.

3. Make a line graph showing the variations of low and high temperatures in Melbourne during the year, labeling the axes in the same way.

Refer to your line graphs to determine which city has

4. the least variation in temperature during the year.

5. the greatest variation in temperature during the year.

6. What advantage do your line graphs have over the tables in presenting the information about the temperature variations in the three cities?

7. Graph the Melbourne information again, using a temperature scale ranging from 40 to 90 as shown at the right.

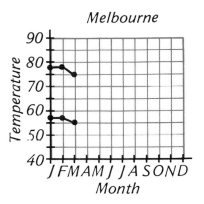

Compare this graph with the other Melbourne graph.

8. In what way does changing the temperature scale from 0°– 110° to 40°–90° change the impression given of the variation in temperatures?

9. Why does the change in the temperature scale do this?

This bar graph appeared in an ad of an insurance company.

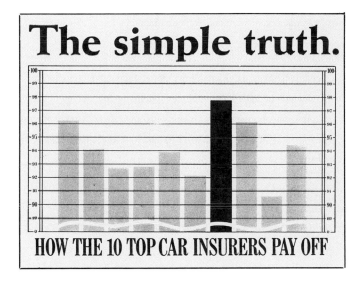

10. Does it really show the simple truth?

The bars represent the "payoffs" of 10 insurance companies, the tallest bar representing that of the company running the ad.

11. Roughly how many times as tall as the shortest bar in the graph is the tallest one?

12. Make a list of the "payoff" numbers shown for the 10 companies, rounding each one to the nearest whole number. For example, the rounded "payoff" numbers for the first three bars are 96, 94, and 93.

13. Graph this information to give a more honest picture of the situation. Number the vertical scale in 10's from 0 to 100.

14. Why do you suppose the more honest graph was not used in the ad?

The shaded slices of these circle (or "pie") graphs show the parts of the population of the United States in 1890 and 1990 who were less than 20 years old.

Part of the U.S. population
less than 20 years old

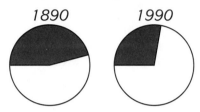

15. In which year was a greater percentage of the population less than 20 years old?

16. Does it follow that there were more people in the United States less than 20 years old in that year than in the other? Explain.

This graph appeared in some advertising for a program to help people withdraw from drugs.

80%	
70%	
60%	*Graduates*
50%	*Remaining*
40%	*Drug Free*
	One Year
30%	*After the*
20%	*Program*
10%	
0%	

17. What is strange about the scale used in this graph?

18. Although the program seems fairly effective, the graph exaggerates its success. What is it about the graph that makes it misleading?

SET II

By permission of Johnny Hart and Field Enterprises, Inc.

Picture graphs, frequently used to make comparisons, can be easily misinterpreted. Suppose, for example, that Peter has twice as many clams as B. C. and that we decide to show this with two bags of clams, one drawn so that it is twice as tall as the other.

1. Does the larger bag look as though it contains exactly twice as many clams as the smaller one?

Rather than comparing the heights of the two bags, someone might compare the *areas* covered by them. Notice that the larger bag is twice as wide as the smaller one, in addition to being twice as tall.

> 2. How do you think their areas compare?
> *Hint:*

The number of clams that each bag contains is determined by the *volume.*

> 3. How do you think their volumes compare?
> *Hint:*

A different picture graph could be drawn, also using bags of clams, which makes it perfectly clear that Peter has just twice as many clams as B. C.

> 4. What would it look like?

The USA cable television network wanted to compare the relative size of its daytime audience with that of two competing networks:

> USA: 849,000 households
> TBS: 722,000 households
> Lifetime: 221,000 households

In an ad, the company used three pyramids to illustrate the relative sizes of these three numbers.

The figure below shows the pyramids drawn so that their relative *heights* represent the numbers.

Correct heights

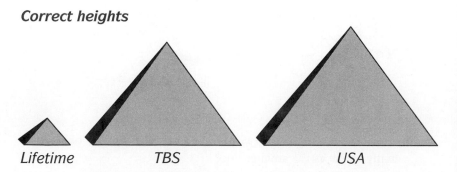

Lifetime   *TBS*   *USA*

These figures show the pyramids so that their relative *areas* represent the numbers and so that their relative *volumes* represent the numbers.

Adapted from USA Cable Network advertisement.

5. Which one of these three sets of pyramids do you suppose the cable company used in its ad?

6. Does that set of pyramids give the correct impression of the relative sizes of the daytime audiences? Why or why not?

This graph appeared several years ago in a bulletin for the employees of a large company. It compared the number of absences on four consecutive Mondays, the last one preceding a legal holiday.

7. What impression does the graph seem intended to convey?

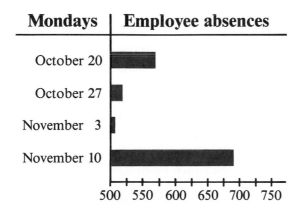

8. The actual numbers of absences on these four days were

566, 517, 501, and 689

Do the lengths of the four bars make this obvious?

9. Why does the graph give the impression that it does?

10. Graph the number of absences again, renumbering the scale along the bottom as follows: 0, 150, 300, 450, 600, 750.

This graph shows the change in the U.S. population per square mile of land as recorded by each census since 1790. Although almost every census has shown an increase in the population per square mile, three censuses have shown a decrease.

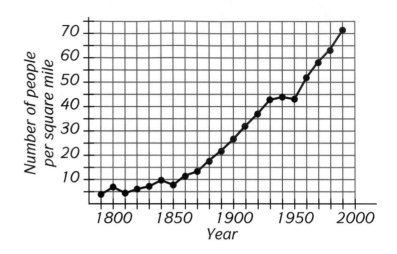

11. Which years are they?

Actually, the *population* increased in those years as well.

12. Explain how the *population per square mile* of the U.S. could decrease even though the population continues to increase.

This graph was used by the sugar industry to show what happened to the sales of Classic Coke during the five-year period (1979–1984) when the company changed from using sugar to using corn syrup as a sweetener.

13. What should you notice when interpreting this graph?

The "height" of the graph in 1984 is about one-fourth of its height in 1979.

14. Does it follow that Coca-Cola's share of the soft drink market in 1984 was about one-fourth of what it was in 1979?

In 1979, Coca-Cola's share of the soft drink market was about 24.4%.

15. What was it in 1984?

16. Sketch a graph that more honestly illustrates the change in sales of Coca-Cola during these years.

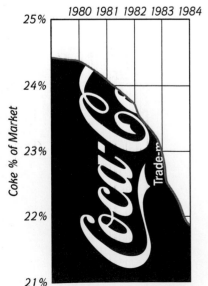

Adapted from The Sugar Association, Inc., advertisement.

SET III

Which face looks happier? Your answer to this question may depend on your age.

The graph below shows the results of an experiment in which people of different ages were shown drawings of two faces, identical except for the size of the pupils, and asked to choose the happier one.* The dark-brown bars represent the percentages of people who chose the face with the large pupils and the light-brown bars represent the percentages of people who chose the face with the small pupils.

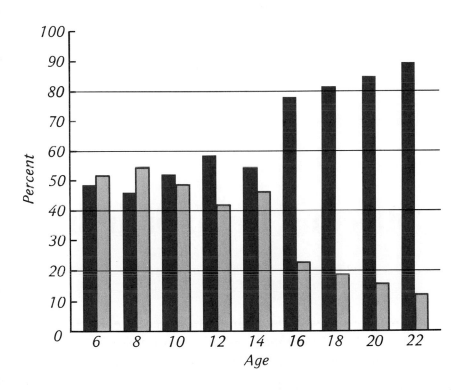

Courtesy of Scott Lewis; photography by Robert Ishi

1. Do you notice any pattern in the graph? If so, what is it?

2. Is there a connection between the height of each light-brown bar and the height of the corresponding dark-brown bar? If so, what is it?

3. Why do you suppose both bars are shown for each age?

4. What conclusion do you suppose was drawn from this experiment?

*The Role of Pupil Size in Communication," by Eckhard H. Hess. Copyright © 1975 by Scientific American, Inc. All rights reserved.

'I can't say that I go for this kind of TV poll . . .'

6

Collecting Data: Sampling

The popularity of television programs in the United States is determined by sampling. The Nielsen ratings are based on a sample of 4,000 homes out of the approximately 92,000,000 homes with television sets. According to the Nielsen Company,

> To do a proper job, a sample needs to reflect, with a proper degree of accuracy, the universe it measures. It is, in effect, a scale model of this universe. The Nielsen TV samples, for example, include all types of households and neighborhoods.*

| A **sample** is a group of items chosen to represent a larger group called the *population* or *universe*.

---

*"Audience Research," *T.V. Guide Almanac,* compiled and edited by Craig T. and Peter G. Norback and the editors of *TV Guide* magazine (Ballantine Books, 1980).

The word *population* in statistics does not refer to people inhabiting a place but rather to a group of things (people or animals or objects) being studied.

Samples are used because they can supply accurate information at less cost and in less time than would be required by a survey of the entire population. For very large populations, such a survey may be impossible. For example, the average annual rainfall in New York City is determined from samples recorded at weather stations because it would be impossible to collect and measure all the rain that fell in that city in a year.

Samples are sometimes the only way that information can be obtained about even a limited population. If a fireworks company tested every one of the firecrackers that it produced to find out whether it would explode, it would not have any firecrackers left to sell.

For a sample to accurately represent a population, it must be *sufficiently large*. If the sample is too small, there is a good chance that the items included in it are not typical of the population. Statisticians have methods of determining just how large a sample should be to be reasonably dependable.

It is also important that the sample be a *random* one.

In a **random sample,** the items are chosen in such a way that every item in the population has an *equal chance* of being included in the sample.

A sample that is not chosen in this way may be biased and give a distorted picture of the population that it is supposed to represent.

There is another reason for wanting samples to be random. It is impossible to be absolutely certain of the nature of a population on the basis of samples of it because the samples will vary. If the samples are random, however, then the methods of probability can be applied to them to get an idea of how accurately they represent the population. Probability theory, then, is fundamental to the sampling process and, hence, a very important part of statistics.

Exercises

Set I

W. E. Smith, Life Magazine © 1948

Before each presidential election, public opinion polls are conducted to try to predict who will win. In 1948, the polls incorrectly predicted that Thomas Dewey would defeat Harry Truman.

The people conducting such a poll cannot find out how everyone will vote. Instead, they have to settle for a *sample.*

1. What is meant by a "sample"?

2. In a poll to predict the winner of an election, what group of people should the sample accurately represent?

3. What are some things that could go wrong with the sample?

Public opinion polling is sometimes done by telephone surveys.

4. Why isn't a telephone survey in a given city a random sample of the people living in that city?

5. If the sample is chosen at random from the telephone directory, what people having telephones will be omitted?

6. How could a random sample be made that would include all people having telephones?

Many years ago, the Chesapeake and Ohio Railroad Company made a study of sampling. From a *sample* of freight bills for a six-month period, the company estimated that another railroad company owed it $64,568. By checking *all* of the freight bills of that period, the company found that the other company actually owed it $64,651.

7. How much money would the Chesapeake and Ohio Railroad Company have lost by trusting the number from the sample?

Checking the sample of freight bills cost the company $1,000; checking all the freight bills cost the company $5,000.

8. How much money would the Chesapeake and Ohio Railroad have saved by relying on the sample rather than checking all of the bills?

9. If these results are typical, which method would benefit the company more: taking a sample of the freight bills for each six-month period or checking all of them?

Composers of music belong to one of two organizations, ASCAP and BMI, which pay them royalties according to how frequently their music is played on radio or television.

10. Why do you suppose that it is impossible for these organizations to keep track of every time that every song is broadcast?

11. How do you suppose these organizations figure out how to distribute the royalties to their members?

An article in the magazine *California Highways** reported that a survey of the clothing worn by pedestrians killed in traffic at night revealed that about 80% of the victims were wearing dark clothes and 20% light-colored clothing. The conclusion was drawn that pedestrians are safer at night if they wear something white so that drivers can see them more easily.

12. What conclusion could be drawn instead?

13. What additional survey would be helpful in deciding which conclusion is correct?

Set II

The effectiveness of the Salk vaccine against polio was tested in 1954 with approximately 2 million children. In one of the towns in which it was tested, there were approximately 1,000 children, half of whom were given the vaccine and half of whom were not.

A polio epidemic passed through the state in which the town was located, and not one of the children inoculated with the vaccine caught the disease.

---

*Cited in *How to Take a Chance,* by Darrell Huff (Norton, 1959).

1. Did this prove that the vaccine was effective?

None of the children who had not been inoculated caught the disease either.

2. What do you suppose this indicated about the size of the sample?

A method sometimes used to estimate the number of fish in a lake is to catch a sample of the fish, tag them, and throw them back. Later, another sample is caught and the number of tagged fish counted.

Suppose that the second sample consisted of 50 fish, of which 10 were tagged.

3. What fraction of the fish in this sample were tagged?

4. If we assume that this sample is typical of the population of fish in the lake, what fraction of the fish in the lake are tagged?

Suppose that the first sample consisted of 100 fish.

5. How many fish do the samples suggest are in the lake?

In using this method, it is assumed that each fish has an equal chance of being caught and that this is true for both of the samples.

6. Give a reason why this may not be a good assumption.

Suppose that, to find out how many owners of cars are making payments on their cars, a group of poll-takers station themselves at the exits of the parking lot of an amusement park. After questioning 800 drivers leaving the lot, they report that 92% of the owners of cars had paid for them in full.

7. Why might it be likely that some of the drivers would not tell the poll-takers the truth?

8. In what way do you think this might affect the figure reported?

Pepsi-Cola once based an advertising campaign on a "blind taste test" comparing Pepsi and Coca-Cola.

A sample of Coca-Cola drinkers was served Pepsi in a glass marked M and Coke in a glass marked Q. More than half of them preferred the cola in the glass marked M.

9. What do you think Pepsi-Cola claimed that this test proved?

10. What do you think Coca-Cola claimed that this test proved about the two letters used?

Coca-Cola then did a test in which Coca-Cola was served in glasses marked M and in glasses marked Q.

11. What do you think they discovered?

After Coca-Cola advertised the results of this test, Pepsi changed the letters on its glasses to L and S, with Pepsi in the glasses marked L and Coke in the glasses marked S. More than half of the people tested preferred the cola in the glass marked L.

12. What would have been a more convincing way of changing the test?

The managers of a summer camp were curious about the numbers of children in the families of the children that attend it. They asked each of the 100 children attending the camp one week how many children were in his or her family. The results are summarized in the frequency distribution below.

Number of children in family	Number of children reporting
1	17
2	23
3	22
4	15
5	9
6	8
7	0
8	6

From this information, they computed the mean number of children per family.

13. What did they get?

The actual mean number of children in the families of the 100 children attending that week was 2.4.

14. What was wrong with the procedure used by the management to find the average?

SET III

Two psychologists in Los Angeles did an experiment to test people's honesty. They used 375 envelopes that were addressed to one of the psychologists at his home. Of these envelopes:

75 were empty and had typed on them: "This is a research study. Drop this envelope in the nearest postbox. Thank you for your cooperation."
150 contained blank sheets of folded paper but did not have a message typed on the outside.
150 seemed to contain money (two coins and a bill that were actually fake.)

The psychologists dropped each envelope on a sidewalk near a mailbox. A third of the envelopes were left in wealthy sections of the city, a third in middle-income areas, and a third in poor neighborhoods. Each one was marked on the inside to show the area in which it had been left.

The numbers of each type of envelope that were returned by the people who found them are shown in the table below.

Type of envelope	Number dropped on sidewalks	Number returned
Empty envelopes labeled "research study"	75	68
Envelopes containing blank sheets of paper	150	120
Envelopes seeming to contain money	150	102

Which type of envelope had

1. the highest rate of return?

2. the lowest rate of return?

A purpose of the experiment was to find out how many people would do something dishonest, such as taking a letter or money that did not belong to them, if there was very little chance of being caught.

3. Why, then, do you think empty envelopes labeled "This is a research study" were used in the experiment?

Some of the envelopes that contained either blank sheets of paper or the fake money were opened and resealed by the people who found them before they were returned.

4. How do you think these envelopes were counted?

The table below shows the number of each type of envelope returned from each area.

	Area of city		
Type of envelope	**Poor**	**Middle**	**Wealthy**
Empty envelopes labeled "research study"	22	24	22
Envelopes containing blank sheets of paper	43	37	40
Envelopes seeming to contain money	28	33	41

There seems to be one significant pattern in this table.

5. What is it?

Although the sample used in this experiment was both sufficiently large and random, the design of the experiment prevents any conclusion about people's honesty from being drawn from the pattern.

6. Why? (*Hint:* The problem concerns temptation.)

CHAPTER

Summary and Review

© 1993 Sidney Harris

In this chapter we have become acquainted with the branch of mathematics that deals with the collection, organization, and interpretation of numerical facts. We have studied:

Organizing data: frequency distributions *(Lessons 1 and 2).* A frequency distribution is a convenient way of organizing data to reveal what patterns they may have. The data can be condensed by grouping them together in intervals.

Measures of central tendency *(Lesson 3).* Three measures of central tendency are commonly used:

The *mean,* or average, of a set of numbers is found by adding them and dividing the result by the number of numbers added.

The *median* is the number in the middle when the numbers are arranged in order of size.

The *mode* is the number that occurs most frequently, if there is such a number.

Measures of variability *(Lesson 4)*. The *range* of a set of numbers is the difference between the largest and the smallest numbers in the set.

The *standard deviation* of a set of numbers is determined by finding:

1. the mean of the numbers,
2. the difference between each number in the set and the mean,
3. the squares of these differences,
4. the mean of the squares, and
5. the square root of this mean.

The standard deviation is useful because of its relation to the normal curve.

A *normal curve* is a bell-shaped curve that closely matches the distribution of many large sets of numbers. For such sets of numbers,

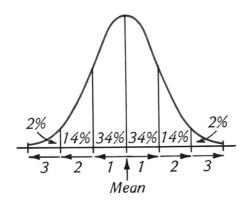

68% of the numbers are within one standard deviation of the mean, 96% are within two standard deviations of the mean, and nearly 100% are within three standard deviations of the mean.

Displaying data: statistical graphs *(Lesson 5)*. Statistical graphs are sometimes drawn in a way that gives a false impression. In bar and line graphs, it is especially important to look at the numbering of the scales. Picture graphs can be misleading if it is not clear whether heights or areas or volumes should be compared.

Collecting data: sampling *(Lesson 6)*. A *sample* is a group of items chosen to represent a larger group called the *population*. For a sample to be representative, it is important that it be sufficiently large and that it be random. In a *random sample*, the items are chosen in such a way that every item in the population has an equal chance of being included in the sample.

EXERCISES

SET I

Florence Nightingale, famous as a nurse, was also a pioneer in the application of statistics to the medical profession.

This bar graph is adapted from a graph by Nightingale showing deaths due to contagious diseases in British military hospitals during the early part of the Crimean War.

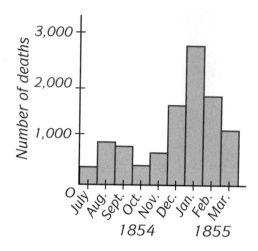

1. In which month was the number of deaths the greatest?

2. During which months did the numbers of deaths increase?

3. For how many months was the number of deaths per month more than 1,000?

Nightingale's use of graphs helped to dramatize the extent to which many deaths could be prevented and led to important reforms in health care.

In the introduction to their book *American Averages,** Mike Feinsilber and William B. Mead wrote:

> Did you know that about half the people in America are below average?
> Not you. Not us.
> Them.

4. Which measure of central tendency is being referred to?

---

*Dolphin Books, 1980.

5. For which measure(s) of central tendency is it possible that more than half of the people in America could be "below average"?

A drug company once included a graph like the one shown here in one of its ads claiming that its product acts "twice as fast as aspirin."

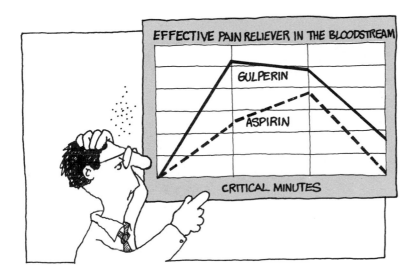

6. Do the lines in the graph seem to show that the drug acts twice as fast as aspirin?

7. What is missing from the graph?

8. Is it possible to tell from the graph how long the effects of the drug will last?

The graph does not seem to be accurately drawn over the time shown.

9. What is strange about its shape?

Petitions are often circulated to force a public vote on a city or state ordinance. A standard procedure is to check out a random sample of the signatures collected, perhaps only 5 out of every 100, to see if they are valid.

10. Why do you think only a sample of the signatures is checked rather than all of them?

There is a legend that many years ago someone wanted to find out how tall the emperor of China was. To ask to measure the emperor was out of the question, so this person decided to poll a million citizens to learn what they thought the height of the emperor was.

After the poll was completed the investigator based his guess on the mean height named, even though not one of the people asked had ever seen the emperor.

Suppose that a graph of the guesses looked like this.

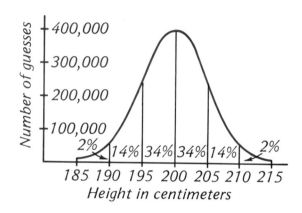

11. What kind of curve does this appear to be?

12. What was the mean, or average, guess of the emperor's height?

13. What was the standard deviation from the mean?

What percentage of the guesses were within

14. one standard deviation of the mean?

15. two standard deviations of the mean?

16. three standard deviations of the mean?

Approximately how many of the million people polled thought that the emperor was

17. at least 200 centimeters tall?

18. more than 215 centimeters tall?

Federally owned land

California   *Oregon*

The shaded slices of the circle graphs at the left show the parts of California and Oregon owned by the federal government.

19. In which state does the government own a greater percentage of the land?

20. Does it follow that the government owns more land in that state than in the other? Explain.

Set II

By permission of Johnny Hart and Field Enterprises, Inc.

If a runner had a race with a roller skater, the skater would probably win. The world record times for three different distances are shown in the table below.

Distance	Runner	Roller skater
400 meters	44 seconds	35 seconds
800 meters	102 seconds	73 seconds
1,500 meters	210 seconds	135 seconds

1. Which one of these three bar graphs comparing the times for the distances is misleading? Why?

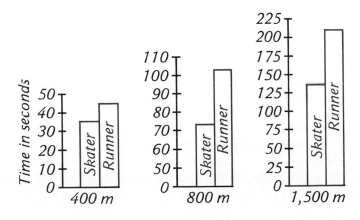

2. What should be noticed in comparing the graphs for 400 meters and 1,500 meters?

Seven people were asked to measure the length of a dollar bill in centimeters. The answers that they reported are listed here:

15   15.2   15.2   15.3   15.3   15.3   153

3. Which answer does not seem to be in centimeters?

4. Judging from the answers, what do you think the length of a dollar bill in centimeters is?

5. Which measure of central tendency did you use? Explain.

6. Find the mean of the seven measurements.

7. Which measures of central tendency are the least affected by the fact that one person's answer is so far off?

Alfred Hitchcock directed 30 films in America, varying in length from *Rope* (80 minutes) to *North by Northwest* (136 minutes). The lengths of the 30 films are listed below.

130	108	80	105	105	120
120	96	117	112	120	128
95	111	110	97	136	127
99	101	101	99	109	116
108	116	95	120	120	120

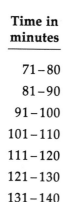

**Time in
minutes**

71–80
81–90
91–100
101–110
111–120
121–130
131–140

8. Make a frequency distribution of these numbers. Use the intervals shown at the left.

9. Make a histogram to illustrate your frequency distribution.

The mean length of Hitchcock's American movies is 111 minutes and the standard deviation is 12 minutes. (Both these numbers have been rounded.)

10. What percentage of the 30 movies are within one standard deviation of the mean?

11. What percentage are within two standard deviations of the mean?

12. Are these close to the percentages that would be expected for a normal curve? Explain why or why not.

Voting records for a presidential election showed that 54% of those old enough to vote did so. In a survey of eligible voters the following year,

59% of the people in the sample said they voted.

13. Why do you suppose the figure obtained from the sample was higher than the figure obtained from the voting records?

14. What other properties of a sample might cause the information obtained from it to be inaccurate?

This picture graph accompanied a newspaper ad with the headline: "The Times has 2,244,500 readers every weekday—more than the next four area newspapers *combined!*"

One weekday issue	The Times	The Examiner	The Register	The Daily News	The Press-Telegram
Number of readers	2,244,500	624,000	485,500	350,500	310,500

15. How do the numbers of readers of the five papers seem to have been used in determining the sizes of the trucks?

16. Is the graph misleading? Explain.

Set III

A well-known signal of distress is · · · ––– · · · , which is Morse code for the letters S O S. In the transmission of Morse code, each dash takes three times as long as each dot. The figure below shows the relative amounts of time for each part of the S O S signal. To send an S takes 5 units of time and to send an O takes 11 units of time.

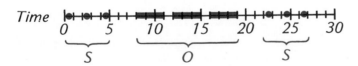

The table below shows the Morse code symbol for each letter of the alphabet.

Letter	Symbol		Letter	Symbol
A	.‐		N	‐.
B	‐...		O	‐‐‐
C	‐.‐.		P	.‐‐.
D	‐..		Q	‐‐.‐
E	.		R	.‐.
F	..‐.		S	...
G	‐‐.		T	‐
H	....		U	..‐
I	..		V	...‐
J	.‐‐‐		W	.‐‐
K	‐.‐		X	‐..‐
L	.‐..		Y	‐.‐‐
M	‐‐		Z	‐‐..

1. Which letter takes just 1 unit of time to send?

2. Which letters take 3 units of time to send?

3. Which letters take 5 units of time to send?

Look at the graph of the frequencies of the letters in ordinary English on page 540.

4. What do you notice about the frequencies of the letters that you named in exercises 1 through 3?

5. Which letters take the longest time (13 units) to send?

6. What do the frequencies of these letters have in common?

7. Which letters take 11 units of time to send?

The symbol for one of these letters seems like a very poor choice.

8. Which letter?

9. Why?

Further Exploration

LESSON 1

1. If three cards are dealt from a shuffled deck, which of the following is more likely: that all three are of different suits, that two cards are of the same suit, or that all three cards are of the same suit?

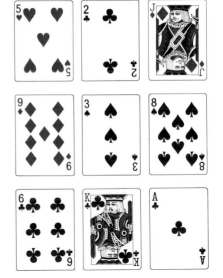

a. Take a deck of playing cards, shuffle it, and deal the cards out in groups of three. Make a frequency distribution of the 17 groups that result, labeling the first column as shown below.

**Greatest number of
cards in any suit**

3
2
1

Reshuffle the cards and deal them out again, adding the results of the 17 newly formed groups to the distribution. Continue to do this until you have recorded the results for exactly 100 groups of three. (This means that you will shuffle the deck six times since 17 + 17 + 17 + 17 + 17 + 15 = 100.)

Use what you have learned about probability to calculate the probability in percent of dealing three cards from a complete deck and getting

b. three cards of the same suit.
c. three cards of different suits.
d. two cards of the same suit. (*Hint:* Subtract the sum of the other two probabilities from 100%.)
e. How do the numbers in your frequency distribution compare with the probabilities that you calculated?

2. In 1787–1788, a series of essays was published to persuade the citizens of New York to ratify the proposed constitution of the United States.* These essays, called the *Federalist* papers, were written anonymously by Alexander Hamilton, John Jay, and James Madison. It is now known that 51 of them were written by Hamilton, 14 by Madison, 12 by either Hamilton or Madison, and 5 by Jay.

To determine the authorship of the 12 disputed papers, historians have studied the frequency with which certain words appear in them. One word used was *by*. The numbers of times that *by* was used in 48 essays by Hamilton, 50 essays† by Madison, and the 12 disputed papers are summarized in the table below.

Number of times per 1,000 words	48 essays by Hamilton	50 essays by Madison	12 disputed papers
1.0–2.9	2	0	0
3.0–4.9	7	0	0
5.0–6.9	12	5	2
7.0–8.9	18	7	1
9.0–10.9	4	8	2
11.0–12.9	5	16	4
13.0–14.9	0	6	2
15.0–16.9	0	5	1
17.0–18.9	0	3	0

From this table, we see that *by* was used between 1.0 and 2.9 times per 1,000 words in 2 of the 48 essays by Hamilton:

$$\frac{2}{48} \times 100\% = \frac{200\%}{48} \approx 4\%$$

a. Copy the table, replacing each number of essays with the corresponding percentage.

b. Refer to your table to make three histograms to represent the appearance of *by* in the essays by Hamilton, the essays by Madison, and the disputed papers. Label the axes of each histogram as shown at the top of the next page.

---

*The information in this exercise is from the article "Deciding Authorship," by Frederick Mosteller and David L. Wallace, in *Statistics: A Guide to the Unknown*, edited by Judith M. Tanur (Holden-Day, 1972).

†This number includes essays in addition to the *Federalist* papers.

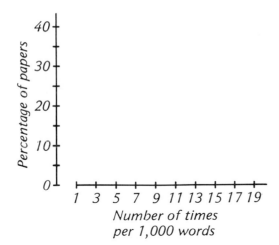

Refer to your histograms to answer the following questions.

c. Who used the word *by* more frequently: Hamilton or Madison?

d. What do the histograms seem to indicate about the authorship of the disputed papers? Explain your reasoning.

LESSON 2

1. The arrangement of the alphabet on the typewriter keyboard, designed by Christopher Sholes in 1872, is shown in the figure below.

Left hand ●
Right hand ●

The row of keys outlined in black is the "home row" on which the fingers rest.

The table at the right shows the typical number of times each key is used in typing 100 letters. Refer to this table and the picture of the keyboard above to determine, in typing 100 letters, the number of times each of the following would be likely to occur.

a. Using a key in the home row.

b. Using a key in the row above the home row.

c. Using a key in the row below the home row.

d. Using a finger of the left hand.

e. Using a finger of the right hand.

A	8	N	7
B	1	O	8
C	3	P	2
D	4	Q	0
E	13	R	7
F	2	S	6
G	1	T	9
H	6	U	3
I	7	V	1
J	1	W	1
K	1	X	1
L	3	Y	2
M	3	Z	0

The American Simplified Keyboard, designed by August Dvorak in 1936, is illustrated below.

Left hand
Right hand

Refer to the table on page 593 and this picture to determine, in typing 100 letters on the American Simplified Keyboard, the number of times each of the following would be likely to occur.

    f. Using a key in the home row.

    g. Using a key in the row above the home row.

    h. Using a key in the row below the home row.

    i. Using a finger of the left hand.

    j. Using a finger of the right hand.

In what way is the American Simplified Keyboard an improvement over the keyboard designed by Sholes with respect to the frequency with which

    k. the rows are used?

    l. the left and right hands are used?

2.   Cryptograms, ciphers that have the original punctuation and spacing between words, often appear in puzzle magazines. Their popularity dates back to 1841, when Edgar Allan Poe wrote an article titled "Secret Writing" for a magazine of the time.

You may enjoy trying to decipher the following cryptograms, each of which is written in a different cipher. Because they are so brief, one word is given in each to help you get started.

    a. NO BAGS ITSRUYE XNXU'Y CTLR TUB
        PCNQXSRU, YCRSR'E T MAAX PCTUPR
        YCTY BAG DAU'Y CTLR TUB.

                                        —PQTSRUPR XTB

        (The fourth word is DIDN'T.)

    b. A HARD HPTE SCHRF, DFIDXACHHO
        SVDT MVDO CUD MCRDT GO IDPIHD
        SVP CTTPO ZD.

                                        —JUDY CHHDT

        (One of the words is ANNOY.)

c. EVO CODE BNJ EY WOOL AVSRFTOZ
VYUO SD EY UNWO EVO VYUO
NEUYDLVOTO LRONDNZE — NZF ROE
EVO NST YHE YM EVO ESTOD.

— FYTYEVJ LNTWOT

(One of the words is PLEASANT.)

LESSON 3

1. One way in which languages differ is in the numbers of syllables in their words. The histograms below show the frequencies of the numbers of syllables per word in English, German, Japanese, and Arabic.

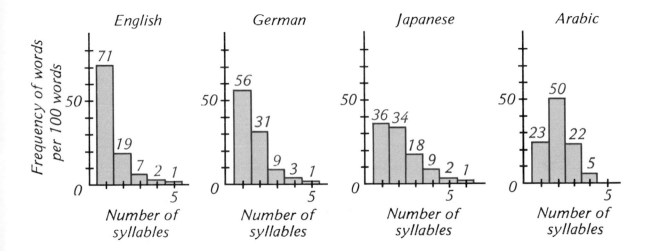

From the histogram for English, we see that the most frequent number of syllables per word is one.

a. What is the most frequent number of syllables per word for each of the other languages?

From the histogram for English, we see that, if a typical sample of 100 words is arranged in order of increasing numbers of syllables, the 50th (or 51st) word would have one syllable.

b. What is the median number of syllables per word for each of the other languages?

c. Find the mean number of syllables per word for English. Show your work. (*Hint:* Use the information contained in the sample of 100 words. For example, 71 words have one syllable each, 19 words have two syllables each, and so on.)

d. Find the mean number of syllables per word for each of the other languages. Show your work in each case.

2. Experiment: *Rolling a Die to Get All Six Numbers*

How many times, on the average, must a die be rolled so that each of its six numbers comes up at least once?

One way to get an idea of the number of times is to take a die and try it. For example, successive rolls of a die might turn up the following numbers:

3   5   2   2   1   2   3   6   4

Number on die	Example trial
1	|
2	|||
3	||
4	|
5	|
6	|_
	9

In this case, it took nine tosses to get all six numbers. A convenient way to keep track of the numbers is with a frequency distribution like the one at the left. The outcomes of the rolls are tallied in a column until there is at least one mark for each number; the total number of rolls is then written at the bottom of the column. The outcomes of additional trials can be recorded in additional columns.

a. Take a die and carry out 25 trials of this experiment, recording the results as shown above.

b. Make a frequency distribution of the results; label the first column "Number of rolls" and tally the numbers of rolls made in each of the 25 trials.

c. Use your frequency distribution to determine the median number of rolls necessary to get all six numbers.

d. Calculate the mean number of rolls necessary.

LESSON 4

1. A letter once appeared in Dear Abby's column from a woman who said she had been pregnant for 310 days before giving birth to her baby. This is considerably longer than the typical pregnancy.

Lengths of pregnancy of women having children are normally distributed, with a mean of 266 days and a standard deviation of 16 days.

Between what two lengths of time would you expect

a. about 68% of the pregnancies to last?

b. about 96% of the pregnancies to last?

c. almost 100% of the pregnancies to last?

The table at the left shows the percentages of the numbers in a normal distribution at various distances from the mean.

Number of standard deviations from mean	Percentage of numbers
0.25	19.7
0.50	38.3
0.75	54.7
1.00	68.3
1.25	78.9
1.50	86.6
1.75	92.0
2.00	95.5
2.25	97.6
2.50	98.8
2.75	99.4
3.00	99.7

d. What percentage of children are born from pregnancies lasting 310 days or more?

e. About how many such children would be expected in 1,000 births?

f. Do you think it is reasonable to assume that the woman was pregnant as long as she claimed?

2. A bakery sells small loaves of bread that supposedly weigh 250 grams each. A statistics student bought three of the loaves, took them home, and weighed them. Their weights were 247 grams, 252 grams, and 239 grams.

a. What was their mean weight?

Wondering about this result and about the variation in the weight, the student decided to buy one loaf of bread at the bakery each morning and weigh it. The weights of the loaves bought on 25 successive days are listed below.

$$
\begin{array}{ccccc}
238 & 242 & 249 & 236 & 253 \\
258 & 234 & 245 & 249 & 244 \\
247 & 248 & 244 & 242 & 239 \\
240 & 246 & 251 & 249 & 254 \\
244 & 239 & 243 & 247 & 244 \\
\end{array}
$$

Weight in grams

230–234

235–239

240–244

245–249

250–254

255–259

b. Make a frequency distribution of these weights by grouping them in intervals of 5 grams. Number the first column in your table as shown at the right.

c. Make a histogram of the weights, labeling the axes as shown here. After you have made the histogram, mark the midpoint of the top of each bar with a dot and connect the dots with a smooth curve.

d. What does the curve seem to be?

e. Use the curve to estimate the mean value of the weights of the loaves of bread.

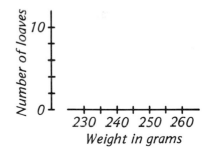

On discovering this, the student complained to the baker, who said he would correct the situation. The weights of the loaves bought on the 15 days that followed the complaint are listed below.

$$
\begin{array}{ccccc}
251 & 254 & 250 & 252 & 255 \\
253 & 251 & 257 & 250 & 250 \\
251 & 250 & 255 & 252 & 250 \\
\end{array}
$$

f. Make a frequency distribution of these weights, using the same intervals as in part b.

g. Make a histogram of the weights, labeling the axes as in part c.

h. On the basis of this evidence, how do you think the baker "corrected the situation"?

i. Explain.

LESSON 5

1. A common marketing practice is to offer a new product at an "introductory low price." In one experiment to find out if this increases sales in the long run, rolls of aluminum foil were introduced in seven stores of a large discount chain at a price of 59¢ and at seven other stores of the chain at a price of 64¢. After three weeks, the price in all the stores became 64¢.*

A graph of the sales for the first eight weeks is shown here.

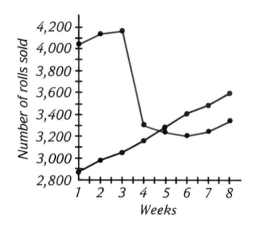

a. Which line do you think represents the sales of the foil introduced at 59¢?
b. What happened to its sales in the course of the eight weeks of the experiment?
c. What happened to the sales of the foil introduced at 64¢ in the course of the eight weeks of the experiment?

Similar patterns in the sales of other items were observed in additional experiments. In each case, the prices were not advertised and the customers did not know that the product was being sold in other stores at a different price. Those buying it at the lower price were not told that it was an "introductory price" and did not know that the price would go up after three weeks.

d. Why do you suppose that the item that started out with the lower sales each time did better in the long run?
e. Judging from these experiments, does the practice of introducing a new product at less than its normal price seem like a good idea?

---

*"Effect of Initial Selling Price on Subsequent Sales," by Anthony N. Doob, J. Merrill Carlsmith, Jonathan L. Freedman, Thomas K. Landauer, and Soleng Tom, Jr., *The Journal of Personality and Social Psychology*, 11(4):1969.

2. *Discrimination in the Wistful Vista School District**

The following table shows the numbers of men and women who applied for teaching positions one year in the Wistful Vista School District and the numbers who were hired.

Grade level	Men who applied	Men hired	Women who applied	Women hired
K–6	40	2	120	12
7–9	50	7	20	6
10–12	180	72	10	9
Total	270	81	150	27

From it, we see that 2 of the 40 men who applied for teaching positions in grades K–6 were hired:

$$\frac{2}{40} \times 100\% = \frac{200\%}{40} = 5\%$$

a. Make similar calculations and copy and complete the following table.

Grade level	Percentage of male applicants hired	Percentage of female applicants hired
K–6	5	▓▓▓
7–9	▓▓▓	▓▓▓
10–12	▓▓▓	▓▓▓
Total	▓▓▓	▓▓▓

b. Copy and complete the bar graph at the right comparing the percentages of men and women hired at each grade level.

This seems to show that the Wistful Vista School District practices discrimination in hiring its teachers.

c. Explain.

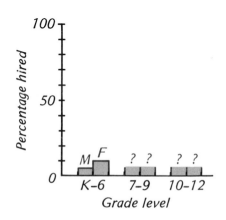

*This exercise is adapted from an example in *Winning with Statistics,* by Richard P. Runyan (Addison-Wesley, 1977).

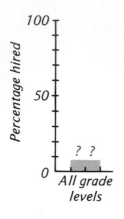

d. Copy and complete the bar graph at the left comparing the percentages of men and women who were hired at all grade levels combined.

e. What does this graph seem to show?

Lesson 6

1.   How many words do you suppose are in this book? The only way to find the *exact* number would be to count them all. If we are content with knowing the *approximate* number, however, we might make an estimate of it on the basis of a sample.

a. Determine, as accurately as you can, the number of words in this book.

b. Explain the procedure that you used to make your estimate.

2.   *The Case of the Missing Stockings**

The owner of a company that manufactured nylon stockings had reason to believe that about $1,000,000 worth of the stockings were being stolen from the factory each year. He brought in detectives, put recording devices on the machines, and had the supervisors questioned, but without success.

Finally, a psychologist hired by the owner discovered the following facts:

1. The amount of material used in making the stockings varied from one worker to another.
2. The annual output of the company was estimated from a test run by one of the machine operators.
3. This operator was the company's best worker.

On the basis of these clues, what do you suppose was happening to the missing stockings? Explain your reasoning.

---

*From *Sampling* (originally titled *Sampling in a Nutshell*), by Morris James Slonim (Simon & Schuster, 1960).

Topics in Topology

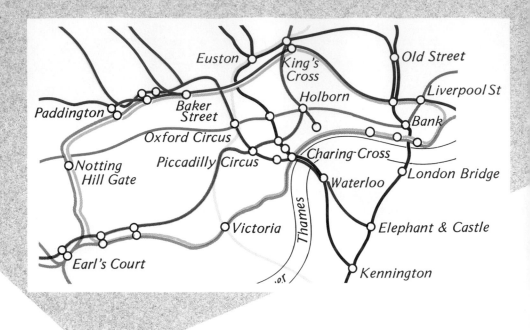

1

The Mathematics of Distortion

Beneath the city of London is an elaborate network of subways. Called the Underground, it consists of 279 stations connected by 410 kilometers of tracks.

Although the map above is an accurate representation of the subway lines and stations in central London, it is not the map used by passengers. For someone traveling on the Underground, all that matters is the way in which the stations are connected by the various lines. The map on the facing page, which shows these connections in a more straight-forward way, is the one posted in each Underground station.

It is easy to see how the second map was made. The artist treated the lines of the original map as if they were elastic bands and stretched and bent them into simpler shapes. The result is a figure having some properties different from the original one and some properties that have remained unchanged. The properties that have remained unchanged are *topological*.

602   Chapter 10: Topics in Topology

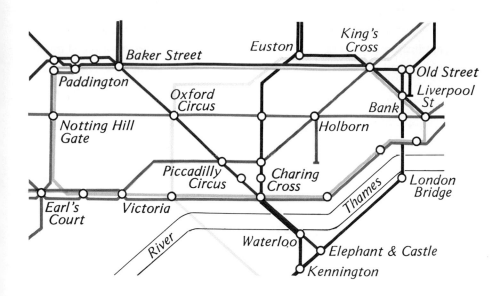

Topology, sometimes called "the mathematics of distortion," deals with very basic properties of geometric figures: properties that remain unchanged no matter how those figures are stretched or bent. Look, for example, at the figures below.

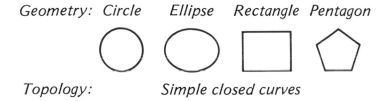

From the standpoint of geometry, each figure is different and has its own name. From the standpoint of topology, they are all alike. The figures are examples of *simple closed curves*.

A **simple closed curve** is a curve on which it is possible to start at any point and move continuously around the curve, passing through every other point exactly once before returning to the starting point.

The figures are also *topologically equivalent.*

Figures are **topologically equivalent** if they can be stretched and bent into the same shape without connecting or disconnecting any points.

Because topology deals with very basic ideas, its influence on other areas of mathematics has been immense. Even though it is a comparatively young branch of mathematics, it has proved to have many practical applications and, because of its many surprises, is often a source of much enjoyment.

EXERCISES

SET I

The figures below are from a book on fingerprints.

Ridge types

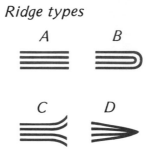

Two of these figures can be stretched and bent into the same shape without connecting or disconnecting any points.

   1. Which two figures are they?

   2. What are two figures that possess this property called?

The figures below were devised by Marvin L. Minsky and Seymour A. Papert of the Massachusetts Institute of Technology as a test of visual perception.

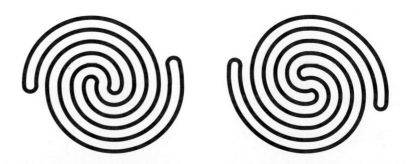

   3. Place tracing paper over this page and trace each figure. Are the figures topologically equivalent?

   4. What does each one consist of?

The first neon sign made its appearance at the Paris Motor Show in 1910. The symbols that are the easiest to make in neon are those that are topologically equivalent to a line segment. Examples are shown at the right.

5 7

Use the forms of the capital letters shown below to answer the following questions.

ABCDEFGHIJK
LMNOPQRSTUVWXYZ

5. Twelve of these letters are topologically equivalent to a line segment. Which letters are they?

Which letters are topologically equivalent to each of the following shapes?

6. (Two letters)   7. (Two letters)   8. (Two letters)   9. (Two letters)   10. (Four letters)

The following topological problem was presented in a lecture by the German mathematician August Ferdinand Moebius in 1840.*

A king with five sons stated in his will that after his death his land was to be divided into five regions so that each region shared some of its border (more than just a point) with each of the others. Can the will be carried out?

Look at the adjoining map.

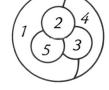

11. Does region 1 share some of its border with each of the other regions?

12. Does region 2 share some of its border with each of the other regions?

13. Does region 3 share some of its border with each of the other regions?

14. Does the map solve the problem? Explain why or why not.

15. Can you draw a different map that solves the problem?

Famous Problems of Mathematics, by Heinrich Tietze (Graylock Press, 1965).

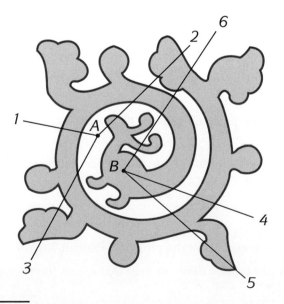

SET II

Even though this figure looks quite complex, in topology it is called a *simple closed curve.*

1. Use the definition of simple closed curve in this lesson to explain why.

2. Which of the figures below are simple closed curves?

Cardioid   Spiral   Octagon   Lemniscate

A basic idea of topology is that a simple closed curve in a plane divides the plane into exactly two regions: an inside and an outside. This fact is called the *Jordan Curve Theorem*, after the nineteenth-century French mathematician Camille Jordan.

3. Which of the four figures in exercise 2 divide the plane into exactly two regions?

This simple closed curve appeared on a medieval wall tile.*

Handbook of Regular Patterns, by Peter S. Stevens (MIT Press, 1980).

4. Is point A inside or outside the curve?

5. How many times does each straight line from point A to the outside cross the curve?

6. Is point B inside or outside the curve?

7. How many times does each straight line from point B to the outside cross the curve?

Whether a point is *inside* or *outside* a simple closed curve can be found from the number of times that a straight line joining that point to the outside of the curve crosses the curve.

8. How?

9. Determine whether each of the points C, D, and E is inside or outside the curve below.

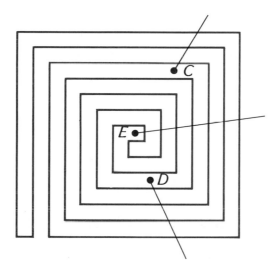

An old puzzle titled "Water, Gas, and Electricity" is to draw lines connecting the three utilities to each of three houses without any of the lines crossing each other.*

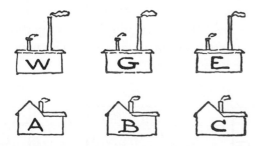

Amusements in Mathematics, by Henry Ernest Dudeney (Nelson and Sons, 1917; Dover, 1958).

10. Copy the figure below showing the utilities and houses and try to connect them as indicated.

The figure below illustrates a partial solution to the puzzle.

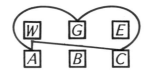

11. What do the four connections shown form?

It is now impossible to connect one of the utilities to one of the houses without crossing one of these connections.

12. Which utility cannot be connected to which house?

13. Why is this impossible?

Regardless of how the connections are made, a situation of this sort always arises.

14. What does this indicate about the puzzle?

Topology is used in the design of the electric circuits used in microelectronics. The connections between the parts of the circuits cannot cross each other or the circuits will short.

The following figures represent electric circuits. Copy each one and then try to find a way to connect, within each figure, each pair of squares labeled with the same letter so that none of the connections cross. If you think a figure does not have such a circuit, explain why not.

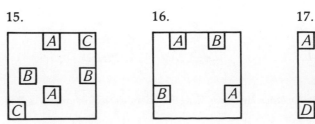

15.

16.

17.

SET III

Copyright Universiteitsmuseum, Utrecht; photograph by Jac. P. Stolp

This strange picture from the eighteenth century is titled *Two Men Boxing.* It consists of a painting with a mirror in the shape of a cone placed at its center. What you see in the center of the picture is the reflection of the painting in the mirror.

1. What topological relation do the picture and its image in the mirror seem to have?

2. As the artist distorted the picture to produce the intended image, were all parts of it stretched and bent by the same amount? Explain.

3. What does the mirror do to the picture in addition to removing the distortion?

More examples of topological distortions in art are included in the book *Hidden Images* by Fred Leeman (Abrams, 1976) and in "Anamorphic Art," Chapter 8 of *Time Travel and Other Mathematical Bewilderments* by Martin Gardner (W. H. Freeman, 1988).

KONINGSBERGA

A Das Schlofs.   E Saghumfse Kirch.   I Das Closter.
B Alt Stener Kirch. F Die Dom kirch.   K Haberbergsche Kirch.
C S. Niclaus.   G Das Collegium.   L Haber kirch.
D S. Barbara.   H Rathhaus im Knypho M Hospital.

From *Topographia Prussiae et Pomerelliae* by M. Zeiller. Frankfurt. c. 1650.

The Seven Bridges of Königsberg: An Introduction to Networks

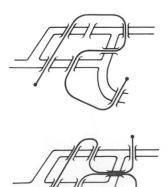

One of the problems that led to the development of topology is about seven bridges in the city of Königsberg in old Germany. The drawing above, from a book published in about 1650, shows that the center of Königsberg was on an island in the middle of a river. The river flowed around the island from the left and, on the right, separated into two branches. Seven bridges made it possible to travel from one part of the city to another.

The citizens of Königsberg wondered whether it was possible to travel around the city and cross each of the seven bridges exactly once. Everyone who tried it ended up either skipping or recrossing at least one bridge. Most people came to the conclusion that a path crossing every bridge exactly once did not exist, but they did not know why.

Eventually, the problem of the Königsberg bridges came to the attention of the Swiss mathematician Leonhard Euler. Euler found the puzzle interesting and, in an article published in 1736, proved that it could not be solved. He began by representing each area of land with a capital letter and each bridge with a small letter. The map can be simplified even further by representing the four areas of land by four points and the seven bridges by seven lines connecting the points. This is shown in the diagram at the right, which we will call a *network.*

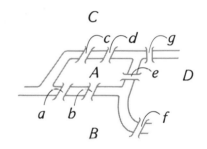

| A **network** is a figure consisting of points, called *vertices,** connected by lines, called *edges.*

The problem of the Königsberg bridges is equivalent to the problem of drawing this network without retracing any edge or taking the pencil off the paper. Euler showed that whether or not this problem has a solution depends on the *degree* of each vertex of the network.

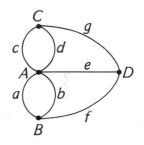

| The **degree** of a vertex of a network is the *number* of edges that meet at it.

In the network for the bridge problem, five edges meet at vertex A and so the degree of vertex A is 5. Three edges meet at each of the vertices B, C, and D, and so the degree of each of these vertices is 3. Euler's results showed that this network, which has four vertices of an odd degree, cannot be drawn with one continuous stroke of a pencil.

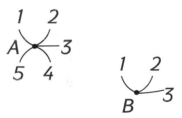

EXERCISES

SET I

A network can be "traveled" if it is possible to draw it without retracing any edge or taking your pencil off the paper. It is easy to see that the first network below can be traveled by starting at any vertex. The second network can be traveled only if you start on one of the vertices marked with an arrow.

Networks   *Example paths*

---

*Vertices is the plural of vertex.

Each of the networks below is topologically equivalent to a simple closed curve.

1. Do you think that every network topologically equivalent to a simple closed curve can be traveled?

All the vertices of such networks have the same degree.

2. What is meant by the *degree* of a vertex?
3. What is the degree of each vertex of the three networks above?

Each of these networks is topologically equivalent to a line segment.

4. Do you think that every network topologically equivalent to a line segment can be traveled?
5. How many vertices of degree 1 do such networks have?
6. What is the degree of the other vertices in such networks?

Each of these networks has four vertices. Notice that the degree of each vertex of network A is 2, so it has 4 "even" vertices.

7. Copy and complete the following table for these networks.

Network	Number of even vertices	Number of odd vertices	Can the network be traveled?
A	4	0	Yes
B	▨	▨	▨
C	▨	▨	▨

(Continue the table to show all nine networks.)

8. What do all the vertices of the networks that cannot be traveled have in common?

Each of the following networks also has four vertices.

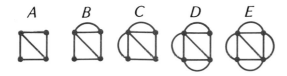

A   B   C   D   E

9. Which networks have two even vertices and two odd vertices?

10. Which have four even vertices?

11. Which have four odd vertices?

12. Which network(s) cannot be traveled?

The maps below show some different possible nonstop flights between the Hawaiian Islands. Place your paper over each map and try to travel each flight network. (Whether or not you are able to do this may depend on where you start.)

13.   *Route map A*

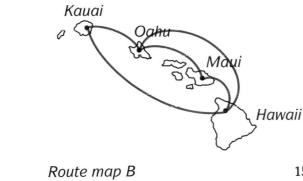

14.   *Route map B*   15.   *Route map C*

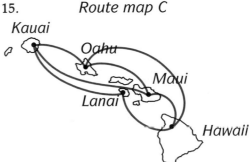

16. How does route map B differ from route map A?

17. How does route map C differ from route map B?

18. Which maps show flight networks that can be traveled?

19. Does adding a vertex to a network seem to have any effect on whether it can be traveled?

20. Does adding an edge to a network seem to have any effect on whether it can be traveled?

Set II

In 1809, the French mathematician Louis Poinsot noted that the figure shown at the left cannot be drawn in one continuous stroke of a pencil.

1. What do the numbers at the vertices of this network represent?

2. How many vertices of even degree does it have?

3. How many vertices of odd degree does it have?

Place tracing paper over the networks below and try to travel each one. (Some of the networks can be traveled only if you start from certain vertices.) Keep a record of those that can be traveled and those that cannot.

4.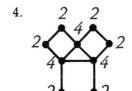

5.

6.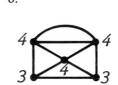

7.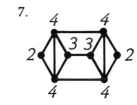

8.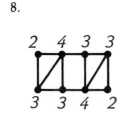

9.

10.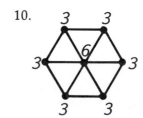

11.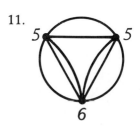

12.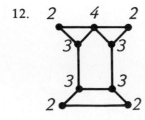

13.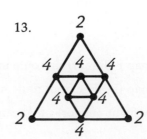

14. Copy and complete the following table for these networks.

Network	Number of even vertices	Number of odd vertices	Can the network be traveled?
4	9	0	Yes
5	▓▓▓	▓▓▓	▓▓▓

(Continue the table to show all ten networks.)

Of the ten networks, six can be traveled. If your table shows fewer than six, try again to draw each of the networks that you have marked "no."

On the basis of the information in your table, does it seem that a network can be traveled if

15. all of the vertices are even?

16. it has more even vertices than odd vertices?

17. it has two odd vertices?

18. it has more than two odd vertices?

19. Write a conclusion about whether a network can or cannot be traveled based on your answers to exercises 15–18.

The brown dots on the map at the left below show the locations of parking meters along several city streets. The person who has the job of collecting money from these meters has to travel each edge of the network shown at the right of the map.

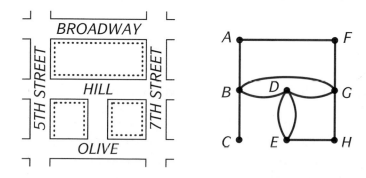

20. Place tracing paper over the network and try to travel it.

21. Where do you think would be the best places for the route to begin and end?

22. In what way do these vertices of the network differ from the rest of the vertices?

Set III

An amusement park has a fun house in the form of a maze with the floor plan shown below. The maze contains a room with trick mirrors, labeled M in the figure, together with seven smaller rooms.

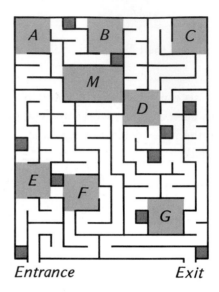

Entrance   *Exit*

1. To show how the eight rooms of the maze are connected, draw a network in which the entrance, exit, and the rooms are represented as vertices and the possible paths between rooms are represented as edges. For example, there is one path connecting rooms A and B, and so an edge would be drawn between vertices A and B of your network, as shown here.

Use your network to answer the following questions.

2. Is it possible to enter the maze, go to the room of mirrors, and exit without traveling any of the corridors more than once and without entering any room more than once? If it is, through what sequence of rooms could the path go?

3. Is it possible to travel the maze as described in exercise 2 in more than one way?

After Topographia Prussiae et Pomerelliae by M. Zeiller. Frankfurt. c. 1650.

KONINGSBERGA

Euler Paths

It was in 1736 that Leonhard Euler proved that it was impossible to walk through the city of Königsberg and cross each of its seven bridges exactly once. In 1875, an eighth bridge was built. It is not difficult to show that the addition of this new bridge made it possible to travel the network.

Networks representing the city before and after the new bridge was built are shown below. The first network contains four odd vertices.

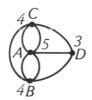

Adding an edge between vertices B and C to form the second network changes two of these vertices to even. One way to travel the new network is to start at vertex D and follow the path shown in the adjoining figure, ending at vertex A. Such a path is called an *Euler path.*

> An **Euler path** is a continuous path that passes along every edge of a network exactly once.

Notice that the Euler path for the network with eight bridges begins at one of the odd vertices and ends at the other one. To see why, look at what would happen if we did not start at vertex D. In traveling the network we must go along each edge, including the three that end at D, exactly once. The first time we come to D, it will be along one of these three edges and we can leave along either of the other two. This leaves one edge untraveled and when we cover it we will be "stuck" at D because there are no edges remaining along which we can leave. In other words, if our trip does not *begin* at vertex D, it must *end* there.

Reasoning in the same way, we can show that the path must also either begin or end at the other odd vertex, A. Because a path has exactly one beginning and one ending, it follows that it must begin at one of the odd vertices and end at the other. It also follows that a network with more than two odd vertices cannot be traveled in a single trip. Euler proved that every network having no more than two odd vertices has an Euler path.

EXERCISES

SET I

Two of the networks below have Euler paths and one does not.

1. What is meant by *Euler path?*

2. Which network has an Euler path that ends at the same vertex from which it begins?

3. Which network has no odd vertices?

4. Which network has an Euler path that ends at a different vertex from which it begins?

5. Which network has two odd vertices?

6. Which network does not have an Euler path?

7. How many odd vertices does it have?

On the basis of your answers to exercises 2 – 7, write rules for identifying each of the following. State each rule as a complete sentence.

8. A network that has an Euler path that returns to its starting point.

9. A network that has an Euler path that does not return to its starting point.

10. A network that does not have an Euler path.

This figure is the symbol for the Olympics.

11. Does it have an Euler path?

12. How can you tell without trying to draw it?

13. Place tracing paper over the symbol and try to travel it in a single trip.

14. Does whether or not you are able to travel the symbol depend on the vertex at which you start?

This figure represents the floor plan of a small art museum consisting of a foyer and three exhibit rooms. Pictures are hung on all four walls of each exhibit room.

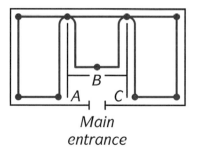

Main
entrance

15. Is it possible to start at door A and walk along each of the walls with the paintings exactly once? If so, which door — A, B, or C — would you end at?

16. Is it possible to start at door B and walk along each of the walls with the paintings exactly once? If so, which door — A, B, or C — would you end at?

17. How can you tell from the vertices of the network whether or not the walks described in exercises 15 and 16 are possible?

To draw network A above, two trips are necessary. In other words, the figure can be drawn without retracing any edge if the pencil is removed from the paper once. One way of doing this is shown in the second figure. The leap of the pencil is equivalent to adding an edge. By adding the edge in color, as in the third figure, we can connect two odd vertices and thus make both of them even.

18. Find the fewest number of trips necessary to travel each of the networks below by drawing them.

B

C

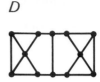

D

Network	Number of odd vertices	Least number of trips
A	4	2
B		
C		
D		

19. Copy and complete the table at the left for these networks.

20. State a rule for how the number of trips necessary to travel a network is related to the number of odd vertices that it has.

Network A at the left separates the paper into three regions: two are inside the network and the third is the region outside. Network B has just one region: the one outside.

The number of regions of a network is related to the numbers of vertices and edges that it contains.

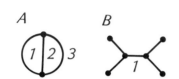

A   *B*

21. Copy and complete the following table for the nine networks at the left and below.

Network	Number of regions	Number of vertices	Number of edges
A	3	2	3
B	1	6	5

(Continue the table to show all nine networks.)

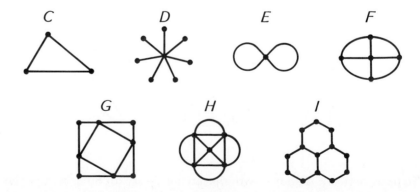

C   *D*   *E*   *F*

G   *H*   *I*

22. Write a formula relating the number of regions, R, in a network, to the number of vertices, V, and the number of edges, E. (*Hint:* Compare $R + V$ with E.)

This photograph shows cells of soap film between two sheets of glass. Four of the cells are pictured in the network at the right.

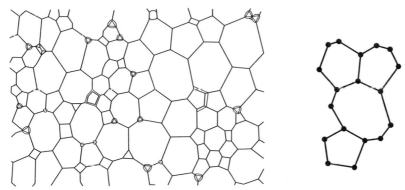

Institut fur leichte Flachentragwerke, Stuttgart, Professor Frei Otto.

23. Count the numbers of regions, vertices, and edges of this network.

24. Does the formula you wrote for exercise 22 fit this network? Show why or why not.

Set II

In his article on the Königsberg bridges, Euler considered another bridge problem, which is illustrated below.*

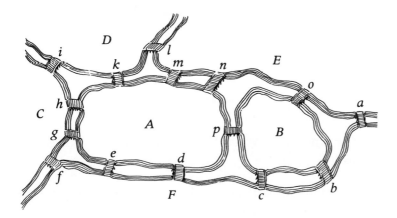

Two islands, A and B, are surrounded by water that leads to four rivers. Fifteen bridges cross the rivers and the water surrounding the islands. Is it possible to make a trip that crosses each bridge exactly once?

*Euler's article is included in *Mathematics: An Introduction to Its Spirit and Use,* with introduction by Morris Kline (W. H. Freeman and Company, 1979).

1. Draw a network in which six vertices represent the regions of land and 15 edges represent the bridges.

2. What do you notice about the vertices of the network?

3. Can a trip be made that crosses each bridge exactly once? If so, can it begin and end on any of the regions of land?

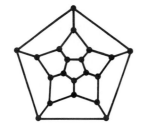

In 1857, the Irish mathematician William Rowan Hamilton invented a game based on the network at the left. One object of the game was to find a path that goes through every *vertex* of the network exactly once and ends at the vertex at which it started. Such a path is called a *Hamilton path,* in contrast with an *Euler path,* which goes along every *edge* of a network. The figures below show a network that has both an Euler path and a Hamilton path.

Network

An Euler path

A Hamilton path

4. Which of the networks below have Euler paths?

A

B

C

D

5. Which networks have Hamilton paths?

6. Can a network have an Euler path without having a Hamilton path?

7. Can a network have a Hamilton path without having an Euler path?

Map A

These maps show streets of a city in which there is a newspaper rack at every intersection.

8. Would someone responsible for keeping the racks filled with newspapers be more interested in an Euler path or a Hamilton path connecting them?

Map B

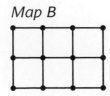

9. Why?

10. Does either map have such a path? If so, sketch the path(s).

Look again at the network for Hamilton's game.

11. Does it have an Euler path? Explain why or why not.

12. Place tracing paper over the figure and see if you can discover a Hamilton path for it.

Set III

This advertisement of the Bell Telephone Laboratories concerns finding the shortest Hamilton path for a given network. According to the advertisement, this problem is "important in many areas of modern business and technology, where 'shortest path' may really mean the least hook-up wire, travel time, or transmission power."

Report from
**BELL
LABORATORIES**

Found: A rapid route to the shortest path

The critical feature of Shen Lin's method is its speed; it makes many good approximations in a reasonable time and selects the best.

To make one approximation, the computer chooses a "starting path" at random. It removes three links of this path (thus breaking it into three sections — see figures) and connects the sections differently to see if a shorter path results. If not, it systematically removes other combinations of three links in the original random path, until all combinations have been tried. But, whenever such a reconnection *does* produce a shorter path, it takes this as a *new* starting path, and begins the series of breaks again. One "approximation" is completed when no further improvement results from such breaking and reconnection.

In the same way — beginning each time with a new and different "starting path" — many additional approximations are found. They usually have some path sections in common; it simplifies the problem to assume that these are part of the absolute minimum path. So, they are routinely incorporated into every new starting path and no longer broken. This speeds computation and the time that's saved is used to find even more approximations.

In general, using a high-speed digital computer, 100 approximations take about $0.75n^3$ milliseconds (n = number of points). For a typical 40-point problem, experiments indicate that about one out of 16 approximations will be the actual minimum solution; for 60 points, about one out of 64. So, if we find 300 approximations in a 60-point problem (roughly eight minutes on a computer) there is a high probability that one of these is the shortest possible.

Start with a random path...

Break it into three sections...

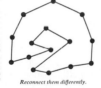

Reconnect them differently.

What is the shortest path through a number of points, touching each just once and ending at the starting point? This "traveling salesman problem" is important in many areas of modern business and technology, where "shortest path" may really mean the least hook-up wire, travel time, or transmission power.

It might seem that the problem could be solved by measuring all paths and taking the shortest but, even with a computer, this is a colossal task. At a million paths per second, for instance, it would take several billion years to compute and compare all paths in a 25-point problem! Shortcut methods have been devised, but they are still too slow when, say, 60 points are involved. In practice, approximate solutions (almost-shortest paths) are found largely through the educated judgments of engineers looking at graphs or maps . . . or for certain limited problems, through special computer programs.

Now, mathematician Shen Lin of Bell Telephone Laboratories has developed a new way of getting good approximate solutions to problems of up to 145 points. Because his method is fast, it is possible to find many such approximations. It is then easy to pick the shortest of these. Often (see left), this is the absolute minimum. If not, it is at least short enough for most engineering purposes.

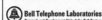

Bell Telephone Laboratories
Research and Development Unit of the Bell System

By permission of Bell Telephone Laboratories, Inc.

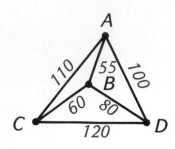

The four points in the figure at the left represent four vertices of a network that have to be connected.

1. Place tracing paper over the figure and connect the vertices with edges to form a Hamilton path.

2. How many different Hamilton paths connecting the four vertices do you think are possible? Draw each one.

Suppose the distances between the pairs of vertices are as follows:

$$AB = 55 \quad AD = 100 \quad BD = 80$$
$$AC = 110 \quad BC = 60 \quad CD = 120$$

3. What is the length of the *shortest* Hamilton path connecting the four vertices? Show how you found it.

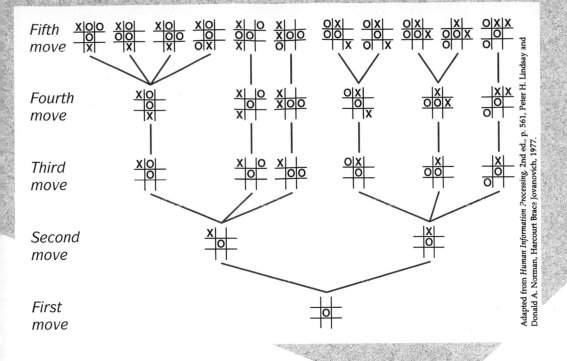

Fifth
move

Fourth
move

Third
move

Second
move

First
move

Adapted from *Human Information Processing*, 2nd ed., p. 561, Peter H. Lindsay and Donald A. Norman, Harcourt Brace Jovanovich, 1977.

LESSON

4

Trees

One of the simplest of games is tic-tac-toe. Because of its simplicity, a child who plays the game for awhile usually learns how to avoid losing.

Although there are nine spaces in which to make the first move, the number of reasonable choices for successive moves rapidly becomes smaller. For example, if the first player chooses the center space, the second player has only two choices that are actually different: either a corner space or a side space. Possibilities for the first five plays in the game are shown in the diagram above.*

The basic structure of this diagram is shown in the network at the right. Mathematicians call such a network a *tree*.

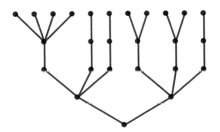

| A **tree** is a network that does not contain any simple closed curves.

*Adapted from a figure in *Human Information Processing*, by Peter H. Lindsay and Donald A. Norman (Harcourt Brace Jovanovich, 1977.)

A real tree (the kind birds build nests in) is like a topological tree in that it consists of a trunk, branches, and twigs that ordinarily do not grow back together. A cat on one branch of a tree cannot climb around the rest of the tree and come back to where it started without returning along the same branches.

The examples below include some networks that are trees and some that are not.

Trees   *Not trees*

There is just one tree that contains two vertices and one tree that contains three. They are shown as figures A and B below. There are two different trees that contain four vertices, and they are shown as figures C and D.

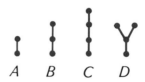

A   B   C   D

The longest path on any of these trees is clearly on the one labeled C: it consists of three edges.

> The **diameter** of a tree is the largest number of edges that a path on the tree can have.

The diameters of the four trees above are 1, 2, 3, and 2, respectively.

Tree networks appear in a wide variety of subjects. You have already seen trees used in mathematics in solving problems in counting and in probability.* Mapmakers draw complicated trees to represent rivers and their tributaries. The map on the next page shows the Mississippi River and some of the other rivers that flow into it. Genealogists use trees to show the relationships between members of different generations of a family. The process used by the postal service to sort mail can be represented by a tree: the ZIP code identifies its main branches. Tree diagrams are drawn by chemists to show the arrangements of atoms in molecules.

---

*See, for example, the trees on pages 403 and 458.

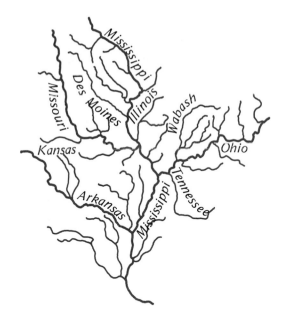

EXERCISES

SET I

Four generations of a person's family tree are shown in this figure.

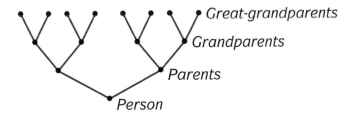

1. How many vertices does this family tree have?

2. How many edges does it have?

This figure illustrates a typical arrangement of stems of a cactus of the genus *Hatiora*.

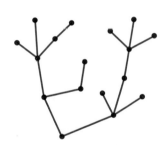

3. How many vertices does this cactus tree have?

4. How many edges does it have?

5. How is the number of edges in each of the trees for exercises 1–4 related to the number of vertices?

The 10 digits are shown below as they appear on digital clocks.

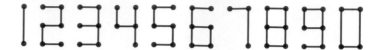

6. Which of these figures are trees?

7. Copy and complete the following table for the 10 digits as they appear on digital clocks.

Digit	Number of vertices	Number of edges	Is it a tree?
1	2	1	Yes

(Continue the table to show all ten digits)

8. Does the pattern you noticed in exercise 5 apply to all of the digits whose digital clock symbols are trees?

9. Write a formula for the pattern, letting V represent the number of vertices of a tree and E represent the number of edges.

10. Try to draw an example of a tree for which this formula is not true.

These networks show the arrangement of the atoms in molecules of nitroglycerin and saccharin.

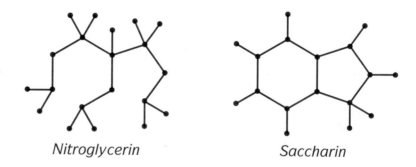

Nitroglycerin   *Saccharin*

11. Is the formula you wrote for exercise 9 true for the nitroglycerin molecule? Show why or why not.

12. Is it true for the saccharin molecule? Show why or why not.

13. How many edges of the network for the saccharin molecule would have to be removed in order for it to become a tree?

14. If these edges were removed, would the formula be true for the resulting tree?

The diameter of a tree is the largest number of edges that a path on the tree can have. For example, the diameter of this tree is 3 because no path on it has more than three edges.

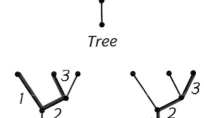

Tree

15. Are there any paths for this tree other than the ones shown in the examples that also contain three edges?

Many snowflakes have the shapes of trees. Below is a photograph of one taken through a microscope, together with a tree diagram of its structure.

Example paths

16. What is the diameter of this tree?

The trees below are adapted from figures in *Patterns in Nature* by Peter S. Stevens.* Notice that each tree has the same number of vertices.

"Spiral"　　　*"Explosion"*　　　*"Branching"*

17. How many vertices does each tree have?
18. What is the diameter of "Spiral"?
19. What is the diameter of "Explosion"?
20. What is the diameter of "Branching"?

---

*Little, Brown, 1974.

The Bettman Archive.

Arthur Cayley.

SET II

In 1875, the English mathematician Arthur Cayley wrote a paper on trees that proved to be particularly useful in chemistry.

In it, he counted the numbers of possible trees having different numbers of vertices. All of the different trees having five or fewer vertices are shown in the table below. An example of a molecule having each structure is also shown.

	Tree	Example molecule
A	•—•	Oxygen
B	•—•—•	Water
C	•—•—•—•	Hydrogen peroxide
D		Ammonia
E	•—•—•—•—•	Calcium hydroxide
F		Nitric acid
G		Carbon tetrafluoride

Notice from the table that there is only one tree having two vertices. How many different trees are there that have

1. three vertices?

2. four vertices?

3. five vertices?

Although the tree shown here may seem different from any of the trees in the table above, it is considered to be the same as tree B. This is because it and tree B both have three vertices and are topologically equivalent (they can be stretched and bent into the same shape).

Each of the following trees has four vertices.

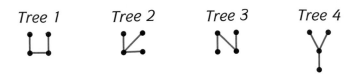

Tree 1   *Tree 2*   *Tree 3*   *Tree 4*

Which of these trees are topologically equivalent to

4. tree C?

5. tree D?

Each of the following trees has five vertices.

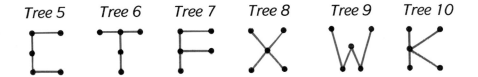

Tree 5   *Tree 6*   *Tree 7*   *Tree 8*   *Tree 9*   *Tree 10*

Which of these trees are topologically equivalent to

6. tree E?

7. tree F?

8. tree G?

Cayley showed that there are six different trees that have six vertices.

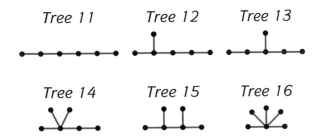

Tree 11   *Tree 12*   *Tree 13*

Tree 14   *Tree 15*   *Tree 16*

To which of these six trees is each of the following trees topologically equivalent?

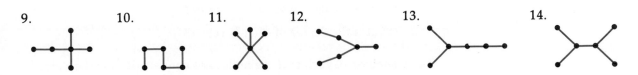

9.   10.   11.   12.   13.   14.

Suppose that two trees have the same number of vertices.

15. If they have the same diameter, are they necessarily topologically equivalent?

16. If they are topologically equivalent, do they necessarily have the same diameter?

Cayley showed that there are eleven different trees that have seven vertices.

17. Make drawings of as many of the eleven trees as you can.

SET III

Chemists use networks called *structural formulas* to show the ways in which atoms are linked together in molecules. Photographs of models of several hydrocarbon molecules, along with trees showing their structures, are shown below. The vertices of the trees labeled C represent carbon atoms and the vertices labeled H represent hydrogen atoms.

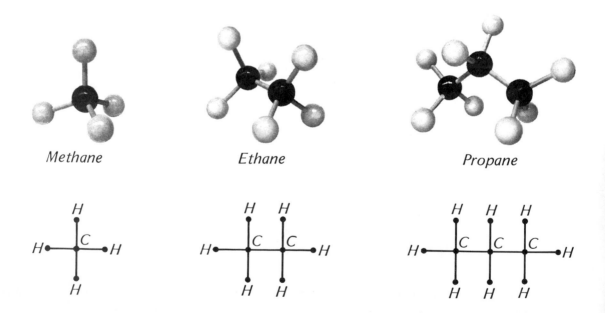

Methane   Ethane   Propane

1. What is the degree of each C vertex?

2. What is the degree of each H vertex?

The "skeleton" of each of these molecules consists of its carbon atoms and the bonds between them. Here is the skeleton of the propane molecule.

C   C   C

There are two different butane molecules containing four carbon atoms each. Their skeletons are:

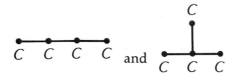

and

3. Copy these figures and then add enough edges and vertices representing hydrogen atoms to them so that the degrees of the vertices are the numbers you gave as the answers to exercises 1 and 2.

There are three different pentane molecules containing five carbon atoms each.

4. Draw tree diagrams of their skeletons and then complete each tree to show the rest of the molecule.

5. Does each of your three drawings for exercise 4 contain the same number of hydrogen atoms?

6. Refer to the figures on the previous page and your drawings to copy and complete the following table.

Hydrocarbon	Number of carbon atoms	Number of hydrogen atoms
Methane	1	4
Ethane	▨	▨
Propane	▨	▨
Butane	▨	▨
Pentane	▨	▨

Notice that a *pentane molecule* contains five carbon atoms and that a *pentagon* contains five vertices.

7. How many carbon atoms do you think an octane molecule has?

8. How many hydrogen atoms do you think an octane molecule has?

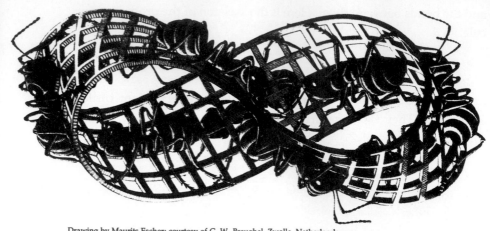

Drawing by Maurits Escher; courtesy of G. W. Breughel, Zwolle, Netherlands

5

The Moebius Strip and Other Surfaces

There is something rather remarkable about the band pictured in the woodcut by Maurits Escher shown above. Look carefully at the procession of ants crawling around it. Escher wrote:

> An endless ring-shaped band usually has two distinct surfaces, one inside and one outside. Yet on this strip nine red ants crawl after each other and travel the front side as well as the reverse side. Therefore the strip has only one surface.*

This band, called a *Moebius strip,* is named after one of the pioneers in topology, the German mathematician and astronomer August Ferdinand Moebius. The idea of the Moebius strip is so simple that it is rather surprising that it did not become widely known until 1865, when Moebius wrote a paper about its properties.

---

*Escher's description of *Moebius Strip II* in *The Graphic Work of M. C. Escher* (Meredith Press, 1967).

The two adjoining figures show an ordinary "belt-shaped" loop and a Moebius strip. The belt-shaped loop has two sides and two edges; it is impossible to travel on this loop from one side to the other without crossing over an edge.

The Moebius strip contains a "half-twist," so-called because it can be made by turning one end of a rectangular strip through an angle of 180° (half of 360°) before taping it to the other end. As a result of this half-twist, the Moebius strip has only one side and one edge. The strip's one-sidedness gives it a number of strange properties that are easily discovered by experiment.

Some of these properties have been put to practical use. Physicists have used the Moebius strip to help explain the interactions of atomic particles. The B. F. Goodrich Company has patented a conveyor belt in the shape of a Moebius strip — the belt lasts longer because both "sides" are actually one and receive equal wear. The advertisement below reveals that the Moebius strip has been put to use in the design of electronic resistors.

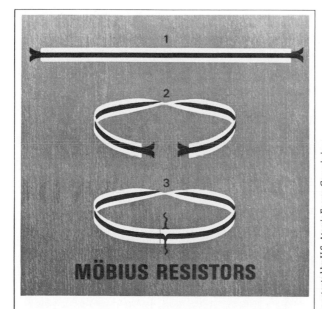

MÖBIUS RESISTORS

The Möbius loop has fascinated mathematicians and parlor magicians for years. At Sandia, researchers recently discovered that it also forms the basis for excellent nonreactive resistors. An elementary Möbius resistor can be made by adhering two 1-foot strips of aluminum tape to opposite sides of a dielectric ribbon (1), twisting these a half turn (2) and joining their ends. Current leads soldered to opposite surfaces of the loop (3) complete the unit. The design formulas governing this basic configuration permit endless variations in performance, size, and shape. Besides being nonreactive, Möbius resistors will not couple to metallic objects, external fields, or themselves, nor will handling, folding or winding disturb the balance.

SANDIA CORPORATION

ALBUQUERQUE, NEW MEXICO / LIVERMORE, CALIFORNIA

Courtesy of Sandia Laboratories; Moebius resistor patented by U.S. Atomic Energy Commission

The properties of the Moebius strip also make it a source of amusement in a wide variety of entertainment, including magic tricks, science fiction, and even sports. An acrobatic trick performed by free-style skiers who make a twist while doing a somersault is now commonly known as the Moebius flip!*

EXERCISES

SET I
EXPERIMENT: THE MOEBIUS STRIP

The best way to understand the properties of the Moebius strip is to make one and discover them yourself. Before making a Moebius strip, however, we will experiment with an ordinary loop.

Cut two strips, each as long as possible and 1 inch or 2 centimeters wide, from graph paper. Take one of the strips, make it into an ordinary "belt-shaped" loop and tape the two ends together. Do the same thing with the other strip so that you have two ordinary loops.

Put your pencil down midway between the edges of one of the loops and draw a line down its center, continuing the line until you return to the point at which you started.

1. How many sides of the loop is the line on?

2. How many sides does the loop have?

Cut the loop along the line you have drawn.

3. What is the result?

Cut the other loop that you made parallel to and about one-third of the way from the edge. Cut all the way around the loop.

4. What is the result?

5. How do they compare in width?

6. How do they compare in length?

Next, make two Moebius strips by doing the following.

---

*Examples of recreational applications of the Moebius strip can be found in the chapter "Moebius Bands" in *Mathematical Magic Show*, by Martin Gardner (Knopf, 1977).

Cut two more strips of the same size that you made at the beginning of this experiment. Take one of the strips, make it into a loop and turn one end over before taping the two ends together. The result, a loop with a half-twist, is a Moebius strip. Do the same thing with the other strip so that you have two Moebius strips.

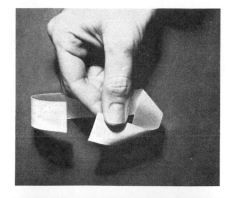

Put your pencil down midway between the edges of one of the Moebius strips and draw a line down its center, continuing the line until you return to the point at which you started.

7. How many times does the line go around the strip before returning to the original point?

8. How many sides of the strip does the line seem to be on?

9. In drawing the line, you never crossed over the edge to get to the other side; so how many sides does the strip have?

10. If you tried to paint just one side of the strip, what would happen?

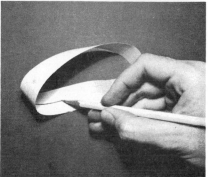

Cut the band along the line you have drawn.

11. What is strange about the result?

Put your pencil down midway between the edges of the resulting strip and draw a line down its center, continuing the line until you return to the point at which you started.

12. How many sides of the strip is the line on?

13. How many sides does the strip have?

14. Is it a Moebius strip? Explain why or why not.

Cut the new band along the line that you drew.

15. What is the result?

Cut the other Moebius strip that you made parallel to and about one-third of the way from the edge. When you have cut all the way around the loop you will find that you are across from the point at which you started. Continue cutting, staying the same distance from the edge as before, until you come back to where you began.

16. What is the result?

17. How do they compare in width?

18. How do they compare in length?

19. Which one is a Moebius strip?

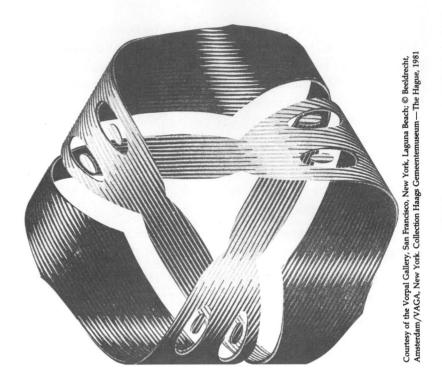

Courtesy of the Vorpal Gallery, San Francisco, New York, Laguna Beach; © Beeldrecht, Amsterdam/VAGA, New York. Collection Haags Gemeentemuseum—The Hague, 1981

This woodcut of a Moebius strip was created by Escher several years before the one shown at the beginning of this lesson.

20. What property of a Moebius strip that you discovered in the preceding exercise does it illustrate?

Set II
Experiment: Other Surfaces

By doing the Set I exercises, you discovered some of the properties of the Moebius strip, a band that contains a half-twist. The experiment that follows will reveal some of the properties of surfaces with more than one half-twist.

Cut from graph paper six more strips of the same size as those made for Set I. Make two of the strips into bands having two half-twists. Draw a line down the center of one of the bands.

1. Would it be possible to paint just one side of this band?

2. How many sides does a band with two half-twists in it have?

Cut the band along the line.

3. What is the result?

4. How do they compare in width and length?

Cut the other band one-third of the way from an edge.

5. What is the result?

6. How do they compare in width and length?

Make two additional strips into bands having three half-twists. Draw a line down the center of one of the bands.

7. How many sides does a band with three half-twists in it have?

Cut the band along the line.

8. What is the result? (Look carefully. It is unlike any of the previous results.)

Cut the other band one-third of the way from an edge.

9. What is the result? (The result will be so tangled up that you will have to pull at it a bit and study it closely in order to figure out what it is.)

Make another strip into a band having four half-twists.

10. How many sides does it have?

Cut the band down the center.

11. What is the result?

Make the last strip into a band having five half-twists. Cut the band down the center.

12. What is the result?

Answer the following questions on the basis of the results of both experiments.

13. How is the number of sides that a band has related to the number of half-twists in it?

14. What do you think would be the result if a long band with 100 half-twists in it were cut down the center?

15. What do you think would be the result if a long band with 101 half-twists in it were cut down the center?

Set III
Experiment: An Antitwister Principle

Tape one end of a strip of paper of about the same dimensions as those used in Sets I and II of this lesson to your desk. Then, beginning with the free end level, give the free end four half-twists counterclockwise as shown in this diagram. Now hold the free end down on the desk with your finger as shown in the figure at the left below.

Free end   *Taped-down end*

1. Do you think that it is possible to untangle the strip as long as both ends remain fixed in the positions shown?

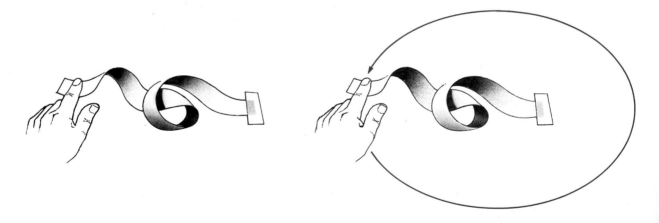

Keeping the free end pointing to the left, slide it with your finger in a circle counterclockwise around the taped end as shown in the second figure above.

2. What happens to the strip?

This principle has been applied in an invention that makes it possible to transmit electricity from a fixed source to something that is turning.*

---

*The device, patented by D. A. Adams, is described and illustrated in "The Amateur Scientist" by C. L. Stong, *Scientific American*, December 1975.

The Moebius strip on a Brazilian postage stamp

Summary and Review

In this chapter we have studied some ideas in topology.

The mathematics of distortion *(Lesson 1).* A *simple closed curve* is a curve on which it is possible to start at any point and move continuously around the curve, passing through every other point exactly once before returning to the starting point.

Figures are *topologically equivalent* if they can be stretched and bent into the same shape without connecting or disconnecting any points.

The Jordan Curve Theorem says that a simple closed curve in a plane divides it into exactly two regions: an inside and an outside.

Networks *(Lessons 2, 3, and 4).* A network is a figure consisting of points, called "vertices," connected by lines, called "edges."

The *degree* of a vertex of a network is the number of edges that meet at it.

An *Euler path* is a continuous path that passes along every edge of a network exactly once. A network has an Euler path only if it has no more than two vertices of odd degree.

A *tree* is a network that does not have any simple closed curves.

The *diameter* of a tree is the largest number of edges that a path on the tree can have.

The Moebius strip and other surfaces *(Lesson 5).* A Moebius strip is a band that contains a half-twist. It has only one side and one edge.

EXERCISES

SET I

The ten figures shown below are used in the Gesell Figure Copying Test, usually given to children between the ages of 3 and 12. The figures are shown in order of difficulty as determined by the percentage of children who are able to copy them at each age level.

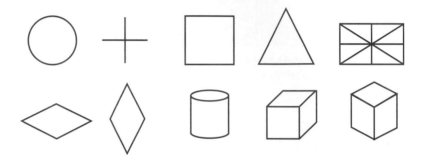

Draw the figure(s)

1. that are simple closed curves.

2. that, other than the simple closed curves, can be drawn without taking the pencil off the paper or retracing any part.

3. that, other than the simple closed curves, are topologically equivalent to each other.

The floor plan of a house can be represented by a network called an access diagram.

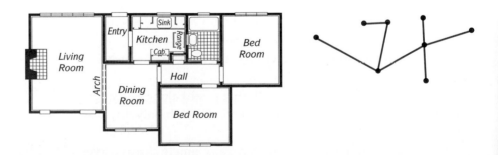

In the network, what represents

4. each room of the house?

5. the door connecting each pair of rooms?

Which room of the house is represented by

6. the vertex of highest degree?

7. the vertex with degree 2?

8. Is the access diagram a tree? Explain why or why not.

9. What is its diameter?

These figures illustrate a double-stranded molecule with cross-links that might be joined as shown in part A to form the arrangement shown in part B.*

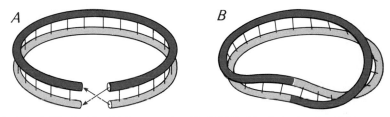

10. What does the molecule shown in part B look like?

11. Thought of as a surface, how many edges does it have?

12. What would be the result if all of the cross-links were broken?

The maps below show the streets in two hilly regions in which a truck delivers newspapers. The papers are thrown from both sides of the truck so that it does not retravel any street unless it is necessary.

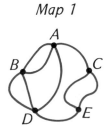 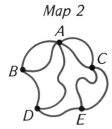

Map 1   *Map 2*

13. Is it possible to plan a route on map 1 that covers each street exactly once?

14. Is it possible to plan a route on map 2 that covers each street exactly once?

---

*"Chemical Topology" by Edel Wasserman, *Scientific American*, November 1962.

15. Explain how it is possible to tell, without resorting to trial and error, that one of the maps does not have such a route.

16. On the map that has such a route, can the route begin at any intersection? If not, where must it begin?

Drawing by Roger Hayward; reproduced in Mathematical Games by Martin Gardner, Scientific American, May 1970, p. 124.

Set II

This drawing by Roger Hayward is based on the network below.

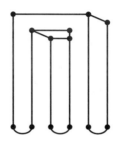

1. How many vertices does it have?

2. How many edges does it have?

3. How many regions does it have? (Remember that one of the regions of a network is the region outside it.)

In doing the Set I exercises in Lesson 3, you discovered a formula relating the number of regions, R, of a network, to its number of vertices, V, and number of edges, E. One way to write the formula is

$$R + V = E + 2$$

4. Does this formula apply to the peculiar network above?

An old puzzle requires drawing a continuous path that crosses each edge of the network at the left exactly once, without going through any of its vertices. A couple of unsuccessful attempts to solve the puzzle are shown below. In the first figure, one of the edges has not been crossed, and in the second one, an edge has been crossed more than once.

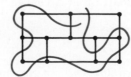 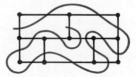

To discover whether the puzzle can be solved, place tracing paper over the networks below and try to draw such a path on each. The path should cross each edge exactly once and should not pass through any vertex. Keep a record of which networks have such a path.

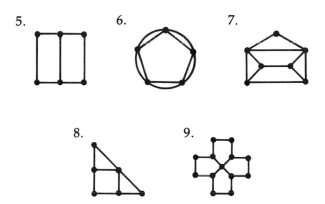

5.   6.   7.

8.   9.

Each network above contains two or more internal regions. A region is considered to be of an even or odd degree according to whether it is surrounded by an even or odd number of edges. For example, both regions in the network in exercise 5 are of even degree because each is surrounded by four edges.

10. Copy and complete the following table for the five networks.

Network	Number of even regions	Number of odd regions	Does the network have a path?
5	2	0	Yes
6	▓▓▓▓	▓▓▓▓	▓▓▓▓

(Continue the table to show all five networks.)

11. What seems to determine whether or not a continuous path can be drawn that crosses each edge of a network exactly once?

12. Do you think the original puzzle has a solution? Explain why or why not.

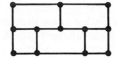

5		6	
3	1		2
	4		

5		6	
3	1		2
6	4		5

Set III

In 1976, it was proved that every map on a plane can be colored in four or fewer colors, so that no two regions sharing a border have the same color. The adjoining figure, from a paper written in 1910, shows that more than four colors may be needed to color a map on a Moebius strip.*

First, imagine that the figure is flat as shown.

1. Are there any pairs of numbered regions that do not share part of their borders? If so, which pairs are they? (For your convenience, every pair of numbered regions in the map is given below.)

$$1-2, \quad 1-3, \quad 1-4, \quad 1-5, \quad 1-6, \quad 2-3, \quad 2-4, \quad 2-5, \quad 2-6,$$
$$3-4, \quad 3-5, \quad 3-6, \quad 4-5, \quad 4-6, \quad 5-6$$

Imagine that the rectangle was cut out and made into a Moebius strip by giving its left edge a half-twist and bringing it around and taping it to the right edge. Because of the half-twist, the unnumbered region at the lower right would be connected to region 5, becoming part of that region. The unnumbered region originally at the lower left would be connected to region 6, becoming part of that region. The adjoining figure has been numbered to indicate this.

Also think of the Moebius strip as being made of transparent material so that the regions show through it.

2. Are there any pairs of regions on the Moebius strip map that do not share parts of their borders? If so, which pairs are they?

Because the Moebius strip is transparent, color in any region will also show on "the other side."

3. How many colors are needed to color this map so that no two regions sharing a border have the same color?

*The map-coloring problem on a plane was discussed in Chapter 1, Lesson 3. The above figure is from "Some Remarks on the Problem of Map-Coloring on One-Sided Surfaces" by Heinrich Tietze, a paper in the Annual Report of the German Mathematics Association, 1910.

Further Exploration

LESSON 1

1. Experiment: *The Borromean Rings*

The first figure below shows two rings linked together; they cannot be separated without cutting one of the rings. The second figure shows two rings that are not linked at all.

Strange as it may seem, *three* rings can be linked together *without any two of the rings being linked with each other.* In other words, the three rings cannot be separated without cutting one of them, yet if any one of the rings were to disappear, the other two would immediately come apart.

Use a compass to draw three identical rings of about the size shown in the first figure at the right on a large file card. Cut the rings out and then cut one of them as shown in the second figure.

In the diagram below, the three rings are linked together in such a way that the brown ring is not linked to either white ring. The white rings, however, are linked to each other so that, if the brown ring were cut and removed, they would not come apart.

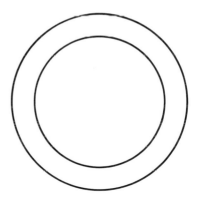

a. Can you link the three rings that you have made together so that *no pair* of them is linked? If so, tape the result to your paper.

Three rings linked in this way are called Borromean rings after an Italian family of the Renaissance named Borromeo, in whose coat of arms they appeared.

The adjoining photograph shows three lizards carved from wood.

b. What is interesting about the way in which they are linked together?

c. How do you suppose that these "linked" lizards were made?

2. The fingerprint classification system used by the FBI uses letters and numbers to identify the topological patterns in fingerprints.* Each fingerprint is assigned a letter according to its basic pattern.

Pattern	Letter
Arch	A
Tented arch	T
Radial loop	R
Ulnar loop	U
Whorl	W

The figure below illustrates the steps in classifying a person's fingerprints: first with letters and then numbers to determine the file classification.

Example 1

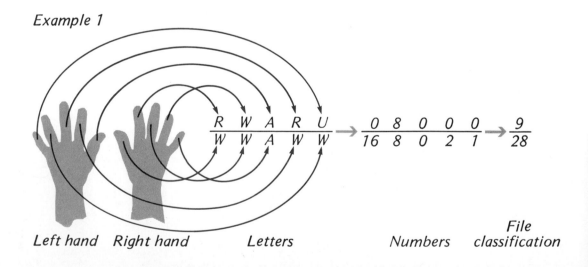

Left hand   Right hand   Letters   Numbers   File classification

---

*The examples in this exercise are taken from *Fingerprint Techniques*, by Andre A. Moenssens (Chilton, 1971).

Additional examples are shown below.

Example 2

$$\frac{T\ U\ U\ T\ W}{W\ W\ W\ W\ U} \rightarrow \frac{0\ 0\ 0\ 0\ 1}{16\ 8\ 4\ 2\ 0} \rightarrow \frac{2}{31}$$

Example 3

$$\frac{W\ W\ T\ W\ W}{R\ U\ U\ W\ W} \rightarrow \frac{16\ 8\ 0\ 2\ 1}{0\ 0\ 0\ 2\ 1} \rightarrow \frac{28}{4}$$

a. How do the numbers seem to be assigned?
b. How do the file classifications seem to be determined? What can you conclude about a person's fingerprints if his or her file classification is

c. $\frac{1}{1}$?

d. $\frac{17}{9}$?

Lesson 2

1. A letter carrier is supposed to deliver mail to all the buildings on both sides of the streets shown on the map at the right. The carrier starts from the post office at the upper left corner and would like to travel each side of each street exactly once before returning to the post office. Find out whether this is possible by doing each of the following.

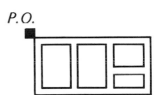

P.O.

   a. Draw a network to represent the letter carrier's territory. Label the corner representing the location of the post office P. (Note that each street must be represented by two edges because the carrier must deliver mail on both sides.)
   b. Can the network be traveled in a path that begins and ends at P? If it can, trace the map and draw such a path.
   c. What do you notice about the vertices of the network?
   d. Do you think every letter carrier's route can be traveled in a path that ends where it begins? Explain.

2. In 1967, John Horton Conway and Michael Stewart Patterson, two mathematicians at Cambridge University in England, invented a topological game called Sprouts.* The game is played by two people who take turns drawing a network.

Mathematical Carnival, by Martin Gardner (Knopf, 1975).

First, several points, which serve as the original vertices of the network, are marked on a piece of paper. The players then take turns drawing edges, following these rules:

1. Each edge either must connect two vertices or must connect one vertex to itself.
2. When an edge is drawn, a new vertex must be chosen somewhere on it.
3. No edge may cross itself, cross another edge, or pass through any vertex.
4. No vertex may have a degree of more than 3.

The last person able to play wins the game. The figures below show one way in which a game that starts with two points might be played.

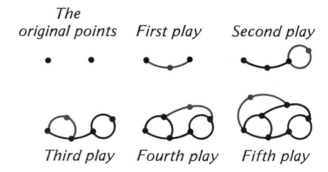

The original points *First play* *Second play*

Third play *Fourth play* *Fifth play*

This game ended on the fifth play because no more edges can be added without breaking the rules.

Is there a limit to the number of moves that a Sprouts game can last if the players want to continue it as long as possible? To find out, try playing it with yourself or with someone else. Play several games starting with just two points and see if you can draw any conclusions. Then try playing games that start with three points and games that start with four points.

a. Is it possible for a game that starts with two points to go on indefinitely? If not, what is the greatest number of moves it can last?

What conclusions can you draw about a game that starts with

b. three points?
c. four points?
d. ten points?

LESSON 3

1. In his book *On the Trail of the Bushongo,* Emil Torday wrote of encountering a circle of African children playing with sand:

The children were drawing, and I was at once asked to perform certain impossible tasks; great was their joy when the white man failed to accomplish them.*

One task was to trace the figure at the right in the sand with one continuous sweep of the finger.

a. Draw an outline of the figure on graph paper and then see if you can figure out how to draw it.
b. Why is it unlikely that someone without any knowledge of networks would be able to do this on the first attempt?
c. What was the children's secret for drawing the network?

2. *The Air-Conditioning Inspector*†

To locate the source of an obnoxious odor in an office building, an air-conditioning inspector is called in and asked to inspect the air-conditioning ducts thoroughly. Because the ducts are cramped, the inspector prefers not to go through any section more than once if it can be avoided. A scale map of the system is shown at the right with the access points shown in color. An access point can be used either as an entrance to the system or as an exit from it.

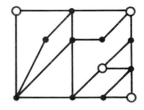

a. Can the inspector check all the ducts without crawling through any of them more than once? Explain.
b. Place tracing paper over the figure and, by drawing several possible paths, try to find the one that you think is best.

LESSON 4

1. A family gathering consists of father, mother, son, daughter, brother, sister, cousin, nephew, niece, uncle, and aunt. But only two men and two women are present. They have a common ancestor and there has been no marriage between relatives.‡

Explain by drawing a tree diagram how the four people are related.

2. Seven secret agents need to be able to communicate with each other either directly or indirectly.§ These communications can involve a certain

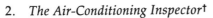

*Related by Claudia Zaslavsky in *Africa Counts* (Prindle, Weber, and Schmidt, 1973).

†From *A Sourcebook of Applications of School Mathematics,* by Donald Bushaw, Max Bell, Henry O. Pollak, Maynard Thompson, and Zalman Usiskin (N. C. T. M., 1980).

‡This puzzle, by Pierre Berloquin, is from his book *One Hundred Games of Logic* (Scribner's, 1977).

§Adapted from a problem by Gary Chartrand in *Graphs as Mathematical Models* (Prindle, Weber, and Schmidt, 1977).

amount of risk. The following table lists the "risk factors" of direct communication between certain pairs of agents.

Agent pair	Risk factor	Agent pair	Risk factor
A–C	7	B–E	5
A–D	4	B–G	6
A–E	2	C–E	8
A–F	4	D–E	3
A–G	5	D–F	5
B–D	6	E–G	4

All other direct communications are either impossible or too dangerous.

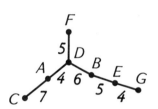

The tree at the left, in which each agent is represented as a vertex and each direct communication as an edge, shows one possible communications network for the agents. It has a total risk of 31.

a. Try to discover a communications network with as small a total risk as possible.
b. Does it have to be a tree? Explain.
c. What is the total risk of the network that you discovered?
d. What method did you use to find it?

LESSON 5

1. Experiment: *A Pair of Topological Surprises*

Part 1. Cut out a strip of paper 1.5 inches wide and as long as possible. Fold the strip in half and cut two slits into both ends as shown in the first figure below.

Unfold the strip and number the ends as shown in the second figure.

Put three short pieces of tape on ends 4, 5, and 6, and then make a loop as shown in the figure below.

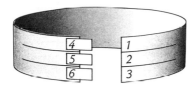

Now tape end 4 to end 1.

Pass end 5 *over* end 4, and end 2 *under* end 1 and tape together.

Pass end 6 *between* ends 4 and 5; pass end 3 *over* end 1 and tape together.

Finish both cuts so that they go all the way around the band.

   a. What is the result?

Part 2. Prepare another strip in exactly the same way as described in the first paragraph of Part 1.

Then turn end 6 over (give it a half-twist) and tape it to end 1.

Turn end 4 over and tape it to end 2.

Turn end 5 over and tape it to end 3.

Finish both cuts so that they go all the way around the band.

   b. What is the result?

2.   Experiment: *More Topological Surprises**

Cut out four large crosses like the one at the right from graph paper. Each arm should be 1 inch wide and extend 3 inches from the center of the cross.

Part 1. Tape the ends of one cross together to make two connected bands as shown in the figure below. Cut along the center of each band.

   a. What is the result?

---

*Most of this experiment is derived from material in *Mathematical Magic Show,* by Martin Gardner (Scribner's, 1977).

Part 2. Repeat the directions of Part 1 with another cross, but put a half-twist in one of the bands as shown in the figure below.

    b. What is the result?

Part 3. Repeat the directions of Part 1 with another cross, but put a half-twist in *each* band as shown in the figure below.

    c. What is the result?

Part 4. Tape the ends of the last cross together to make two connected bands, one with a half-twist, as shown in the first figure on this page. Then cut the twisted band *one-third of the way from the edge* and finally cut the other band along its *center*.

    d. What is the result?

Appendix

Basic Ideas and Operations

1   ANGLES AND THEIR MEASUREMENT

Imagine a large pie that has been divided by a very sharp knife into 360 equal slices. Viewed from above, one of the slices would look like the figure below. The slice's two straight edges form an *angle* and the point at which they meet is called the *vertex*. The straight edges lie along the *sides* of the angle.

The *measure* of an angle gives the size of the "opening" between its sides. The measure of the angle of the slice shown above is one degree, which is written as 1°. If all 360 slices were left together, they would completely surround the center of the pie, and so the number of degrees about a point is 360.

The degree, then, is the basic unit used to express the measures of all angles. Here are other examples. Consider the angle formed by the sides of one-half of a pie. Because one-half of a pie is equal to 180 of the small slices, its angle has a measure of 180°. Such an angle is called a *straight angle* because its sides lie along a straight line.

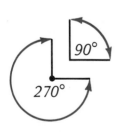

Next consider a pie from which a quarter slice has been cut. The angle of the quarter slice has a measure of 90° and is called a *right angle.* The rest of the pie can be thought of as another slice; the measure of its angle is $360° - 90° = 270°$.

Angles are usually measured with a protractor. Although some protractors are circular in shape, most are semicircular like the one pictured here. Protractors usually have two scales so that they can be used to measure angles in either a clockwise or a counterclockwise direction.

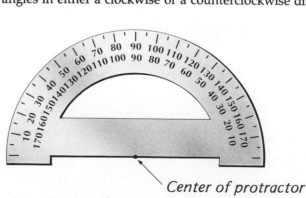

Center of protractor

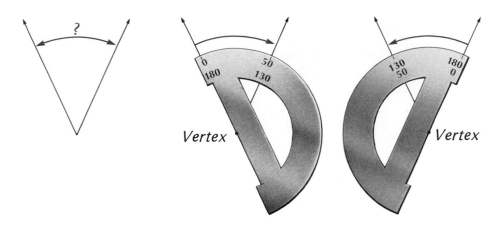

To measure an angle with a protractor, first place the center of the protractor on the vertex of the angle. Next, line up the edge of the protractor with one side of the angle (which side does not matter; both possibilities are shown above). Finally, read the measure of the angle by looking at where its other side falls under the scale; in the example, it is 50°.

For practice, measure the angles below. The measures of the angles are given at the bottom of this page so that you will know whether or not you are measuring them correctly.

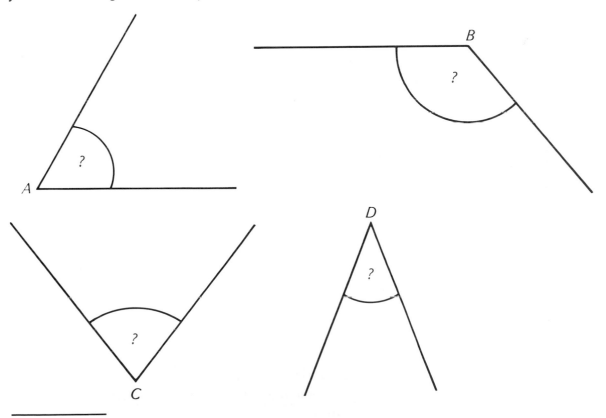

Angle A = 60°, Angle B = 130°, Angle C = 75°, Angle D = 42°.

2 THE DISTRIBUTIVE RULE

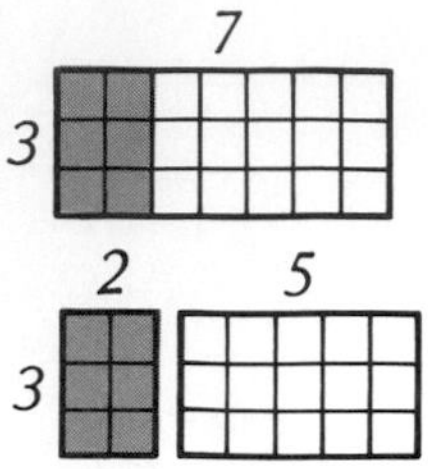

The distributive rule relates multiplication and addition. It is illustrated by the adjoining figure. The number of small squares in the figure can be expressed as either

$$3 \times (2 + 5) \quad \text{or} \quad (3 \times 2) + (3 \times 5).$$

Notice that

$$3 \times (2 + 5) = 3 \times 7 = 21$$

and

$$(3 \times 2) + (3 \times 5) = 6 + 15 = 21.$$

In general,

$$a \times (b + c) = a \times b + a \times c.$$

Look at how this applies to the number trick on page 41.

Choose a number.	n
Add 5.	$n + 5$
Double the result.	$2 \times (n + 5) = (2 \times n) + (2 \times 5)$
	$\qquad\qquad = 2n + 10$

Dividing 21 by 3 is the same as changing the fraction $\dfrac{21}{3}$ into decimal form or as multiplying 21 by $\dfrac{1}{3}$. It is therefore not surprising that a similar distributive rule applies to fractions:

$$\frac{b + c}{a} = \frac{b}{a} + \frac{c}{a}$$

For example,

$$\frac{6 + 15}{3} = \frac{6}{3} + \frac{15}{3}$$

because

$$\frac{6 + 15}{3} = \frac{21}{3} = 7 \quad \text{and} \quad \frac{6}{3} + \frac{15}{3} = 2 + 5 = 7$$

Continuing with the number trick on page 41,

	$2n + 10$
Subtract 4.	$2n + 6$
Divide by 2.	$\dfrac{2n + 6}{2} = \dfrac{2n}{2} + \dfrac{6}{2}$
	$\qquad\qquad = n + 3$
Subtract the number first thought of.	3
The result is 3.	

Check your understanding of the distributive rule by trying these problems. The answers are given on page 662.

1. Multiply by 2: $n + 3$.
2. Multiply by 3: $n + 5$.
3. Divide by 2: $2n + 10$.
4. Divide by 4: $4n + 12$.

3 SIGNED NUMBERS

Everyone knows that the whole numbers have a definite order. After you learned to count, you were able to say, for example, what number comes after 12 or what number comes before 8. The order of the whole numbers can be shown by representing them as evenly spaced points along a line.

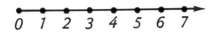

$$0 \quad 1 \quad 2 \quad 3 \quad 4 \quad 5 \quad 6 \quad 7$$

If the line is extended beyond 0 in the opposite direction, how should the points on the other side be numbered? The customary way of doing it looks like this:

$$-4 \quad -3 \quad -2 \quad -1 \quad 0 \quad 1 \quad 2 \quad 3 \quad 4$$

The numbers of the points on the line left of the zero are *negative* and the numbers on the right of the zero are *positive*. Because they are identified by the signs − and +, these numbers are often called *signed numbers*. (A number without a sign is assumed to be positive.)

The rules for calculating with signed numbers that are needed for the exercises in this book are reviewed here.

Addition An easy way to add two signed numbers is to think of gains and losses. You might think of a football field to help in picturing the situation. For example, what is the sum of 2 and −5? A gain of 2 yards followed by a loss of 5 yards is the same as a loss of 3 yards:

$$2 + -5 = -3$$

What is the sum of −3 and 7? A loss of 3 yards followed by a gain of 7 yards is equivalent to a net gain of 4 yards:

$$-3 + 7 = 4$$

What is the sum of −8 and −4? A loss of 8 yards followed by a loss of 4 yards is the same as a net loss of 12 yards:

$$-8 + -4 = -12$$

Try the following exercises. The answers are on page 662.

5. $-6 + 5$
6. $9 + -2$
7. $-3 + -15$
8. $12 + -27$
9. $-64 + 10$

Subtraction To subtract a positive number, think of it as a loss. For example, what is 10 subtracted from 3? A gain of 3 yards followed by a loss of 10 yards is the same as a net loss of 7 yards:

$$3 - 10 = -7$$

What is 6 subtracted from 0? A gain of 0 yards followed by a loss of 6 yards is the same as a loss of 6 yards:

$$0 - 6 = -6$$

Try these exercises. The answers are on page 662.

10. $8 - 9$
11. $1 - 12$
12. $5 - 20$

Multiplication The multiplication of signed numbers is easily illustrated by a multiplication table.

$\times$	3	2	1	0	-1	-2	-3
3	9	6	3	0	-3	-6	-9
2	6	4	2	0	-2	-4	-6
1	3	2	1	0	-1	-2	-3
0	0	0	0	0	0	0	0
-1	-3	-2	-1	0	1	2	3
-2	-6	-4	-2	0	2	4	6
-3	-9	-6	-3	0	3	6	9

This table shows, for example, that

$$3 \times -2 = -6$$

and that

$$-1 \times -3 = 3.$$

In general, the product of two numbers having *opposite* signs is *negative*, and the product of two numbers having the *same* sign is *positive*.

Try these exercises. The answers are on page 662.

13. -4×3
14. -1×-8
15. 7×-2
16. $(-5)^2$

Division The rules for dividing signed numbers, or changing signed fractions into decimal form, are the same as those for multiplying them. The quotient of two numbers having *opposite* signs is *negative*; the quotient of two numbers having the *same* sign is *positive*.

For example,

$$\frac{12}{-3} = -4 \quad \text{and} \quad \frac{-20}{-2} = 10.$$

Try these exercises. The answers are on page 662.

17. $\dfrac{-35}{5}$

18. $\dfrac{-24}{-8}$

19. $\dfrac{10}{-4}$

4 PERCENT

The word *percent* literally means *per hundred*. For example, to say that there is a 50% chance of a coin coming up heads when it is tossed means that it will turn up heads about 50 times per 100 tosses. If we change 50% to a fraction, we get

$$\frac{50}{100} = 0.50.$$

To change a fraction into a percentage, *simply multiply it by 100*;

$$0.50 \times 100 = 50\%$$

What is $\dfrac{1}{38}$ expressed as a percentage? Here are the steps in finding out:

Step 1. $\dfrac{1}{38} \times 100 = \dfrac{100}{38}$

Step 2. $\dfrac{100}{38} \approx 2.6$

Step 3. 2.6 rounded to the nearest whole number is 3.

So $\dfrac{1}{38} \approx 3\%$.

Check your understanding of percent by changing each of the following numbers into percentages. (Where necessary, round to the nearest whole number.) The answers are given below.

20. $\dfrac{3}{4}$

21. 1

22. $\dfrac{2}{25}$

23. $\dfrac{1}{36}$

24. $\dfrac{15}{64}$

ANSWERS TO EXERCISES IN THE APPENDIX

1. $2n + 6$.	9. -54.	17. -7.
2. $3n + 15$.	10. -1.	18. 3.
3. $n + 5$.	11. -11.	19. -2.5.
4. $n + 3$.	12. -15.	20. 75%.
5. -1.	13. -12.	21. 100%.
6. 7.	14. 8.	22. 8%.
7. -18.	15. -14.	23. 3%.
8. -15.	16. 25.	24. 23%.

Answers to
Selected
Exercises

Answers to Selected Exercises

CHAPTER 1, LESSON 1 (PAGES 8 – 11)

SET I
2.

SET II
6. $\dfrac{9}{3} = \dfrac{6}{2} = 3.$

CHAPTER 1, LESSON 2 (PAGES 14 – 18)

SET I
7. On a table whose length is even and whose width is odd, the ball ends up in the upper-left corner.

SET II
1. No. **4.** $\dfrac{7}{2}.$ **11.** $\dfrac{95}{85} = \dfrac{19}{17}.$ Upper-right because reduced length and width are both odd.

CHAPTER 1, LESSON 3 (PAGES 20 – 24)

SET I
3. $1 + 3 + 5 + 7 + 9 + 11 + 13 = 7 \times 7.$ **9.** From $7 \times 2.$

SET II
2. 25 units. **4.** The pressure gets smaller. **6.** The pressure is divided by 4.

CHAPTER 1, LESSON 4 (PAGES 27 – 30)

SET I
2. Inductive. **5.** 11. **17.** That it is not prime.

SET II
5. The top of the T has $2 \times 9 = 18$ squares. The bottom has $6 \times 3 = 18$ squares. $18 + 18 = 36.$ **11.** No.
13. 64 square units.

CHAPTER 1, LESSON 5 (PAGES 33 – 38)

SET I
1. If the pennies are placed on squares of the same color, A wins; if they are placed on squares of different colors, B wins. **5.** Seven of the squares must be brown and seven of the squares must be white. **6.** Six are white and eight are brown. **10.** It would have to be white because the label is *wrong.* **13.** No. (It could be brown or white.)

SET II
3. 8. **6.** $5 \times 5 \times 5 = 125.$ **7.** Three.
10. $12 \times 3 = 36.$ **13.** Inside the large cube.
16. $12 \times 12 \times 12 = 1,728.$

CHAPTER 1, LESSON 6 (PAGES 41 – 44)

SET I
3. □○
8. $n + 5.$
11. Choose a number. □ n

Add 3. □○○○ $n + 3$

Multiply by 2. □□888 $2n + 6$

Add 4. □□88888 $2n + 10$

Divide by 2. □○○○○○ $n + 5$

Subtract the number first ○○○○○ 5
  thought of.
The result is 5.

SET II
2. It contains the original three-digit number written twice. **6.** 1,001.

Chapter 2, Lesson 1 (pages $61-66$)

Set I
2. The odd numbers.   **4.** 1, 4, 7.   **9.** That the numbers continue.   **13.** 5, 13, 21, 29, 37.   **19.** -3.
25. 10, 25, 40, 55, 70.

Set II
2. $3 + 7(9) = 3 + 63 = 66$.
5. $8 + 7(10) = 8 + 70 = 78$.   **8.** $55 - 3 = 52$.

Chapter 2, Lesson 2 (pages $69-74$)

Set I
1. By multiplying by 3.   **4.** 3, 6, 12.   **7.** 4, 12, 36, 108, 324.   **11.** 0, 0, 0, 0.   **14.** Arithmetic; difference is 5.   **16.** Neither.

Set II
2. Six. (They are 1, 10, 100, 1,000, 10,000, and 100,000.)   **5.** 5, 10, 20.   **8.** $2 + 3 \cdot 3$ and $2 + 3 \cdot 4$.   **15.** $6 \cdot 4^{99}$.

Chapter 2, Lesson 3 (pages $77-82$)

Set I
2. The binary sequence.   **4.** $4 + 1$.
6. $32 + 8 + 2$.   **9.** 1 0 1.   **11.** 1 0 1 0 1 0.
14. $8 + 4 + 2 + 1 = 15$.   **17.** $16 + 4 = 20$.

Set II
3. They are each 1 less.   **6.** $128 - 1 = 127$.
12. $\$32 - \$31 = \$1$ ahead.

Chapter 2, Lesson 4 (pages $84-91$)

Set I
5. 1, 4, 9, 16, 25.   **9.** 2 meters per second per second.   **10.** 4, 8, 12, 16, 20, 24.   **15.** It is doubled.   **17.** It is multiplied by 4.   **20.** Triangular numbers.

Set II
3. 9.   **8.** No.

Chapter 2, Lesson 5 (pages $94-98$)

Set I
3. 27.   **10.** It is multiplied by 4.   **12.** 81.

Set II
2. The odd numbers (or, an arithmetic sequence).
7. The sequence of cubes.   **14.** $3^4 + 4^4 + 5^4 + 6^4 = 7^4$.

Chapter 2, Lesson 6 (pages $101-106$)

Set I
3. They are every third term.   **6.** 5, 8, and 13.
9. A male bee.

Set II
5. The sums are every other term of the Fibonacci sequence.
7. $1^2 + 1^2 + 2^2 + 3^2 + 5^2 + 8^2 = 104 = 8 \cdot 13$.
11. Pattern B.   **14.** They get closer and closer to a number around 1.618.

Chapter 3, Lesson 1 (pages $124-128$)

Set I
5.

x	1	2	3	4	5
y	12	23	34	45	56

Set II
1. $y = 2x$.   **3.** $y = x - 3$.   **6.** $y = 10x + 1$.
12. By adding 70.   **13.** By multiplying by 7.

Chapter 3, Lesson 2 (pages $131-135$)

Set I
1. C.   **3.** H.   **9.** E.   **11.** A(8, 6), B(4, 5), C(2, 4), D(0, 2), E(-1, 0), F(-2, -4).

Set II
1.

10. A, D, and E.   **15.** A.

CHAPTER 3, LESSON 3 (PAGES 138–144)

SET I
2. Five.   6. At 4.   9. $y = x - 2$.   12. (0, 1).
18. The lines are parallel.

SET II
3. 24 centimeters.
4.

t	0	1	2	3	4	5
h	10	8	6	4	2	0

7. The height of the candle at the beginning.
13. $y = 15x$.
18. It increases the Fahrenheit temperature by 36.

CHAPTER 3, LESSON 4 (PAGES 146–152)

SET I
4. Function A:

x	−3	−2	−1	0	1	2	3
y	11	6	3	2	3	6	11

6. They look alike but one is higher than the other.
12. At 12.   16. At 3.

SET II
2. It is multiplied by 4.   6. The number of subscribers decreases.   10. $180,000.   14. The place where the diver hits the water.

CHAPTER 3, LESSON 5 (PAGES 155–160)

SET I
4. 0.06.
6.

x	−6	−5	−4	−3	−2	−1
y	−1	−1.2	−1.5	−2	−3	−6

11. The curves have the same shape.   14. The exponents in them are even.

SET II
2. It gets shorter.   4. It is divided in half.
8. $\dfrac{275}{2.2} = 125$ times the volume.   13. It is multiplied by 4.

CHAPTER 3, LESSON 6 (PAGES 163–167)

SET I
3. About 40° C.   4. Interpolate. The value estimated

is between values that are known.   7. It slows down.   9. They lie along a line.

SET II
2. 21 centimeters.   7. It gets steeper and steeper.
13. 10° C.

CHAPTER 4, LESSON 1 (PAGES 186–191)

SET I
2. 10^1.   4. 10^{10}.   8. Ten million.
14. $10^{11} \times 10^{12} = 10^{23}$.   18. 10^{20}.   22. 10^{16}.   29. 10^{100}.

SET II
2. One hundred million.   5. 10^{16}.

CHAPTER 4, LESSON 2 (PAGES 194–198)

SET I
3. 3×10^{27}.   7. 2,000,000,000,000,000.
11. 6.35×10^{11}.   17. 200,000,000 and 400,000,000.

SET II
2. 80,000,000,000.   5. 6×10^{15}.
7. $30 \times 10^{11} = 3 \times 10^{12}$.   13. 15.   16. 4×10^2.
20. $\dfrac{3 \times 10^{10}}{7.5 \times 10^9} = 0.4 \times 10^1 = 4$.

CHAPTER 4, LESSON 3 (PAGES 201–206)

SET I
2.

$$128 \rightarrow \quad 7$$
$$32 \rightarrow \underline{+\ 5}$$
$$4{,}096 \leftarrow \quad 12$$

5.

$$65{,}536 \rightarrow \quad 16$$
$$16 \rightarrow \underline{+\ 4}$$
$$1{,}048{,}576 \leftarrow \quad 20$$

11.

$$32{,}768 \rightarrow \quad 15$$
$$2{,}048 \rightarrow \underline{-11}$$
$$16 \leftarrow \quad 4$$

15. 12.

18.

$$4 \rightarrow \quad 2$$
$$\underline{\times\ 5}$$
$$1{,}024 \leftarrow \quad 10$$

SET II
1. 5.   **3.** 12.   **9.** 5.   **12.** $2^5 \times 2^7 = 2^{12}$.
13. 5, 7, and 12.

CHAPTER 4, LESSON 4 (PAGES 211–215)

SET I
1. $0.30 + 0.78 = 1.08$.   **2.** 3 and 4.
5. $0.30 + 0.85 = 1.15$.   **11.** 15.   **13.** 1.60.
15. One thousand.

SET II
1. $0.48 + 6 = 6.48$.   **7.** 2×10^8.
8. $0.30 + 8 = 8.30$.   **12.** 10^2 or 100.   **14.** $6.48 =$
$0.48 + 6$; 3×10^6 or 3,000,000.

CHAPTER 4, LESSON 5 (PAGES 219–223)

SET I
1. .158.   **4.** 3.158.   **7.** 2.2×10^4; 4.342.
11. 1.04.   **14.** 1.04×10^5; 104,000.   **22.** Binary (or geometric).

SET II
3. 100.
8-9. (Beginning of table)

Interval of time	Number of seconds	Log
One second	1×10^0	0
One minute	6×10^1	1.8

12-13. (Beginning of table)

Wave	Frequency in hertz	Log
AC current	6×10^1	1.8
AM radio station at 980	9.8×10^5	6.0

CHAPTER 4, LESSON 6 (PAGES 226–229)

SET I
2. (Beginning of table)

x	0	1	2	3
y	1	10	100	1,000

5. 25 inches.   **8.** Exponent.   **11.** Exponential.

SET II
2. (Beginning of table)

Year	Log
1880	1.5
1890	1.9
1900	2.2

6. (Beginning of table)

x	0	1	2
y	1.8	1.5	1.2

CHAPTER 5, LESSON 1 (PAGES 248–254)

SET I
2. E.   **5.** The point itself.   **7.** C.   **10.** The axis of symmetry.   **17.** 3.   **19.** No.   **22.** 5-fold.
23. Line symmetry (1 line).   **25.** Line symmetry (2 lines) and 2-fold rotational symmetry.

SET II
3. A triangle with 3 lines of symmetry and 3-fold rotational symmetry.   **6.** A square with 4 lines of symmetry and 4-fold rotational symmetry.   **9.** 3.   **10.** 360°.

CHAPTER 5, LESSON 2 (PAGES 258–265)

SET I
1. 90°.   **3.** It decreases.   **7.** $\dfrac{360°}{8} = 45°$.

SET II
2. $\dfrac{360°}{24} = 15°$.   **4.** Line symmetry (24 lines) and 24-fold rotational symmetry.   **11.** A regular pentagon with a five-pointed star inside.   **15.** $\dfrac{360°}{16} = 22.5°$.

CHAPTER 5, LESSON 3 (PAGES 268–275)

SET I
2. Three.   **5.** $90° + 90° + 90° + 90° = 360°$.
10. 135°.   **11.** $135° + 135° + 90° = 360°$.
14. Three triangles and two squares.

SET II

1. 6-6-6.   **7.** A, B, D, and E.   **10.** Yes.   **13.** 3-6-3-6.

CHAPTER 5, LESSON 4 (PAGES 279 – 288)

SET I

2. Octahedron.   **6.** Four.   **9.** Three.   **17.** Six.
25. Thirty.

SET II

1. 3-3-3.   **9.** 180°.   **15.** $6 \times 4 = 24$ corners and
$6 \times 4 = 24$ sides; $\dfrac{24}{3} = 8$ corners and $\dfrac{24}{2} = 12$ edges.
21. An octahedron each of whose six corners touches
one of the six faces of a cube.

CHAPTER 5, LESSON 5 (PAGES 290 – 297)

SET I

1. 3-6-6.   **6.** 3-4-3-4.   **14.** Three.   **15.** An equilateral triangle.   **16.** A regular octagon.   **20.** All of
their faces are equilateral triangles.   **30.** Eight.

SET II

4. 12 squares, 8 hexagons, and 6 octagons.

CHAPTER 5, LESSON 6 (PAGES 300 – 308)

SET I

2. Rectangles.   **3.** A triangular prism.   **7.** Regular
hexagons and rectangles.   **9.** Hexagonal prisms.
13. Six.   **18.** A hexagonal pyramid.   **22.** That an
equilateral triangle and two squares meet at each
corner.   **24.** Cube.

SET II

1. 19.   **8.** $2n$.   **11.** 12.   **14.** $n + 2$.   **18.** 17.

CHAPTER 6, LESSON 1 (PAGES 330 – 337)

SET I

4. Major.   **5.** 20.   **10.** They become more elongated.

SET II

4. 3 cm.   **6.** $AF_1 = 3$ cm and $AF_2 = 7$ cm.
13. 36.   **14.** 6.   **20.** A circle.

CHAPTER 6, LESSON 2 (PAGES 340 – 346)

SET I

1. Both distances are 6.0 cm.   **4.** The distances from
point F are equal to the distances from line ℓ.

SET II

4. The farther the ray, the smaller the angle.
7. Toward the satellite.

CHAPTER 6, LESSON 3 (PAGES 350 – 354)

SET I

2. Two.

SET II

1. 3 cm.   **6.** The distances get smaller and smaller.
8. At 3 and -3.   **9.** $3^2 = (-3)^2 = 9$; 9 appears below
y^2.

CHAPTER 6, LESSON 4 (PAGES 358 – 364)

SET I

2. 135°.   **9.** Rotational symmetry (2-fold).   **13.** 0.

SET II

2. $y = 4$ sine $2x$.   **4.** $y = 2$ sine $5x$.   **11.** The frequency (or wavelength).

CHAPTER 6, LESSON 5 (PAGES 367 – 373)

SET I

3. 6 feet.   **6.** Arithmetic.

SET II

1. 5   10   15   20   25   **3.** Archimedean.   **8.** 32 units.

CHAPTER 6, LESSON 6 (PAGES 376 – 383)

SET I

1. At the top of the wheel.   **4.** No.

SET II

2. One.   **4.** No.   **11.** An epicycloid.   **17.** The
curve is produced by a point on the rim of a wheel rolling around the inside of another wheel.

CHAPTER 7, LESSON 1 (PAGES 404–412)

SET I

1. 3.   3. 18.   4. $2 \times 3 \times 3 = 18$.   8. 3×3.
11. $3 \times 3 \times 3 = 27$.   16. $4 \times 4 \times 1 = 16$.
18. $4 \times 4 = 16$.   22. $2 \times 2 \times 2 \times 2 \times 2 \times 2 = 64$.

SET II

2. $2 \times 2 \times 1 = 4$.   3. $2 \times 2 \times 2 = 8$.
8. $5 \times 5 \times 1 = 25$.   15. $20 \times 20 \times 20 = 8,000$.
19. $6 \times 3 \times 2 = 36$.   21. $8 \times 2 \times 9 = 144$.
27. $2 \times 3 \times 4 = 24$.

CHAPTER 7, LESSON 2 (PAGES 416–419)

SET I

2. 6 gymnasts and 3 medals.   3. $6 \times 5 \times 4 = 120$.
5. $7 \times 6 \times 5 \times 4 \times 3 = 2,520$.   9. $_9P_2$.
10. $9 \times 8 = 72$.   18. 74.   20. $75 \times 74 = 5,550$.

SET II

1. False because $6 + 6 \neq 720$.   6. $8 \times 7 \times 6 = 336$.
9. $\dfrac{40,320}{120} = 336$.   13. $6! = 720$ ways.

CHAPTER 7, LESSON 3 (PAGES 422–427)

SET I

1. 5,040.   4. $\dfrac{11!}{9!}$.   9. $\dfrac{13!}{3! \times 3! \times 4! \times 2!}$.   14. $\dfrac{5!}{2! \times 3!}$.

SET II

2. $\dfrac{10!}{5! \times 5!}$.   3. 252.   10. $\dfrac{10!}{10!} = 1$.   13. $\dfrac{4!}{2! \times 2!} = 6$.

CHAPTER 7, LESSON 4 (PAGES 430–434)

SET I

4. 10.   7. $10 \times 9 = 90$.   8. $\dfrac{10 \times 9}{2!} = 45$.

11. $\dfrac{6}{1!} = 6$.   19. $\dfrac{7 \times 6}{2!} = 21$.

SET II

1. $_9C_2 = \dfrac{9 \times 8}{2!} = 36$.

5. $_{24}C_3 = \dfrac{24 \times 23 \times 22}{3!} = 2,024$.

8. $_{36}C_6 = \dfrac{36 \times 35 \times 34 \times 33 \times 32 \times 31}{6!} = 1,947,792$.

CHAPTER 8, LESSON 1 (PAGES 450–456)

SET I

1. $\dfrac{1}{6}$.   3. $\dfrac{3}{6} = \dfrac{1}{2}$.   4. 0.   7. $\dfrac{1}{5}$.   8. $\dfrac{4}{5}$.

11. $\dfrac{1}{7}$.   15. $\dfrac{2}{8} = \dfrac{1}{4}$.   21. $\dfrac{8}{80} = \dfrac{1}{10}$.   22. 10%.

SET II

2. $\dfrac{5,067}{10,000} = 0.5067$.   5. 90%.   10. $\dfrac{1}{38} \approx 2.6\%$.

11. $\dfrac{18}{38} - \dfrac{9}{19} \approx 47.4\%$.

CHAPTER 8, LESSON 2 (PAGES 459–466)

SET I

2. $6 + 2 = 8$.   3. $\dfrac{8}{26} = \dfrac{2}{9} \approx 22.2\%$.   8. That a 7

will.   14. $\dfrac{4}{36} = \dfrac{1}{9} \approx 11.1\%$.

SET II

2. 3.   8. 6.   10. 3. (The list is: 1, 4, 4; 4, 1, 4; 4, 4, 1.)   13. 1.   15. 6.

CHAPTER 8, LESSON 3 (PAGES 469–475)

SET I

3. $\dfrac{1}{2} \times \dfrac{1}{2} = \dfrac{1}{4}$.   5. $\dfrac{1}{3}$.   7. $\dfrac{1}{3} \times \dfrac{1}{2} = \dfrac{1}{6}$.

9. $0.1\% = 0.001 = \dfrac{1}{1,000}$. One in a thousand.

13. About 17.   14. $\dfrac{3}{10} \times \dfrac{1}{2} = \dfrac{3}{20}$.   18. $\dfrac{1}{37}$.   21. 0.07.

SET II

1. Dependent.   2. $\dfrac{5}{7}$.   4. $\dfrac{4}{6} = \dfrac{2}{3}$.

10. $0.5 \times 0.5 = 0.25$.   **18.** Because there are 10 spades in the remaining 47 cards.

Chapter 8, Lesson 4 (pages 478–485)

Set I
1. $0.50 \times 0.50 = 0.25 = 25\%$.
4. $25\% + 25\% = 50\%$.   **5.** $25\% + 50\% = 75\%$.
7. $0.4 \times 0.4 \times 0.6 = 0.096 = 9.6\%$.   **9.** 3.
10. $3 \times 14.4\% = 43.2\%$.   **14.** 100%.
17. $21.6\% + 43.2\% = 64.8\%$.   **20.** 4.
21. $4 \times 10.29\% = 41.16\%$.   **23.** 6.   **30.** 16.

Set II
1. $\frac{4}{4} = 1 = 100\%$.   **2.** $\frac{0}{4} = 0 = 0\%$.

Chapter 8, Lesson 5 (pages 488–494)

Set I
2. Binary (or geometric).   **4.** $0.07776 \approx 7.8\%$.
6. 10.   **7.** $10 \times (0.4 \times 0.4 \times 0.6 \times 0.6 \times 0.6) = 0.3456 \approx 34.6\%$.   **15.** $0.9 \times 0.9 \times 0.9 \times 0.9 \times 0.9 \times 0.9 = 0.531441 \approx 53.1\%$.   **20.** 20%.
22. $0.8 \times 0.8 \times 0.8 = 0.512 = 51.2\%$.
24. $3 \times (0.8 \times 0.8 \times 0.2) = 0.384 = 38.4\%$.   **29.** 70.

Set II
5. It has a similar shape.   **7.** 10.

Chapter 8, Lesson 6 (pages 497–503)

Set I
1. $\frac{2}{3}$.   **3.** $\frac{1}{3} \times \frac{1}{3} = \frac{1}{9}$.   **4.** $1 - \frac{1}{9} = \frac{8}{9}$.
7. $\frac{2}{3} \times \frac{2}{3} \times \frac{2}{3} = \frac{8}{27}$.   **9.** 70%.

11. $0.30 \times 0.10 = 0.03 = 3\%$.   **14.** $\frac{1}{100}$.   **16.** 97%.

Set II
1. Three out of the ten digits are lucky.
4. $100\% - 6\% = 94\%$.   **11.** Telephone numbers can end in 100 different pairs of digits; the Cracker Jack boxes contain 100 different prizes.

Chapter 9, Lesson 1 (pages 529–536)

Set I
2. 5.   **3.** 8.   **8.**

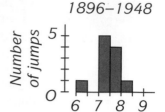

1896–1948

14. The writings of Bacon.

Set II
8. 0 to 1 kilometer.   **18.** That Hawaiian words never begin with A or W.

Chapter 9, Lesson 2 (pages 539–545)

Set I
4. Q and Z.   **5.** 8.   **6.** $3 \times 8 = 24$.   **16.** It appears less frequently.

Chapter 9, Lesson 3 (pages 548–553)

Set I
2. $\frac{48}{9} \approx 5.3$.   **5.** 5.   **8.** 7.   **13.** $5.99.
17. $\frac{192}{24} = 8$ centimeters.   **23.** Each one is 100 more.

Set II
4. The mode.   **8.** Each of them is 2.   **11.** It is not symmetrical.

Chapter 9, Lesson 4 (pages 557–563)

Set I
2. $\frac{1,200}{10} = 120$ seconds.   **5.** $\sqrt{16} = 4$.
6. $\frac{7}{10} = 70\%$.   **13.** It increases.   **18.** 33. (Scores between 12 and 34.)

SET II

1. 33.5 centimeters.   4. 68%.   6. 94%.   13. A normal curve.   15. $800 - 200 = 600$.   17. 34%.
23. 17.   24. $\frac{17}{25} = 68\%$.

CHAPTER 9, LESSON 5 (PAGES 566–573)

SET I

4. Acapulco.   8. It makes the variation look more extreme.   11. Almost four times as tall.   15. 1890.

SET II

2. The area of the larger bag is four times the area of the smaller one.   13. The vertical scale starts at 21% rather than 0.

CHAPTER 9, LESSON 6 (PAGES 576–581)

SET I

2. The people who vote in the election.   5. The people who have unlisted numbers.   7. $64,651 − $64,568 = $83.

SET II

3. $\frac{10}{50} = \frac{1}{5}$.   4. $\frac{1}{5}$.   8. It would make the figure reported too large.   10. That people prefer the letter M to the letter Q.

CHAPTER 10, LESSON 1 (PAGES 604–609)

SET I

1. A and C.   7. K and X.   8. P and Q.   11. Yes.

SET II

2. The cardioid and the octagon.   4. Outside.
11. A simple closed curve.

CHAPTER 10, LESSON 2 (PAGES 611–616)

SET I

3. 2.   9. A, B, D, and E.   11. C.   16. It has an extra vertex (at Lanai).

SET II

1. The degrees of the vertices.   14. (The networks that can be traveled are 4, 6, 7, 9, 11, and 13.)   16. No.

CHAPTER 10, LESSON 3 (PAGES 618–624)

SET I

2. A.   5. C.   8. A network has an Euler path that returns to its starting point if it has no odd vertices.
11. Yes.   18. Network B, 3 trips; network C, 4 trips; network D, 5 trips.

SET II

2. Vertices D and E are odd; the other vertices are even.   5. A and B.

CHAPTER 10, LESSON 4 (PAGES 627–633)

SET I

1. 15.   4. 16.   13. 2.   15. Yes.   16. 6.   19. 4.

SET II

1. 1.   3. 3.   4. 1 and 3.   6. 5 and 9.   9. K.   11. M.

CHAPTER 10, LESSON 5 (PAGES 636–640)

SET I

1. One.   3. Two separate bands.   5. One is twice as wide as the other.   7. Twice.   8. Both.
9. Just one.   10. You would paint the entire strip.
15. Two interlocking bands.

SET II

1. Yes.   2. Two.   7. One.

Index

Index